Y0-CAT-096

Reliability Engineering for Electronic Design

ELECTRICAL ENGINEERING AND ELECTRONICS

A Series of Reference Books and Textbooks

1. Rational Fault Analysis, *edited by Richard Saeks and S. R. Liberty*
2. Nonparametric Methods in Communications, *edited by P. Papantoni-Kazakos and Dimitri Kazakos*
3. Interactive Pattern Recognition, *Yi-tzuu Chien*
4. Solid-State Electronics, *Lawrence E. Murr*
5. Electronic, Magnetic, and Thermal Properties of Solid Materials, *Klaus Schröder*
6. Magnetic-Bubble Memory Technology, *Hsu Chang*
7. Transformer and Inductor Design Handbook, *Colonel Wm. T. McLyman*
8. Electromagnetics: Classical and Modern Theory and Applications, *Samuel Seely and Alexander D. Poularikas*
9. One-Dimensional Digital Signal Processing, *Chi-Tsong Chen*
10. Interconnected Dynamical Systems, *Raymond A. DeCarlo and Richard Saeks*
11. Modern Digital Control Systems, *Raymond G. Jacquot*
12. Hybrid Circuit Design and Manufacture, *Roydn D. Jones*
13. Magnetic Core Selection for Transformers and Inductors: A User's Guide to Practice and Specification, *Colonel Wm. T. McLyman*
14. Static and Rotating Electromagnetic Devices, *Richard H. Engelmann*
15. Energy-Efficient Electric Motors: Selection and Application, *John C. Andreas*
16. Electromagnetic Compossibility, *Heinz M. Schlicke*
17. Electronics: Models, Analysis, and Systems, *James G. Gottling*
18. Digital Filter Design Handbook, *Fred J. Taylor*
19. Multivariable Control: An Introduction, *P. K. Sinha*

20. Flexible Circuits: Design and Applications, *Steve Gurley, with contributions by Carl A. Edstrom, Jr., Ray D. Greenway, and William P. Kelly*

21. Circuit Interruption: Theory and Techniques, *Thomas E. Browne, Jr.*

22. Switch Mode Power Conversion: Basic Theory and Design, *K. Kit Sum*

23. Pattern Recognition: Applications to Large Data-Set Problems, *Sing-Tze Bow*

24. Custom-Specific Integrated Circuits: Design and Fabrication, *Stanley L. Hurst*

25. Digital Circuits: Logic and Design, *Ronald C. Emery*

26. Large-Scale Control Systems: Theories and Techniques, *Magdi S. Mahmoud, Mohamed F. Hassan, and Mohamed G. Darwish*

27. Microprocessor Software Project Management, *Eli T. Fathi and Cedric V. W. Armstrong (Sponsored by Ontario Centre for Microelectronics)*

28. Low Frequency Electromagnetic Design, *Michael P. Perry*

29. Multidimensional Systems: Techniques and Applications, *edited by Spyros G. Tzafestas*

30. AC Motors for High-Performance Applications: Analysis and Control, *Sakae Yamamura*

31. Ceramic Materials for Electronics: Processing, Properties, and Applications, *edited by Relva C. Buchanan*

32. Microcomputer Bus Structures and Bus Interface Design, *Arthur L. Dexter*

33. End User's Guide to Innovative Flexible Circuit Packaging, *Jay J. Miniet*

34. Reliability Engineering for Electronic Design, *Norman B. Fuqua*

35. Design Fundamentals for Low-Voltage Distribution and Control, *Frank W. Kussy and Jack L. Warren*

Additional Volumes in Preparation

Electrical Engineering-Electronics Software

1. Transformer and Inductor Design Software for the IBM PC, *Colonel Wm. T. McLyman*

2. Transformer and Inductor Design Software for the Macintosh, *Colonel Wm. T. McLyman*

3. Digital Filter Design Software for the IBM PC, *Fred J. Taylor and Thanos Stouraitis*

Reliability Engineering for Electronic Design

Norman B. Fuqua

Reliability Analysis Center
IIT Research Institute
Rome, New York

MARCEL DEKKER, INC. New York and Basel

Library of Congress Cataloging-in-Publication Data

Fuqua, Norman B.
Reliability engineering for electronic design.

(Electrical engineering and electronics ; 34)
Includes index.
1. Electronic apparatus and appliances--Reliability.
I. Title. II. Series.
TK7870.F87 1986 621.381'042 86-24046
ISBN 0-8247-7571-6

MARCEL DEKKER, INC.
270 Madison Avenue, New York, New York 10016

Current printing (last digit):
10 9 8 7 6 5 4

PRINTED IN THE UNITED STATES OF AMERICA

TO MY FAMILY

Preface

The purpose of this book is to fill a void between some fine theoretical textbooks on reliability and detailed specifications such as the International Electrotechnical Commission (IEC) specifications, the United States Military Standards (MIL-STDs), and other similar specifications that establish reliability requirements but give little or no detail regarding why and how to accomplish specific reliability-oriented tasks. The book addresses the needs of electronic design engineers, reliability engineers, and their respective managers, stressing a pragmatic viewpoint rather than a vigorous mathematical presentation. To keep the book at a reasonable length I have sometimes had to provide only a brief sketch of topics that may well have been treated in much greater detail. The intent is to give broad coverage of the subject matter without unduly concentrating on a few methods (or tools) to the exclusion of other equally important methodologies.

The principal aim of the book is to provide electronic design engineers with the background necessary to properly interface with their reliability counterpart and to assure that they recognize and adequately address each of the major factors affecting equipment reliability, starting with the initial design effort. One goal of this book is to acquaint the reliability practitioner with modern automated tools and information sources that are available today, and that can be utilized to improve efficiency and effectiveness within the reliability discipline. The book is divided into five major sections: Introduction, Detail Part Considerations, Circuit Design and Analysis, System Design and Analysis, and Management and Testing.

Part I, the introduction, combines basic theory with fundamental definitions and mathematics; that is, it provides the necessary foundation for all the material which follows.

Dealing with piece part considerations, the second part addresses the selection, specification, and control of the individual parts in Chapter 2, their proper application and derating in Chapter 3, and the potential for damage to state-of-the-art parts caused by electrostatic discharge in Chapter 4.

"Circuit Design and Analysis," the third part, discusses the different circuit analysis methodologies and techniques available to the designer in Chapter 5, redundant circuit configurations in Chapter 6, and two complementary reliability analysis techniques, Failure Mode, Effects, and Criticality Analysis and Fault Tree Analysis, in Chapters 7 and 8, respectively.

The fourth part, "System Design and Analysis," includes a variety of subject material. Chapter 9 addresses the quantification of reliability; its specification, apportionment, modeling, and prediction. Chapter 10 deals with the equipment environmental considerations the designer must address. Chapter 11 explores the impact on equipment reliability of specific manufacturing processes that have been selected, and the maintenance of equipment in the field.

"Management and Testing," the fifth and final part, addresses the necessary evaluation and managerial aspects of a successful reliability program. Formal reliability qualification demonstration programs are addressed in Chapter 12, while Chapter 13 considers the equipment's potential for reliability growth under defined operating and test conditions. Chapter 14 discusses environmental stress screening of equipment prior to delivery. In the final analysis, reliability is a management discipline; thus the closing chapter, Chapter 15, deals with the proper management of various types of reliability programs.

This book is an outgrowth of a Design Reliability Training Course that I have developed, directed, and taught over the past nine years. This course is presented under the auspices of the Reliability Analysis Center, operated by IIT Research Institute. As of December 1985 the course has been presented ninety-four times to over 3500 students throughout the United States, England, Denmark, Norway, Sweden, Finland, West Germany, Israel, Canada, and Brazil.

I wish to express my appreciation to my former students in the course, who have contributed many of the pertinent examples; to Mr. Harold Lauffenburger of IIT Research Institute whose vision and challenge made both the course and this book possible; to Mr. James Herz for his helpful critique of the manuscript; and to Mrs. Helen Adsit for the editorial assistance. I am also indebted to my former colleague and fellow instructor in the Design Reliability Training Course, the late Joseph Naresky, principal author of Volume 1 of MIL-HDBK-338, "Electronic Reliability Design Handbook," from which I have borrowed some of the illustrations used in this text.

Norman B. Fuqua

Contents

Reliability Engineering for Electronic Design

Part I
Introduction

1

Basic Theory and Mathematics

1.0 INTRODUCTION

This chapter describes some of the main points of design reliability engineering; the basic concepts and applicable formulas that are required for a better understanding of the underlying principles and design techniques presented in later chapters. It is interesting to note that the same concepts used in reliability theory are also used by actuaries in calculating life insurance premiums, and that the human life span provides a useful analogy for component part reliability. Practical application is emphasized rather than rigorous theoretical exposition. This text embodies a preventive approach to enhancing reliability; it describes the overall approach to reliable design, including both theoretical and practical considerations.

Deliberate and positive measures must be taken during equipment and system design and development for two reasons: to enhance the inherent reliability by forcing the design to be iterated, and to minimize degradation by eliminating potential failures and manufacturing flaws prior to production and operational use.

Reliability Standards and Specifications

Many years ago Lord Kelvin stated, "When you can measure what you are speaking about and can express it in numbers, you know something about it. But when you cannot measure it, when you cannot express it in numbers, your knowledge of the subject is of a meager and unsatisfactory kind." Since we plan to measure reliability we need a standard or a unified series of standards against which we can compare our results. In the field of reliability there are only two such yardsticks presently available. These are the

International Electrotechnical Commission (IEC) specifications and the United States Military Standards (MIL-STDs).

The United States, as well as most other industrial nations, is signatory to the IEC documents. However, these documents are seldom referenced or utilized within the United States. This leaves the United States Military Standards as our primary reference documents. Many industrial and commercial manufacturers are offended by frequent reference to these military documents. However, it appears that we have little other choice, and must out of necessity reference some set of standards, if for no other reason than that of comparison.

A major advantage that the military specifications offer is that they document a selection of practical tools which can be used to enhance equipment reliability, be it military, industrial, or commercial equipment. A recent modification to the philosophy in the military standards is to allow, indeed to require, that reliability tasks be tailored to the unique aspects and characteristics of a given program, rather than requiring slavish adherence to the letter of the specification. This change has greatly enhanced the practicality and usefulness of the military standards in the field of reliability, and has enhanced the possibility of their use in nonmilitary applications.

The Bathtub Curve

Equipment reliability is frequently described by the familiar bathtub curve shown in Figure 1.1. The bathtub curve is actually a composite curve formed by the addition of three separate curves: the first curve or phase is that of infant mortality, the second curve or phase is of useful life, and the third is of wear-out (these three curves are analogous to the three major factors in human lifespan statistics). Each of these curves also relates to that phase during which it is the principle contributor to an individual or an equipment's hazard rate. Distinctly different phenomena dominate during each of these three phases of the equipment's life.

During the first phase, infant mortality, the weaker parts have not yet been removed from the total part population. As these weaker parts are identified and removed from the population, either by a formal screening program or through normal equipment operation, the instantaneous failure rate, or the hazard rate, decreases rapidly to a more or less residual level.

This residual value persists throughout the second and most important phase, the useful life period. During the useful life period, failures still occur occasionally but they are random in nature and are randomly distributed with respect to time. The statistical frequency of these failures can be predicted with fairly good accuracy, but the exact time and location of actual failures cannot be pinpointed and predicted.

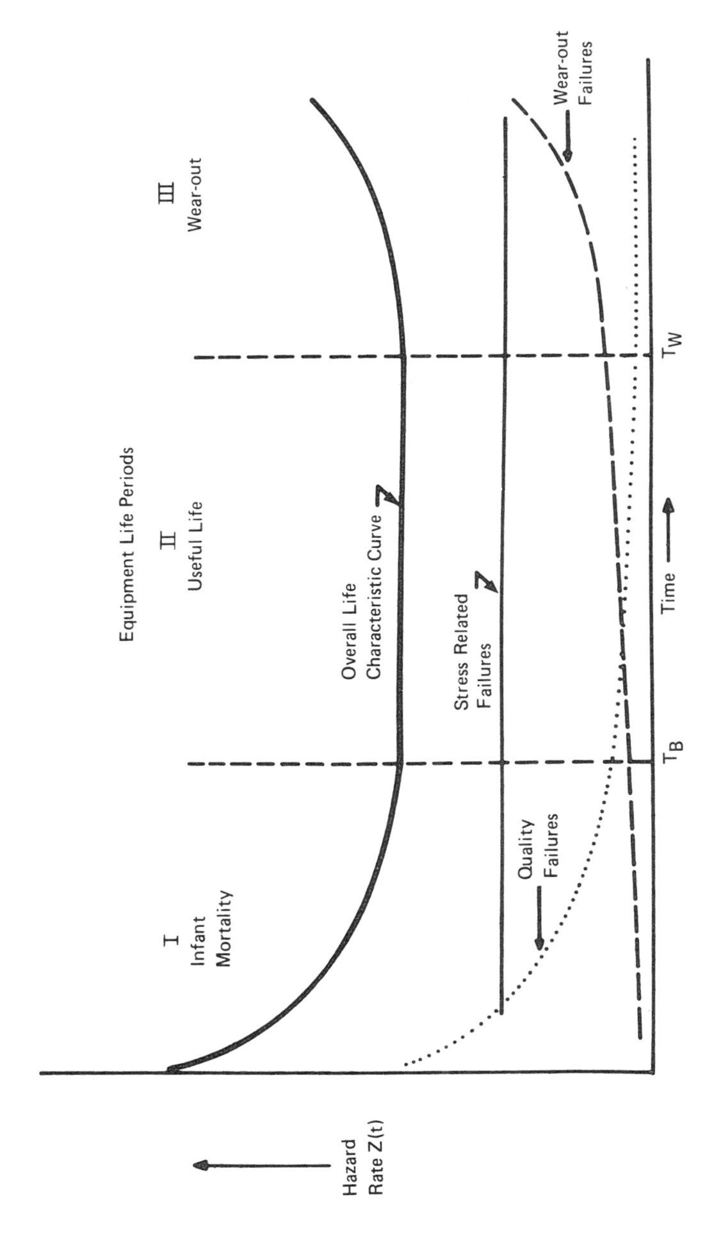

Figure 1.1 The bathtub curve. [From *Reliability Design Handbook* (RDH-376), copyright 1976 by IIT Research Institute, RADC/RAC, Griffiss Air Force Base, N.Y.]

The third and final phase of equipment life is wear-out. During this phase the accumulated damage due to the applied stresses begins to take its toil. The parts generally become weaker, more prone to failure, and thus they fail with increasing frequency. For most types of electronic parts the equipment itself becomes obsolete and is retired long before the parts enter the wear-out period. Thus for the vast majority of electronic parts (but unfortunately not for all) the wear-out period can be safely ignored.

Eliminating Part Failures

The elimination of part failures within the equipment is a major concern of design reliability engineering. As summarized in Table 1.1, design reliability is concerned with the elimination of four different types or classifications of part failures.

Quality Defects

The first type of part failure results from simple quality defects. These failures are the typical dead-on-arrival type. They are not stress or time dependent. They can usually be found and eliminated by a simple quality inspection or an electrical test to ascertain whether the parts are functioning properly and whether they are within the specification requirements at normal room temperature. Since this type of failure is normally handled by existing quality control functions quite adequately, it will not be addressed in this text.

Reliability Failures

The second type of part failure is reliability failure. Reliability failures are stress dependent, and they are not detected by normal quality control functions. Parts exhibiting this type of failure

Table 1.1 Four Types of Part Failures

Type	Dependent upon	Eliminated by
Quality	Unrelated to stress or time	Inspection
Reliability	Stress	Screening
Wear-out	Time	Qualification (life) testing or periodic replacement
Design	May be stress and/or time dependent	Proper application and derating

normally pass the initial inspection test, function properly, and are within the specification requirements at room temperature. Nevertheless they contain some inherent weakness or latent defect, and may fail when they are subjected to stresses that are within the normal equipment stress profile. These weak parts must be stressed during testing to precipitate failures.

Stress testing or screening is normally a two-step process. The part is usually stressed first to activate a failure mechanism and then tested to detect a failure mode. The development of an effective part screening program to eliminate this type of failure is the major thrust of Chapter 2.

Wear-out Failures

The third type of part failure, that of wear-out, is not normally experienced with electronic parts. However, there are some important exceptions which we will later address. The useful life, or the flat exponential failure-rate portion of the life curve of most electronic parts is usually much longer than the actual life of the equipment in which it is installed. Electronic equipment is normally replaced due to obsolescence rather than wear-out. Some important exceptions to this general principle will be identified in Chapter 2.

Wear-out is more characteristic of mechanical parts and equipment than of electronic parts and equipment. Nevertheless, it is good design reliability practice to perform an accelerated life test on samples of all critical electronic parts to qualify those parts for their intended usage and to assure that they do not contain significant wear-out mechanisms.

Design Failures

The fourth type of part failure is related to the design of the equipment. These failures are also stress and/or time dependent. They are usually the result of improper part application or overstressing of the parts. Design-related part failures are eliminated by proper part application and part derating. These topics are studied in detail in Chapter 3.

Reliability Engineering as a Design Discipline

Reliability engineering is the technical discipline of estimating, controlling, and managing the probability of failure in devices, equipment, and systems. In a sense, it is engineering in its most practical form since it consists of two fundamental aspects: paying attention to detail, and handling uncertainties. In reliability engineering a family of design procedures is used to achieve the desired reliability During product development a design is formulated to meet previously defined quantitative reliability requirements. The results of these activities provide inputs for all future actions. The importance of designing in the degree of reliability initially cannot be overemphasized, for once the design is approved, inherent reliability is fixed. Less-than-perfect compliance with the required actions from this point on may result in an achieved reliability level less than the inherent level.

There is a host of design principles and assessment tools of which the designer should be aware, and he should be utilizing these to achieve the required equipment and system reliability. These tools include:

Part selection, specification, and control (Chapter 2)
Part derating (Chapter 3)
ESD damage prevention (Chapter 4)
Circuit analysis (Chapter 5)
Sneak circuit analysis (Chapter 5)
Redundancy (Chapter 6)
Failure mode, effects and criticality analysis (Chapter 7)
Fault tree analysis (Chapter 8)
Reliability apportionment, prediction, and modeling (Chapter 9)
Environmental design (Chapter 10)
Human factors design (Chapter 11)
Reliability demonstration (Chapter 12)
Reliability growth management (Chapter 13)
Environmental stress screening (Chapter 14)
Design review (Chapter 15)

Each of the tools will be discussed in the appropriate chapter in terms of its role in the design of reliable equipment and systems.

Deterministic Versus Probabilistic Considerations

Modern engineering disciplines are based on applied mathematics. An engineer or scientist will observe a particular event, and then will formulate a hypothesis (or a conceptual model) which describes a relationship between the observed facts and the event being studied. In the physical sciences, these conceptual models are usually mathematical in nature. Mathematical models represent an efficient, shorthand method of describing an event and the more significant factors which may cause or affect the occurrence of that event. Such models are useful because they provide the theoretical foundation for the development of engineering disciplines and engineering design principles which can then be applied to either cause or prevent the occurrence of that event.

Mathematical models may be either deterministic or probabilistic in nature. Newton's second law of mechanics, force equals mass times acceleration (F = ma), is a deterministic model. There is nothing indefinite about this model. In contrast, a probabilistic model is one in which the results cannot be determined as exactly as in the deterministic model, but can only be described in terms of probability or of a probability distribution function. Modern atomic theory which defines the exact future location of an electron in terms of a probability function is an example of a probabilistic model. The use of probabilistic as opposed to deterministic models is becoming more common in the modern engineering solutions to problems, and is an important consideration in reliability engineering.

The discipline of reliability engineering is based upon the use of both deterministic and probabilistic, or stochastic, models. Some of the reasons supporting the use of probabilistic models are:

1. It would be extremely difficult to try to identify and quantify exactly all of the variables which contribute to the failure of all of the electronic parts in order to develop exact, deterministic failure models. We are dealing with uncertainty and measured values which cannot be stated with total certainty.
2. Probabilistic models, when applied to large samples, tend to smooth out individual variations. Thus the final average result is simple and accurate enough for engineering analysis and design.

Reliability parameters are defined in probabilistic terms, therefore, probabilistic parameters such as random variables, density functions, and distribution functions are utilized in the development of reliability theory and practice.

1.1 DEFINITIONS

Listings of appropriate reliability definitions can be found in MIL-STD-721C "Definitions of Terms for Reliability and Maintainability" or in IEC STD 271 and 271A "List of Basic Terms, Definitions, and Related Mathematics for Reliability."

For the convenience of the reader, a list of words and terms that are applicable when considering reliability follows.

Reliability: The probability that an item will perform its intended function for a specified time interval under stated conditions.

Apportionment: Assigning reliability requirements to individual items within a system to attain the specified system reliability.

Failure: The termination of the ability of an item to perform a required function within previously specified limits.

Secondary (dependent) failure: Failure of an item caused either directly or indirectly by the failure of another item.

Assessed reliability: The reliability of an item determined by a limiting value or values of the confidence interval associated with a stated confidence level, based on the same data as the observed reliability of nominally identical items.

Inherent reliability: A measure of reliability that includes only the effects of an item's design and its application, and assumes an ideal operation and support environment.

Failure rate: The total number of failures within a population, divided by the total number of life units expended by that population, during a particular measurement interval under stated conditions.

Mean-time-between-failures (MTBF): A basic measure of reliability for repairable items. The mean number of life units (for example, hours $\times 10^6$) during which the component performs to specification during a particular measurement interval under stated conditions.

Mean-time-to-failure (MTTF): A basic measure of reliability for nonrepairable items, and hence more applicable to component reliability than to system reliability. The total number of life units divided by the total number of failures for a population of components operating during a particular measurement interval under stated conditions.

Degradation: A gradual impairment in ability to perform.

Infant mortality: The initial phase of the lifetime of a population of a particular component when failures occur as a result of latent defects, manufacturing errors, and so on.

Burn-in (preconditioning): The operation of an item under stress to stabilize its characteristics and remove infant mortalities.

Useful life: The second phase of the lifetime of a population of a particular component when primarily random failures occur.

Wear-out: The third and final phase of the lifetime of a population of components when failures occur due to wear-out.

Derating: Limiting the stress applied to the parts to levels that are well within their specified or proven capabilities, in order to enhance their reliability.

Random failure: Failure whose occurrence is predictable only in a probabilistic or statistical sense. This applied to all distributions.

Failure mode and effects analysis (FMEA): A procedure by which each potential failure mode in a system is analyzed to determine the results or effects thereof on the system and to classify each potential failure mode according to its severity.

Failure analysis: Subsequent to a failure, the logical systematic examination of an item, its construction, application, and documentation to identify the failure mode, and to determine the failure mechanism and its basic cause.

Failure mechanism: The physical, electrical, thermal, or any other process which results in failure.

Failure mode: The consequence of the mechanism through which the failure occurs, that is short, open, fracture, excessive wear, and so on.

Single-point failure: The failure of an item which results in failure of the system and is not compensated for by redundancy or by an alternate operational procedure.

Failure effect: The consequence(s) a failure mode has on the operation, function, or status of an item. Failure effects may be classified as local effects, next-higher effect and end-item effect.

Latent defect: An inherent weakness that has a high probability of resulting in an early life failure.

Critically: A relative measure of the consequence of a failure mode and its frequency of occurrence.

Demonstrated: That which has been measured by the use of objective evidence gathered under specified conditions.

Reliability growth: The improvement in a reliability parameter caused by the successful correction of deficiencies in item design or manufacture.

Corrective action: A documented design process, procedure, or materials change implemented and validated to correct the cause of the failure or design deficiency.

Redundancy: The existence of more than one means for accomplishing a given function. Each means of accomplishing the function need not necessarily be identical.

Accelerated test: A test in which the applied stress level is chosen to exceed that stated in the reference conditions in order to shorten the time required to observe the stress response in a given interval. To be valid, an accelerated test should not alter the basic modes and mechanisms of failure or their relative prevalence.

Screening: A process for inspecting items to remove those that are unsatisfactory or those likely to exhibit early failure. Inspection includes visual examination, physical dimension measurement, and functional performance measurement under specified environmental conditions.

Environmental stress screening (ESS): A series of tests conducted under environmental stresses to disclose latent part and workmanship defects in order to execute corrective action.

Environment: The aggregate of all external conditions either natural, manmade, or self-induced (such as temperature, humidity, radiation, magnetic and electric fields, shock, vibration, and so on) that influences the form, performance, reliability, or survival of an item.

Maintenance: All actions necessary for retaining an item in or restoring it to a specified condition.

Maintainability: The measure of the ability of an item to be retained in or restored to a specified condition when maintenance is performed by personnel having specified skill levels, who use prescribed procedures and resources at each level of maintenance and repair.

Mean-time-to-repair (MTTR): A basic measure of maintainability: the total number of life units of an item divided by the total number of failures within an item repaired at that level, during a particular interval under stated conditions.

Fault tree analysis: A deductive type of failure analysis that focuses on one particular undesired event at a time, and then provides a method for determining the possible causes of that event.

1.2 STATISTICAL DISTRIBUTIONS

Definitions of Statistical Terms

A variable defined by a probabilistic law is called a random variable, Variables may be either discrete or continuous. The properties of a random variable are specified by the set of possible values it may take, together with associated probabilities of occurrence of these values. The graph of probability against the random variable is called a probability distribution, and the mathematical expression for that distribution is called the probability density function (pdf). Since the relative frequency of occurrence of each value of the random variable is a measure of its probability, the vertical axis usually defines frequency, while the horizontal axis defines the random variable.

The pdf's in this chapter usually have time (or part hours) as the random variable. Some general terms used to quantify a distribution given by f(x), are:

The median, or the middle-ordered value
The mean, or the arithmetic average of x
The mode, or the most common value of x
The variance, or the spread of a set of values
The standard deviation, or the square root of the variance
The skew of a distribution, or a measure of its asymmetry
The kurtosis, or a measure of the length of the tails of a distribution

Cumulative distribution functions, F(t), are also useful. These are simply functions giving cumulative probabilities at the same value of the random variable. They are derived by integrating the pdf over some range of the random variable. Thus, they give the probability that the random variable is in some specified range. Statistical tables give values of the cumulative distribution over specified ranges, typically for the upper tail.

Another important function is the hazard rate function, h(t), or the instantaneous failure rate. The hazard rate represents the probability that a component will fail in the time interval $t + \Delta t$ given that it has survived to time t.

A simple explanation of hazard rate and its relation to failure rate may be shown by analogy to an automobile trip. If a trip of 220 miles is completed in 4 hours, the average rate will be 55 mph although the auto was probably driven faster at some times and slower at others. The rate at any given instant could be determined by radar or by consulting the speedometer. The average speed, 55 mph, is analogous to the failure rate, and the speed at any instant is analogous to the hazard rate.

Reliability studies deal with both discrete and continuous random variables. The number of failures in a given interval of time is an example of a discrete variable. The time from part installation to part failure and the time between successive equipment failures are examples of continuous random variables. The distinction between discrete and continuous variables (or functions) depends upon how the problem is treated and not necessarily on the basic physical or chemical processes involved.

Statistical Distributions Used in Reliability Models

Many standard statistical distributions may be used to model various reliability parameters. However, a relatively small number of statistical distributions satisfies most reliability needs. The particular distribution used depends upon the nature of the data. The most appropriate model for a particular application may be decided either empirically or theoretically or by a combination of both approaches.

A distribution may be chosen empirically by fitting to data. Simple graphical methods have been developed for fitting distributions using probability paper. The method is particularly useful in reliability work using the Weibull distribution because of the Weibull distribution's flexibility.

Goodness-of-fit tests such as Chi-square or Kolmogorov-Smirnov may also be used to decide whether a particular distribution fits a particular set of data well enough. These tests give an unbiased, objective assessment of the goodness-of-fit. Full descriptions and examples of their use are given in any good introductory statistical text.

Following is a short summary of some of the distributions commonly used in reliability analysis, the criteria for their use, and some examples of their application. Figures 1.2 and 1.3 summarize the shape of common failure density, reliability, and hazard rate functions for each of the distributions.

Continuous Distributions

Normal (or Gaussian) distribution There are two principal reliability applications of the normal distribution. The first deals with the

Distribution & Parameters	Probability Density function, f(t)	Reliability Function, R(t) = 1 - F(t)	Hazard function (instantaneous failure rate), $h(t) = \frac{f(t)}{R(t)}$ (general expression)
NORMAL μ = Mean σ = Standard deviation	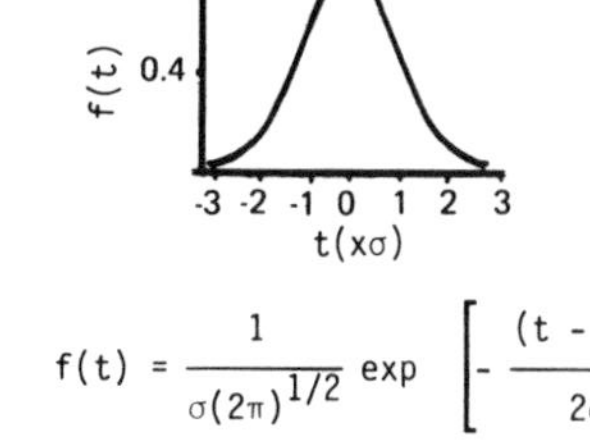 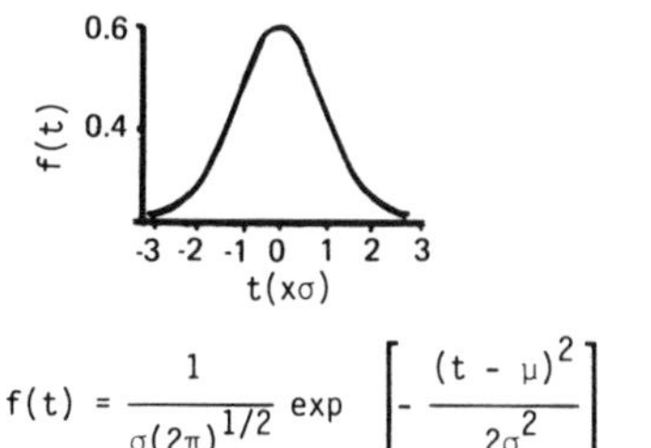 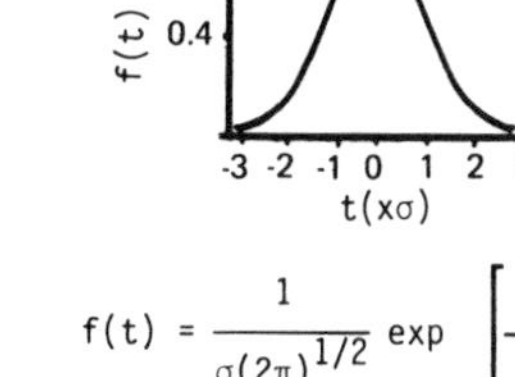		
	$f(t) = \frac{1}{\sigma(2\pi)^{1/2}} \exp\left[-\frac{(t-\mu)^2}{2\sigma^2}\right]$	$R(t) = \int_0^\infty f(t)\,dr$	$h(t) = \frac{f(t)}{R(t)}$ (general expression)
LOGNORMAL μ = Mean σ = Standard deviation	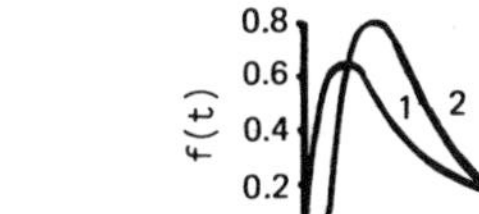 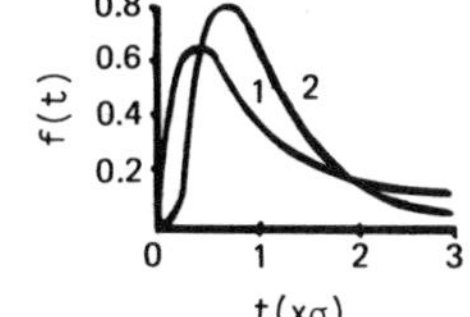 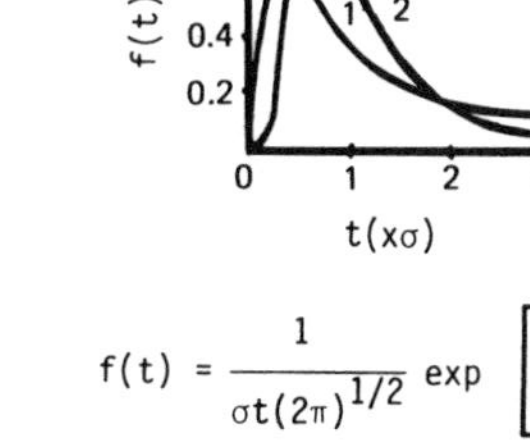		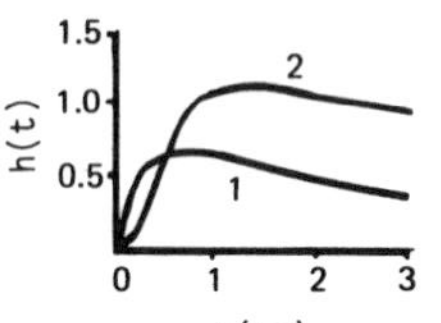 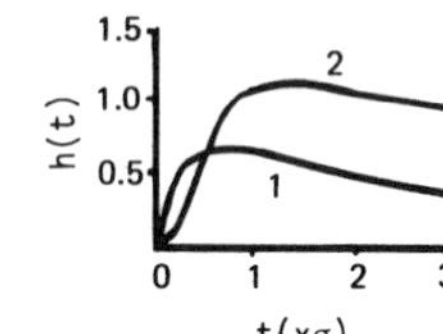 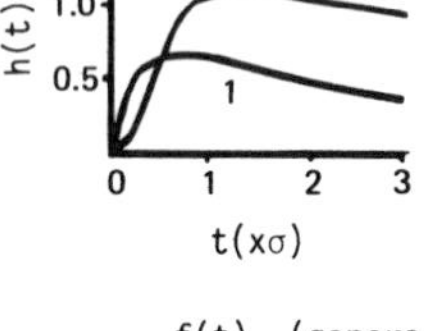 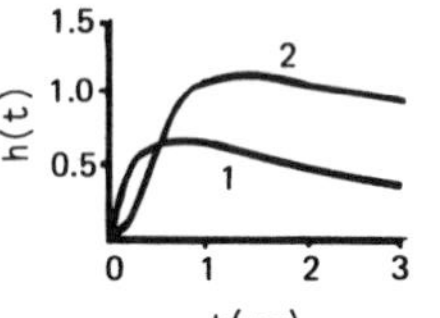
	$f(t) = \frac{1}{\sigma t(2\pi)^{1/2}} \exp\left[-\frac{\ln(t-\mu)^2}{2\sigma^2}\right]$	$R(t) = \int_0^\infty f(t)\,dr$	$h(t) = \frac{f(t)}{R(t)}$ (general expression)

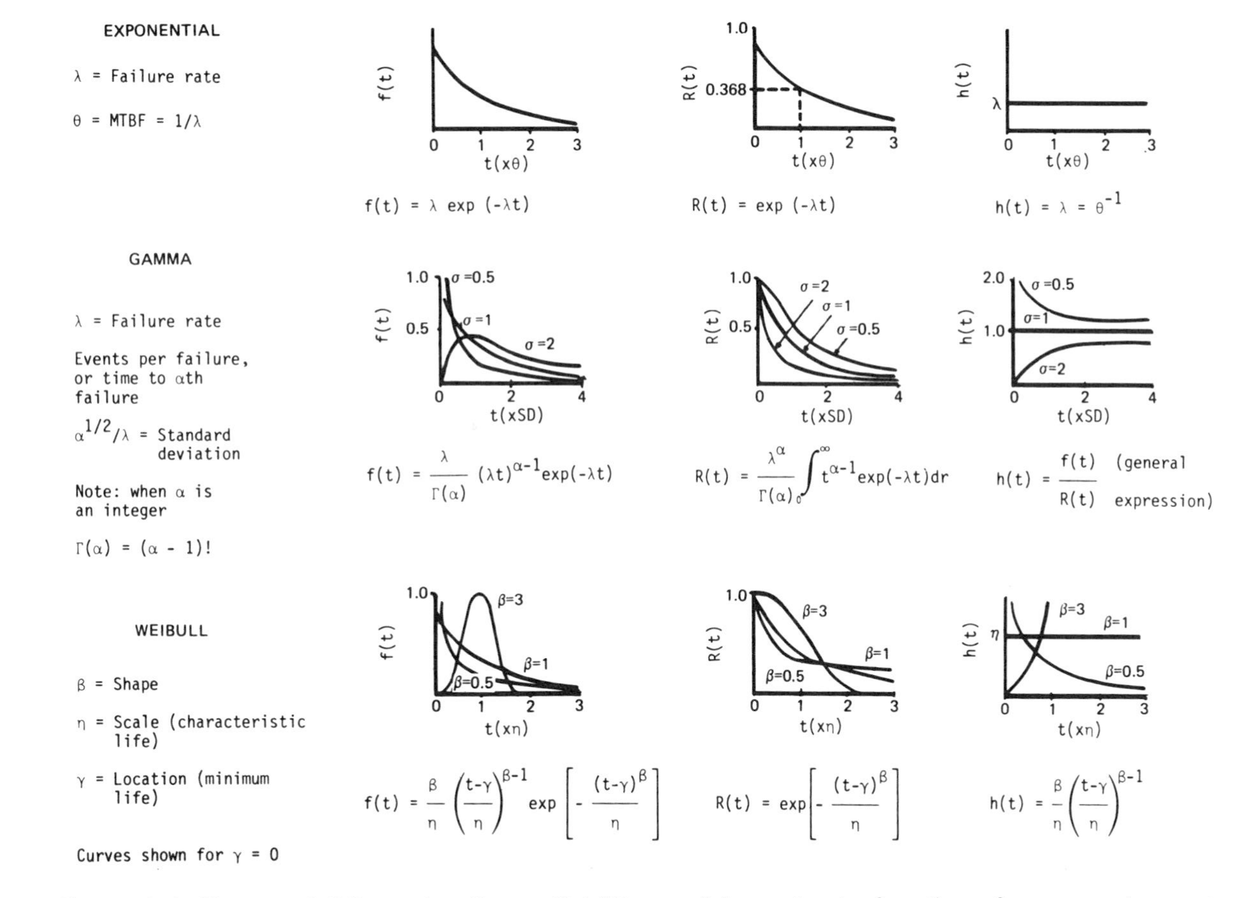

Figure 1.2 Shapes of failure density, reliability, and hazard rate functions for commonly used continuous distributions. (From MIL-HDBK-338, *Electronic Reliability Design Handbook*, Naval Publications and Forms Center, Philadelphia.)

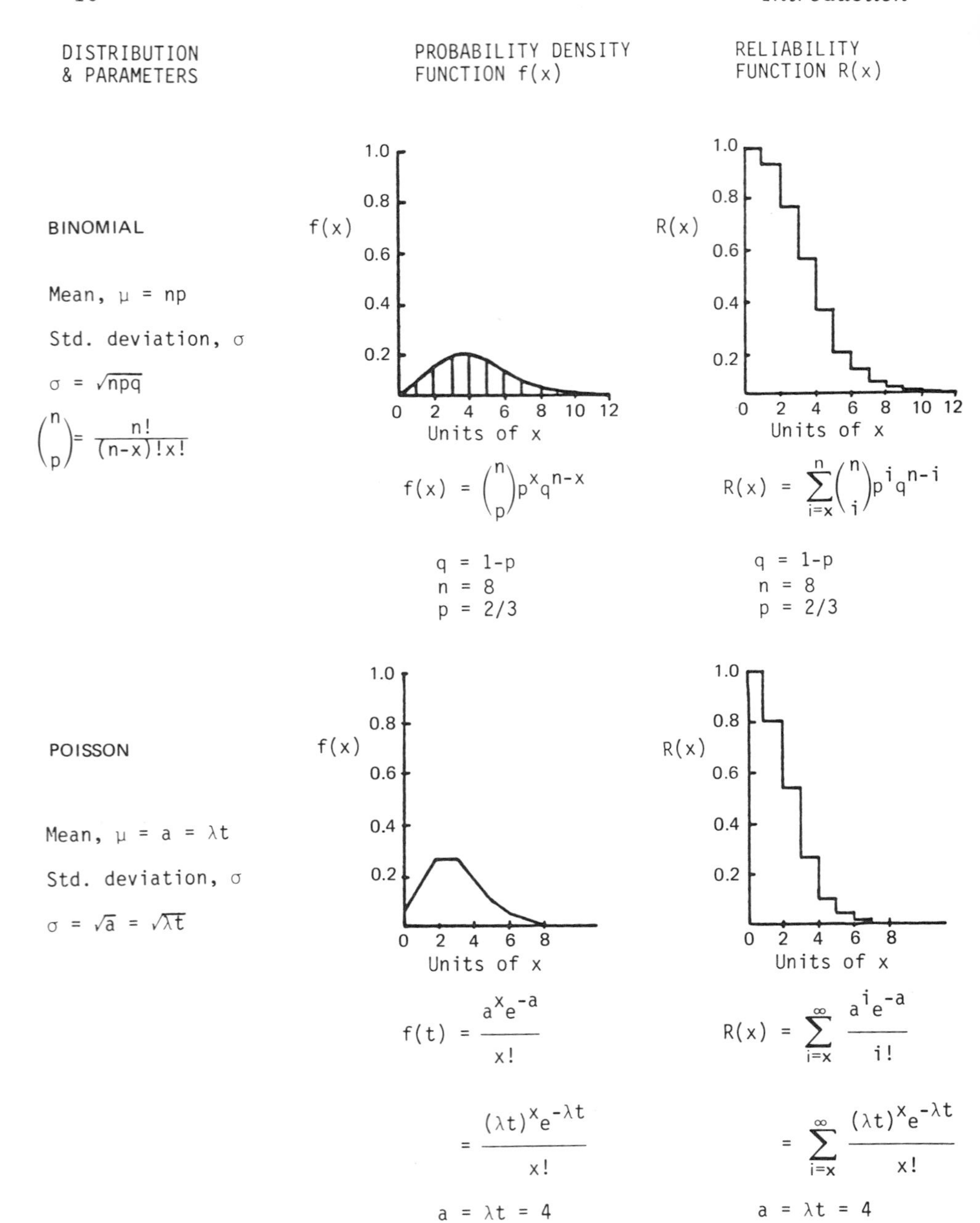

Figure 1.3 Commonly used discrete distributions. (From MIL-HDBK-338, *Electronic Reliability Design Handbook*, Naval Publications and Forms Center, Philadelphia.)

analysis of items which exhibit failure due to wear, such as mechanical devices. Frequently the wear-out failure distribution is sufficiently close to normal that the use of this distribution for predicting or assessing reliability is valid.

The second application deals with the analysis of manufactured items and their ability to meet specifications. No two parts made to the same specification are exactly alike. The variability of parts leads to a variability in hardward incorporating those parts. The designer must take this part variability into account, otherwise the equipment may not meet the specification requirements due to the combined effects of part variability.

Use of the normal distribution in this application is based upon the central limit theorem. It states that the sum of a large number of identically distributed random variables, each with a finite mean, μ, and a standard deviation, β, is normally distributed. Thus, the variations in parameters of electronic component parts due to manufacturing are considered to be normally distributed.

Log-normal distribution If the natural logarithm of a function is found to be distributed normally, then the function is said to be log-normal. As shown in Figure 1.2, μ defines the mean of the distribution and σ defines its standard deviation. A third parameter, t, representing the minimum life, may also be incorporated in the log-normal distribution.

Physical examples of the log-normal distribution are the fatigue life of certain types of mechanical components and incandescent light bulbs. Light bulbs eventually suffer filament deterioration and empirically follow a log-normal distribution. Semiconductor failures may also frequently follow a log-normal distribution.

Exponential distribution This is the most important distribution in reliability work. It is used almost exclusively for reliability prediction of electronic equipment. If failures are random in time, they obey a particular law called the Poisson process, and the hazard rate is constant.

Note the simplicity of the hazard rate model (see Figure 1.2). It is completely defined by a single parameter, λ. This makes estimation from data, and mathematical manipulation relatively easy. Consequently the exponential distribution has become very popular. However, the validity of this assumption should be checked in ever case.

Some major advantages of the exponential distribution are that:

There is single, easily estimated parameter λ.
It is mathematically very tractable.
It has wide applicability.
It is additive, for example, the sum of a number of independent exponentially distributed variables is exponentially distributed.

Specific applications include:

1. Items whose failure rate does not change significantly with age.
2. Complex repairable equipment without excessive amounts of redundancy.
3. Equipments from which the infant mortality or early failures have been eliminated. This is done by "burning-in" the equipment for some time period.

The failure density function for the exponential distribution is:

$$f(t) = \lambda e^{-\lambda t} \quad \text{for } t > 0$$

The reliability function is:

$$R(t) = e^{-\lambda t}$$

where:

λ = the hazard (failure) rate

$\Theta = 1/\lambda$ = mean life = MTBF (for repairable equipment)

The hazard rate for the exponential distribution is constant and equal to λ. This property is the basis for MIL-HDBK-217's comprehensive tabulation of failure rates for various components. They are based on the assumption of a constant failure rate. The underlying assumption for the use of the exponential distribution is randomness. If a component fails purely at random and is not subject to burn-in or wear-out (at least for a specified portion of its lifetime), then the exponential model is justified.

The exponential distribution may also be recognized as a special case of the gamma distribution.

Gamma distribution The gamma distribution is used in reliability analysis for those cases where partial failures can exist, that is, when a given number of partial failures must occur before an item fails for example, redundant systems. Another example could be the time-to-the-second-failure when the time-to-failure is exponentially distributed. If a component fails as a result of some combination or sequence of mechanisms, it may also follow a gamma distribution. The failure density function is as shown in Figure 1.2. Note that the setting $\alpha\alpha = 1$ the time-to-first-failure is considered, and the exponential distribution results.

The gamma distribution can be used to describe either an increasing or a decreasing hazard (failure) rate. When $\alpha > 1$, h(t) increases; when $\alpha < 1$, h(t) decreases. This is also illustrated in Figure 1.2.

Weibull distribution The Weibull distribution, named after the Swedish investigator of metal fatigue problems, W. Weibull, is especially useful in reliability work. It is a very general distribution which, by adjustment of the distribution parameters, can be made to model a wide range of life distribution characteristics for different classes of items. One of the versions of the failure density function is:

$$f(t) = \frac{\beta}{\eta}\left(\frac{t-\gamma}{\eta}\right)^{\beta-1} \exp\left[-\left(\frac{t-\gamma}{\eta}\right)^{\beta}\right]$$

where:

β = shape parameter

η = scale parameter or characteristic life (at which 63.2% of the population will have failed)

γ = minimum life

In most practical reliability situations, $\gamma = 0$ (for example, where failures are assumed to start at $t = 0$). Then the failure density function becomes:

$$f(t) = \frac{\beta}{\eta}\left(\frac{t}{\eta}\right)^{\beta-1} \exp\left[-\left(\frac{t}{\eta}\right)^{\beta}\right]$$

and the reliability and hazard functions become:

$$R(t) = \exp\left[-\left(\frac{t}{\eta}\right)^{\beta}\right]$$

$$h(t) = \left(\frac{\beta}{\eta}\right)\left(\frac{t}{\eta}\right)^{\beta-1}$$

Depending upon the value of β, the Weibull distribution function can also take the form of the following distributions:

Beta value	Distribution type	Hazard rate
<1	Gamma	Decreasing
1	Exponential	Constant
2	Log-normal	Increasing/decreasing
3.5	Normal (approximately)	Increasing

Thus, in addition to its being a distribution in its own right, the Weibull distribution may also be used to identify other distributions from life data. Graphical methods are frequently used to analyze Weibull failure data.

The Weibull probability plot shown in Figure 1.4 is a useful tool in identifying particular distributions. This is a special graph paper whose axes are defined on transformed scales such that the cumulative Weibull distribution function appears as a straight line. Thus, for any set of straight line data plotted on Weibull probability paper, the values of η, β, γ, and the mean life may be read directly from the graph paper. In practice the process is not quite so clean since data rarely fall exactly on a straight line. Statistical tests of linearity are advisable as false conclusions are otherwise easily drawn. Weibull probability paper is particularly useful as an exploratory technique to help in understanding data.

As shown in Figure 1.5, by the appropriate choice of the value of β, the Weibull distribution can be used to describe each of the three regions of the bathtub curve, infant mortality, useful life, and wear-out referred to in the previous portion of this chapter. Extreme value distributions may also be defined by the Weibull distribution, by using a logarithmic transformation on the data.

Discrete Distributions

Binomial distribution The binomial distribution is very useful in both reliability and quality assurance, it is used when there are only two possible outcomes, such as success or failure, and probability remains the same for all trials. The probability density function (pdf) of the binomial distribution and the cumulative distribution function (cdf) are shown in Figure 1.3.

The probability density function $f(x)$ is the probability of obtaining exactly x good items and $n - x$ bad items in a sample of n items where p is the probability of obtaining a good item (success) and q (or $1 - p$) is the probability of obtaining a bad item (failure).

The cumulative distribution function is the probability of obtaining r or fewer successes in n trials.

Computations involving the binomial distribution become rather unwieldy for even small sample sizes. However, complete tables of the binomial pdf and cdf are available in many statistics texts.

Poisson distribution This distribution is used quite frequently in reliability analysis. It can be considered an extension of the binomial distribution when n is infinite. In fact, it is used to approximate the binomial distribution when $n \geq 20$ and $p \leq 0.05$.

If events are Poisson distributed, they occur at a constant average rate, and the number of events occurring in any time interval are independent of the number of events occurring in any other time

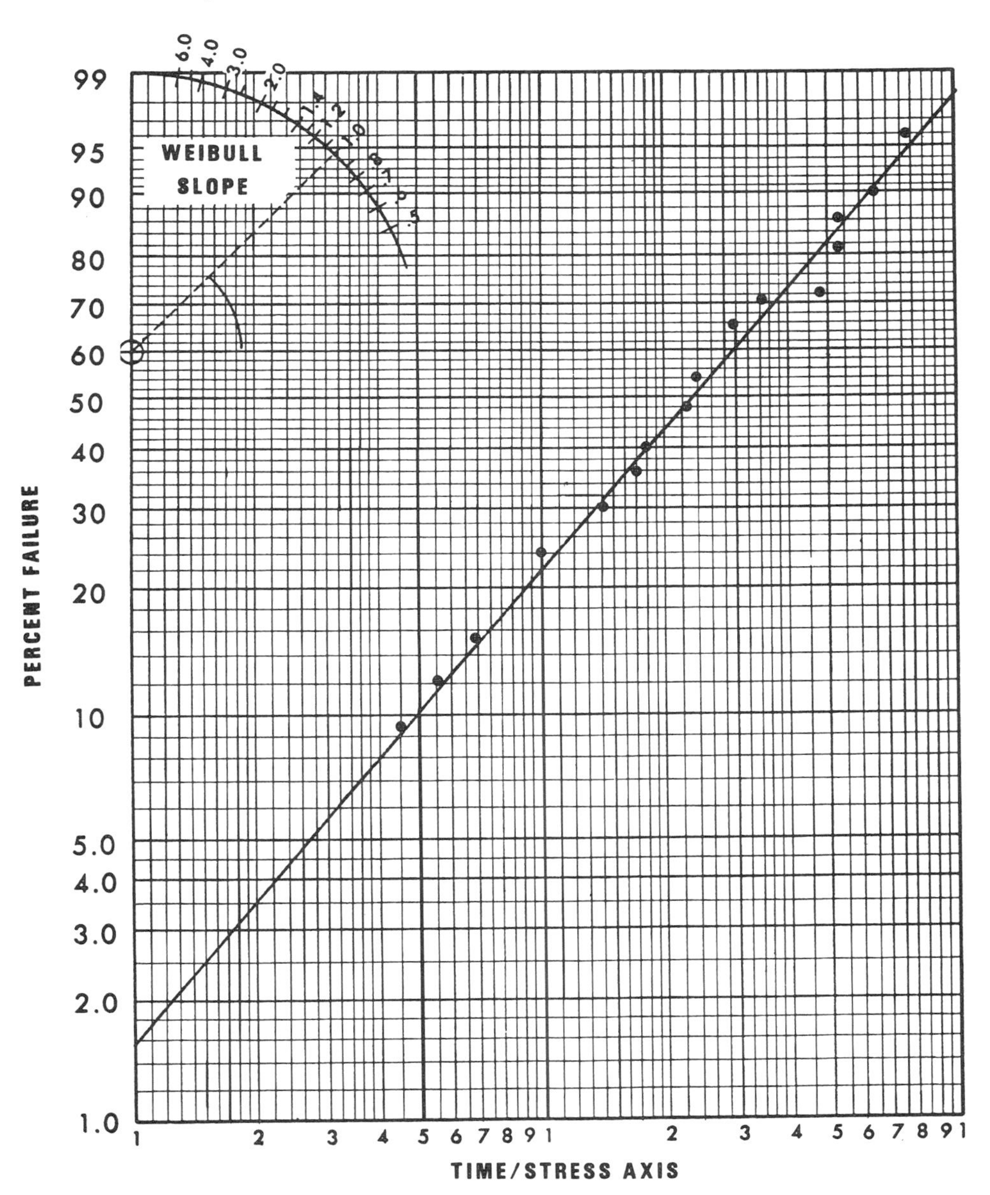

Figure 1.4 Weibull probability plot.

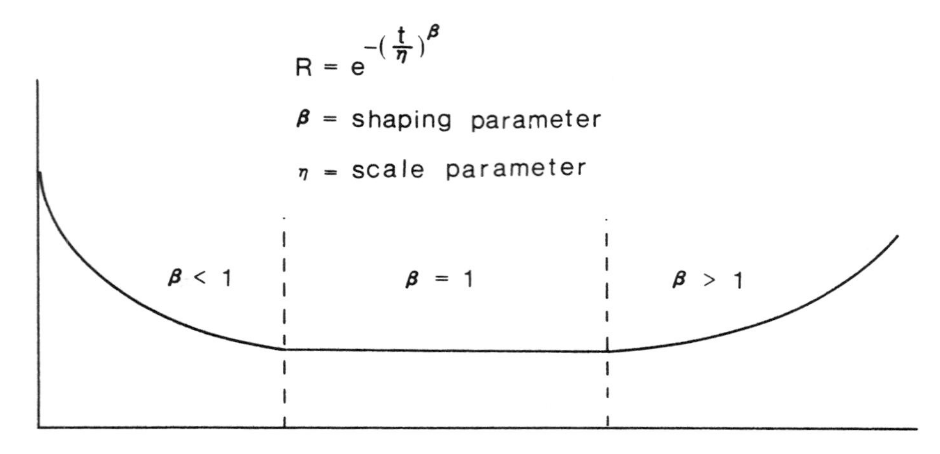

Figure 1.5 Weibull distribution and the bathtub curve.

interval. For example, the number of failures in a given time would be given by:

$$f(x) = \frac{a^x e^{-a}}{x!}$$

where x is the number of failures and a is the expected number of failures. For the purpose of reliability analysis, this becomes:

$$f(x; \lambda, t) = \frac{(\lambda t)^x e^{-\lambda t}}{x!}$$

where:

λ = failure rate

t = length of time being considered

x = number of failures

The reliability function, R(t), or the probability of zero failures in time t is given by:

$$R(t) = \frac{(\lambda t)^0 e^{-\lambda t}}{0!} = e^{-\lambda t}$$

or simply the exponential distribution.

Bayesian statistics in reliability analysis Bayesian statistics, developed by the eighteenth century English clergyman and mathematician, are being used increasingly in reliability analysis. The advantage of Bayesian statistics is that it allows prior information (for example, predictions, test results, engineering judgment, and so on) to be combined with more recent information, such as test or field data, in order to arrive at a prediction or assessment of reliability based upon a combination of all available data. It also permits the reliability prediction or assessment to be updated continually as more and more test data are accumulated.

The Bayesian approach is intuitively appealing to design engineers because it permits them to use engineering judgment, based upon prior experience with similar equipment designs, to estimate the reliability of new systems where only limited field data exists. Bayes' theorem can, for example, combine the data from a reliability test with information available prior to the start of the test, such as component and subassembly test data, data from previous tests on the product, and even intuitive data based upon experience.

The basic difference between Bayesian and non-Bayesian (classical) approaches is that the former uses both current and prior data, whereas the latter uses only current data.

A primary limitation of the Bayesian approach is that one must be extremely careful in choosing the prior probabilities based upon past experience or judgment. If these are capriciously or arbitrarily chosen, the end results of the Bayesian analysis may be inaccurate and misleading. Thus, the key to the successful use of the Bayesian method is in the appropriate choice of prior probability distributions.

Bayes' analysis begins by assigning an initial reliability based on whatever evidence is currently available. The initial prediction may be based solely on engineering judgment, or it may be based on data from other similar types of items. Then, when additional test data is obtained, the initial reliability estimates are revised to reflect this data using Bayes' theorem. The initial reliability estimates are known as *prior* estimates, in that they are assigned before the acquisition of the additional data. The reliability estimates resulting from the revision process are known as *posterior* estimates.

From basic probability theory, Bayes' theorem is given by:

$$\Pr[A/B] = \Pr[A] \frac{\Pr[B/A]}{\Pr[B]}$$

In the specific framework and context of reliability, the various terms in the equation may be notated and defined as follows:

A: A hypothesis or statement of belief. ("The reliability of this component is 0.90.")

B: A piece of evidence, such as a reliability test result that has bearing upon the truth or credibility of the hypothesis. ("The component failed on a single mission trial.")

Pr[A]: The prior probability: the probability assigned to the hypothesis A before evidence B becomes available ["We believe, based on engineering experience, that there is a 50/50 chance that the reliability of this component is about 0.90, as opposed to something drastically lower, for example, Pr(A) = 0.5."]

Pr[B/A]: The likelihood: the probability of the evidence assuming the truth of the hypothesis. ("The probability of the observed failure, given that the true component reliability is indeed 0.90, is obviously 0.10.")

Pr[B]: The probability of the evidence B, evaluated over the entire weighted ensemble of hypothesis A.

Pr[A/B]: The probability of the evidence B, evaluated over the entire weighted ensemble of hypothesis A.

The posterior probability is the end result of the application of Bayes' equation. The reader is referred to standard statistical texts for examples of the use of Bayes' theorem.

Part II
Detail Part Considerations

2

Part Selection, Specification, and Control

2.0 INTRODUCTION

A diversified complement of electronic parts is available as building blocks from which modern electronic systems are fashioned, and, as such, these parts greatly impact hardware reliability. Since the reliability of the end item is dependent upon these building blocks, the importance of selecting and effectively utilizing these parts cannot be overemphasized.

The task of selecting, specifying, and controlling the parts used in complex electronic systems is a major engineering task. Part selection and control is a multidiscipline task involving the best efforts of component engineers, failure analysts, and reliability engineers as well as design engineers. Thus, controls, guidelines, and requirements must be formulated, reviewed, and implemented during the development effort.

Selection and Control Elements

An effective part selection and control process consists of:

Evaluation of the design of the part
Study of the part's reliability history
Analysis of the part construction
Possibly an FMEA of the part
Study of the cost-effectiveness of the use of this specific part
Study of the vendor's production and quality assurance capability

The part control effort will usually include the development of procurement specifications which reflect a balance among design requirements, quality assurance (QA) requirements, and reliability needs consistent with the reliability apportionment (see Chapter 9.2), and the vendor's capabilities. Procurement specification documents should generally include:

Qualification testing (where necessary)
Quality assurance provisions
Lot acceptance testing

An effective parts control program may also involve establishing a vendor control program, audits of the vendor processes, and establishment of source inspection (where applicable), plus the preparation of the associated documentation.

Part control activities comprise a large segment of the total effort for part selection, application, and procurement. This effort encompasses tasks for standardization, approval, qualification, and specification of parts capable of meeting the performance, reliability, and other requirements of the evolving design.

The reliability of the equipment is a function of the reliability of each of its constituent elements (a chain is no stronger than its weakest line). Therefore, we must assure that each element or piece part is properly selected for its intended function, that its reliability parameters are properly specified, and that its procurement is adequately controlled. In order to do this properly we should be able to anticipate how the part is likely to fail. We must understand the basic failure mechanisms and then take steps to guard against these failure mechanisms adequately. This topic will be addressed in general throughout this chapter.

Custom Device and VLSI Microcircuit Considerations

It was once thought that any process change which increased yield of a semiconductor process would also increase device reliability. Serious doubts exist today regarding this premise. High power magnification optical inspection (for a long time a keystone of microcircuit quality assurance programs) is rapidly becoming ineffective. Device geometries are approaching the limit resolvable by optical light. Also, the time required to inspect any given area is directly proportional to the magnification required to resolve critical defects (low power magnification optical inspection of the bonds and the wire dress is still a viable technique and it is not affected by this consideration).

Custom products have historically been among the most difficult products to control because of the high priority for a short design cycle, the relatively low volume, and the often short product life. The design of a high-volume standard product can support experiments with several alternative designs before a final decision is made. Development of a manufacturing process specifically for standard products is not unusual. Also, standard products are able to support engineering studies to refine the process into an easily controlled, well-understood condition.

Custom products rarely allow these luxuries. The development budget is limited, and time is critical. The volume cannot support pulling statistically significant samples for experiments, and the production run of a particular product may be over before the results of many classical engineering studies are available. A custom product must generally fit an existing process and thus must rely on generic quality controls.

The problem is further complicated if the product is designed at one facility and then fabricated at a "silicon foundry." The problems introduced by this arrangement include the fact that the designers are not as likely to be intimately familiar with the foundry's processes and the fact that the foundry will refine its process based on feedback from the customer that protests the loudest. Parameters such as etch temperature or metallization alloy may have a positive effect on one design but may seriously degrade the reliability of another.

Critical Parts

The parts control effort also includes identifying all "critical parts" including equipments or components. Items may be considered critical from any one or more of the following standpoints:

1. Parts which are mission and/or safety sensitive, that is, those parts which are major contributors to the total equipment or system failure rate
2. Parts which are reliability sensitive (as determined from earlier reliability studies, apportionments, and so on)
3. Parts which have a limited life or are required to operate under highly stressed conditions
4. High-cost items
5. Parts which have long procurement lead times
6. Parts which require formal qualification testing
7. Sole source items

A specific type of device which may well meet most or all of the above criteria for criticality is the custom LSI or VLSI microcircuit. MIL-STD-454, "Standard General Requirements for Electronic Equipments," requirement 64 "Microelectronics," defines all custom microcircuits as critical parts. The criticality of such items is not related just to reliability but includes cost and availability considerations as well.

Table 2.1 Parts Selection and Control Checklist

1. Determine the type of part needed to perform the required function and define the environment in which it must operate.
2. Determine the criticality of the part.

   Does this part perform a critical function, i.e., one which is safety-related or mission-critical?
   Does the part have a limited life?
   Does the part have a long procurement lead time?
   Is the part reliability sensitive, i.e., is it a major contributor to the total failure rate?
   Is the part a high-cost item?
3. Determine the availability of the part.

   Does a Preferred Parts List exist and is this part on the list?
   Is the part available from a qualified vendor or does it require formal qualification testing?
   What is the part's normal delivery cycle?
   Will this part continue to be available throughout the life of the equipment?
   Does an acceptable specification procurement document exist for this part?
   Is this part available from multiple sources?
4. Estimate the expected stresses on the part in its circuit application.
5. Determine the reliability level required for the part in this specific application.
6. Select appropriate screen and burn-in methods for reducing the part's failure rate (as required).
7. Prepare an accurate and explicit part procurement specification, where necessary. This should include specific screening provisions to assure adequate part reliability.
8. Determine the actual stresses on the part in its intended circuit application. Perform a detail stress failure-rate calculation per MIL-HDBK-217.
9. Employ appropriate derating factors consistent with reliability prediction studies.

Source: *Reliability Design Handbook* (RDH-376), copyright 1976 by IIT Research Institute, RADC/RAC, Griffiss Air Force Base, N.Y.

Custom LSI or VLSI devices may be divided into three convenient categories:

1. Unique new devices that have not previously been manufactured
2. Devices that have been qualified through previous usage
3. Devices that are customized at the final metallization stage

The unique new devices usually create the most unpredictable cost and schedule problems because they require simultaneous device and system development, thus compounding potential problems. The second category of devices is not usually a risk, unless the sole source vendor should "lose the recipe" or decide to discontinue the production line without adequate warning. Then serious problems can develop. The last category of devices is considered to carry the least risk. Nevertheless a cautious view should still be taken regarding the impact of these items on cost and schedule.

A critical part control plan should include controls for: any special handling that may be required; the identification of critical item characteristics to be inspected or measured during incoming inspection; material review procedures; material traceability criteria; an early procurement schedule; periodic audits and part growth testing if necessary. All items considered safety-critical must be specifically identified. Detailed documentation should then be prepared that describes the procedures, tests, test results, and the efforts to reduce the degree of criticality of each critical item.

A summary of some general ground rules for the selection and control of parts is shown in the checklist in Table 2.1. Key items in the checklist not covered in this chapter will be discussed in more detail in Chapters 3 and 4.

2.1 PART QUALITY GRADES

Component parts may frequently be procured to a variety of different quality grades. Aluminum electrolytic capacitors, for example, are available in a series of different commercial grades: (a) standard commercial grade, (b) computer grade, (c) long-life grade, and (d) long-life computer grade. In addition to the standard commercial grade, discrete transistors and diodes may be procured to military grades of JAN, JAN TX, JAN TXV, or JAN S. Microcircuits may be procured as JAN S or JAN B in addition to the standard commercial grade. There are in essence two commercial microcircuit grades, hermetically sealed or epoxy-encapsulated packages.

These higher quality grades are normally controlled by military specifications such as MIL-M-38510, "General Specification for Microcircuits," and MIL-S-19500, "General Specification for Semiconductor

Devices," for discrete semiconductors. These specifications establish specific requirements for the electrical performance, design, quality, and reliability assurance. They also define the steps necessary to establish vendor certification and vendor qualification.

This chapter dwells heavily on microcircuit examples. This is not intended to imply a lack of concern regarding other types of parts. The author has simply chosen to concentrate on microcircuits, as they appear to offer the most comprehensive, readily recognized example since they, as a group, best represent the leading edge of the state of the art. Most of the concepts and techniques illustrated with microcircuits can be readily adapted to the vast majority of other part types as well.

JAN Microcircuits

The most reliable, highest quality grade of microcircuits available today are JAN S and JAN B microcircuits produced by a Qualified Products List (QPL) Supplier.

The military specification establishes two quality grades for monolithic, multichip and hybrid microcircuits: MIL-M-38510 JAN class S (for space), the highest quality level and class B, the standard military quality level.

Electrical performance tests for the various types of JAN microcircuits are specified in the detailed mircocircuit military specification, and are performed in accordance with the applicable test methods defined in MIL-STD-883, "Test Methods and Procedures for Microelectronics."

Only microcircuits qualified and procured in full accordance with MIL-M-38510 may be marked with the "JAN" or "J" designation. To acquire and maintain listing on the QPL for either classes S or B, MIL-M-38510 device manufacturers must meet rigid qualification requirements. Qualification requires first manufacturer certification (including a government-approved Product Assurance Program Plan) and production-line certification. The second step is formal part qualification testing, and finally quality conformance inspection testing of each lot in accordance with MIL-STD-883, Method 5005, "Qualification and Quality Conformance Procedures" (for monolithic), or Method 5008, "Test Methods for Hybrid and Multichip Microcircuits" (for hybrid microcircuits).

MIL-M-38510 class S and class B microcircuits also require 100% screening of all parts in full accordance with MIL-STD-883, Method 5004, "Screening Procedures" (for monolithic) or Method 5008 (for hybrid devices).

MIL-STD-883 Microcircuits

Microcircuits may also be procured to a MIL-STD-883 class S or class B grade. These devices are subjected to all of the screening requirements of MIL-STD-883, Method 5004 (or Method 5008), and the quality conformance testing of each lot in accordance with Method 5005 (or Method 5008). However, they have not been formally qualified to MIL-M-38510 nor have they had the rigorous in-process controls required by MIL-M-38510. These parts typically exhibit failure rates approximately three times higher than JAN products.

MIL-STD-883 explicitly states that only microcircuits subjected to all of the screening and quality assurance requirements of MIL-STD-883, Method 5004 and Method 5005 (or Method 5008), may be marked with the "883" designation.

"Vendor's Equivalent" Microcircuits

In addition to JAN classes S and B and 883 classes S and B grade microcircuits, there are also various *vendor equivalent* grades and lower commercial grade parts. Vendor's equivalent screening is not performed in full accordance with MIL-STD-883, Method 5004 (or Method 5008). The vendor has elected to take exception to one or more of the specific requirements of MIL-STD-883. Therefore, MIL-STD-883 prohibits the marking of vendor's equivalent grade parts with the 883 designation. (A summary of the product assurance screening required by MIL-M-38510 and MIL-STD-883, Method 5004 (or Method 5008), will be shown later in Table 2.4.) These parts will typically exhibit failure rates six times higher than JAN products.

Quality and Reliability Levels of Microcircuits

The reliability of the parts is determined to a large extent by their quality grade as shown in Table 2.2 extracted from MIL-HDBK-217. The failure rate of the part is a function of π_Q the higher π_Q the higher the failure rate.

The failure rate values calculated using this table in accordance with MIL-HDBK-217 have been found to be quite realistic for military, industrial, and commercial equipments with only one important exception. This exception is what might be called an "ideal computer environment."

An ideal computer environment has three essential characteristics:

1. Operation with continuous temperature and humidity control at moderate levels
2. Continuous equipment operation, for example, power continuously applied (not cycled on and off daily)
3. Use of plastic-encapsulated microcircuits

If all three essential characteristics are true, the failure rates calculated in accordance with MIL-HDBK-217 may be extremely pessimistic for an ideal computer environment. However, if even one of the characteristics is not true, the exception does not apply, and

Table 2.2 Quality Factors for Microcircuits

Quality grade	Description	π_Q
S	Procured in full accordance with MIL-M-38510, class S requirements.	0.5
B	Procured in full accordance with MIL-M-38510, class B requirements.	1.0
B-0	Procured in full accordance with MIL-M-38510, class B requirements, except that the device is not on the Qualified Products List (QPL). The device must be tested to all of the electrical requirements of the applicable MIL-M-38510 slash sheet with no waivers allowed.	2.0
B-1	Procured with all the screening requirements of MIL-STD-883, Method 5004, class B. Electrical requirements in accordance with MIL-M-38510, DESC drawings, or vendor/contractor electrical parameters. Quality conformance requirements must be in full accordance with MIL-STD-883 Method 5005, class B; no waivers are allowed.	3.0
B-2	Procured to the vendor's equivalent of MIL-STD-883, Method 5004, class B screening requirements, and MIL-STD-883, Method 5005, class B quality conformance requirements.	6.5
D	Hermetically sealed parts with no additional screening, or plastic-encapsulated parts with additional screening including burn-in, temperature cycling, and high-temperature continuity test.	17.5
D-1	Standard commercial plastic-encapsulated parts.	35.0

Source: MIL-HDBK-217, *Reliability Prediction of Electronic Equipment*, Naval Publications and Forms Center, Philadelphia.

the MIL-HDBK-217 failure rates are realistic. The reason that this exception exists is due to a weakness in the MIL-HDBK-217 microcircuit failure rate model itself. The model assumes that the quality factor π_Q and the environmental factor π_E are independent variables. Normally they are; nowever, in this unique case they are interrelated, and the model in its present form has no way of handling this interrelationship.

Statistical Quality Control Tools

Various statistical quality control tools are frequently referenced in procurement specifications and other part purchasing documents. They are used to maintain and assure specific quality levels in purchased parts. Three such tools are "AQL," "LTPD," and "PDA."

Acceptable Quality Level (AQL)

Acceptable quality level (AQL) is the maximum percent defective (or the number of defects per hundred units, which may or may not be the same thing) allowed in a given lot. It may be considered to be a satisfactory process average and is best suited to reducing the manufacturer's risk of rejecting a good product. Most large-quantity users of semiconductors now use AQL as the standard by which they judge quality.

Various sampling plans based on the AQL concept are available. The one found in MIL-STD-105, "Sampling Procedures and Tables for Inspection by Attributes," is frequently used. Sampling procedures and sample sizes are defined in this scheme such that the consumer's (buyer's) probability of acceptance of a batch at a specified AQL varies from 0.8 to over 0.99 depending on sample size. The operational curves are also given in the standard for each case so that the associated risks may be assessed.

MIL-STD-105 makes the distinction between percent defective and defects per hundred units since in the latter case more than one defect per unit under inspection may be counted. Thus, MIL-STD-105 is designed to ensure that the great majority of components accepted by the buyer are of quality as good as or better than the specified AQL. The scheme is especially suited for continuous monitoring where batches are inspected one after the other over a period of time. Normal, tightened, and reduced inspection plans are also given for use, depending on how the process is faring (it will, of course, have variations).

Lot Tolerance Percent Defective (LTPD)

The lot tolerance percent defective (LTPD) is also frequently used to specify product quality. It is defined as some chosen limiting value of percent defective in a lot and is favored by some consumers and by the military because it reduces their risk of accepting a bad product. The LTPD is chosen such that components of quality worse than the LTPD are rejected with high probability.

For the purchase of components with very small defect rates the LTPD scheme may be more practical than the AQL. LTPD is utilized in both MIL-M-38510 and in MIL-S-19500.

One significant difference between AQL and LTPD is the fact that for a given quality level, with AQL the sample size is dependent upon the lot size, but with LTPD the sample size (above 200 items) is independent of the lot size.

Percent Defective Allowed (PDA)

This is another statistical tool frequently used to control the quality level in each lot of parts. It is used, for example, in MIL-M-38510 and MIL-STD-883, Method 5004, to define the maximum allowable percentage of failures allowed in a given screen test before the entire lot must be rejected.

2.2 ARRHENIUS REACTION RATE MODEL

Most failure mechanisms involve one or more physical or chemical processes, each of which occurs at a rate which is highly dependent on temperature; chemical reactions and diffusion mechanisms are common examples. Because of this strong temperature dependency, considerable effort has been expended to predict the temperature dependency of various chemical reaction rates by means of mathematical equations. The most widely accepted model in reliability work is the Arrhenius model.

Arrhenius Temperature Dependence Model

The Arrhenius reaction rate model was determined empirically in 1889 to describe the effect of temperature on the rate of inversion of sucrose.

Arrhenius models used in electronics and other areas, while not precise, are widely used because of their simplicity and reasonable accuracy. The Arrhenius model predicts an exponential increase in the rate of a given reaction with temperature.

The form of the model generally used in reliability is:

$$\lambda = A \exp - \left[\frac{E_a}{k}\left(\frac{1}{T} - \frac{1}{T_0}\right)\right]$$

where:

λ = temperature-related failure rate

A = a normalizing rate constant

E_a = activation energy (eV) (a unique constant for each specific chemical reaction or failure mechanism)

k = Boltzman's constant = 8.63×10^{-5} eV/°K

T = ambient temperature in degrees Kelvin (°K)

T_0 = reference temperature (°K) (used for normalization to given temperature)

Activation Energy

The origin Arrhenius model coined the term "experimental activation energy (E_a)," a factor which determines the slope of the reaction-rate curve with temperature. This slope factor describes the accelerating effect that temperature has on the rate of a reaction, and is expressed in the unlikely units of electronvolts (eV). For most applications it is adequate to treat activation energy (E_a) as merely the slope of a curve rather than a basic energy level. A low E_a value indicates a small slope or a reaction that has a small dependence on temperature. A large E_a indicates a high degree of temperature dependence.

Each and every chemical reaction, diffusion mechanism, and so on, has a unique activation energy, E_a, associated with it. In an actual electronic component there are several such reactions proceeding simultaneously, each of which is capable of eventually causing a failure. The cumulative effect of several different Arrhenius reactions will result in a device failure rate which exhibits some complex, non-Arrhenius, temperature-dependent failure rate.

Specific chemical reactions have specific activation energies; components do not. Nevertheless, the activation energy concept is frequently applied to components. For general classes of components, where the distribution of failures remains fairly constant among the various mechanisms or reactions, the cumulative effects of these various reactions is a failure rate whose temperature dependence is approximately exponential over a limited temperature range. Thus, there is a tendency to refer to an activation energy for a given type of component. Even though this is technically erroneous, it does serve a useful purpose, since it indicates that the component is exhibiting a failure rate that has a similar temperature dependence to a component failing because of a single mechanism with the specified activation energy.

In reliability engineering there are two general uses of Arrhenius curves:

1. To show specific time-temperature failure rate effects for particular failure mechanisms
2. To show general time-temperature failure rate effects at the device level

The entire MIL-HDBK-217 prediction methodology is based on a series of different Arrhenius reaction rate models, normally a separate

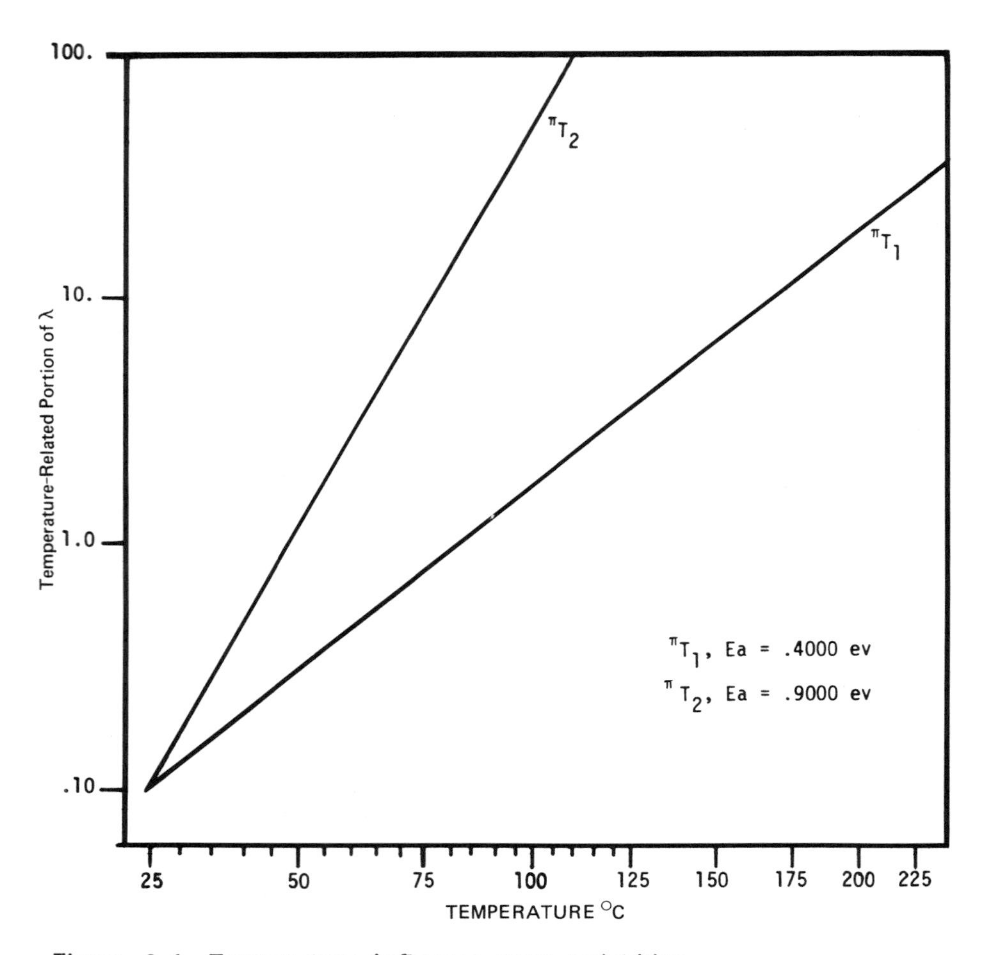

Figure 2.1 Temperature influence upon reliability.

model for each different type of component part. Figure 2.1 graphically depicts the temperature-related portion of the MIL-HDBK-217 predicted failure rate for microcircuits, where π_{T1} and π_{T2} represent the upper and lower activation energies utilized in the prediction model.

2.3 MICROCIRCUIT PACKAGE CONSIDERATIONS

The reliability of a given microcircuit is a function not only of the technology and processing of the die itself but also of the package into which this die is assembled. Some package factors which have a major impact upon microcircuit reliability are:

Whether the package is hermetically sealed or not
How the package is sealed
How the die is attached to its substrate
The style of the package

Hermeticity

Hermeticity is a relative term; everything leaks at some rate. There is no package that can keep a device sealed forever. A package can be considered hermetic if the leak rate is small enough to maintain the desired internal atmosphere for the desired life of the item.

Moisture is the most common contaminant with which to be concerned. The time for moisture to permeate various sealant materials in one defined geometry is shown in Figure 2.2 that illustrates the magnitudes of time involved for different materials. Organics (plastics) are orders of magnitude more permeable than materials typically considered for hermetic seals.

Moisture combines with other contaminants inside the package to cause corrosion and a host of other problems on the surface of the chip. For this reason, hermetic sealing of the package is a very desirable characteristic meant to exclude the entry of moisture contamination.

Thus epoxy packages by definition, are not hermetically sealed. The epoxy itself is not impervious to moisture. The major entry point of moisture, however, is at the interface between the pins and the epoxy. This is a very imperfect seal, and moisture wicks its way into the package following the bond wires from the pins to the surface of the die itself. This is one of the primary reasons why epoxy packages historically have not been recommended for high-reliability applications.

Ceramic-dual-in-line (CER-DIP) packages are hermetically sealed. However, the soft glass seal at the point where the pins egress from

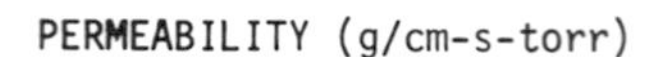

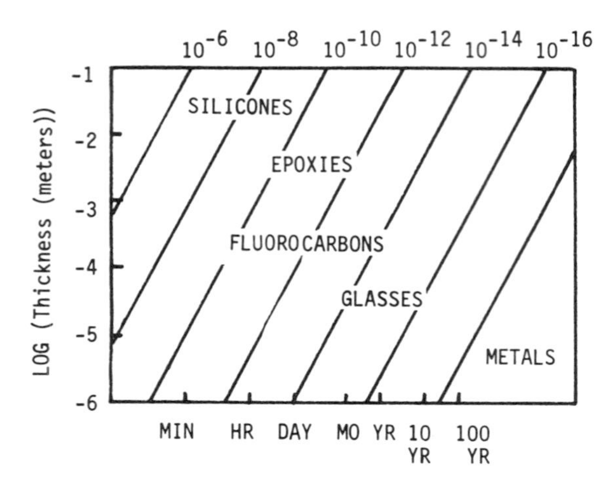

Figure 2.2 Sealant material effectiveness. (From Proceedings of the 26th Annual Electronics Component Conference, copyright 1976 by the Institute of Electrical and Electronic Engineers.)

the package is relatively vulnerable to damage, particularly where automatic insertion equipment is utilized. Fracture of this seal will allow the entry of moisture contamination into the package itself.

For high-reliability applications, packages utilizing a welded, brazed, soldered, or hard glass seal are preferred from the viewpoint of package hermeticity.

Glassivation, passivation, and parylene and other coatings are frequently used on the surface of the die to protect it from moisture, other contaminants, and alpha-particle radiation from the package itself (see Chapter 10.4). Unfortunately, these protective measures can sometimes introduce additional problems. For example, stress cracks in glassivation can act as nucleation points for any moisture which may be present within the package, causing accelerated corrosion at those points. Alpha-particle protective coatings can also introduce problems since they are frequently hydrophilic in nature and can attract moisture to the surface of the chip. Even when effective coatings are used, the bonding pads themselves are still exposed, allowing any moisture present within the package to get a corrosive foothold there.

Die Attachment Methods

The heat generated within the die must be removed. The substrate to which the die is attached is normally the vehicle used for such heat removal. Therefore, the interface between the die and the substrate is a vital reliability concern because it will affect the efficiency with which this heat is removed. Three different attachment mediums are normally employed: eutectic (solder), epoxy, and glass. The eutectic bond offers the best thermal characteristics and thus is preferred.

Fillers are frequently added to die bond epoxies to modify and improve their thermal properties. However, these fillers can sometimes compromise other properties. The epoxies are also subject to outgassing of corrosive contaminants into the package cavity, which can significantly reduce device reliability.

Package Style Considerations

The same microcircuit die is usually available in a variety of different package styles: dual-in-line, TO-5 cans, flat-pack, and leadless chip carrier (LCC), to name a few. Package style may also influence microcircuit reliability.

Dual-in-Lines (DIPs)

Dual-in-lines are the most popular package today. They are ideally suited for mounting on printed circuit boards (PCBs). During production, DIPs can be inserted either manually or automatically into mounting holes on the PCBs and soldered by various mass production techniques.

The DIP, however, is relatively large and heavy, and as pin count increases (as with VLSI devices), these factors are aggravated. DIP packages of high pin-count configurations are subject to excessive breakage due to their long, narrow shape. In addition, the resistance and inductance from the corner pins to the input and output terminals of the die become significant for high-speed functions.

Flat Packs

Ceramic flat packs, although popular with the military in the past, are being phased out by the leadless chip carriers (LCCs). These have the size advantages of flat packs along with much greater interconnection density.

TO-5 Cans

These packages offer good heat conduction, radiation protection, hermeticity, and ruggedness. Their principal disadvantages are their relatively inefficient use of space, and their limited number of leads. Since these packages are mounted cantilevered by their leads, they may also require additional mechanical support to prevent fatigue or damage due to vibration or shock.

Leadless Chip Carriers (LCCs)

Leadless chip carriers are very similar in construction to side-brazed DIP packages. Therefore, in themselves they are highly reliable. They offer some significant advantages over the DIP, for example, they offer a significantly higher pin-out in less space and potentially higher frequency operation. Also they are not subject to the high breakage rate of DIPs.

Unfortunately, LCCs do suffer a serious problem when mounted on a PCB. The thermal coefficient of expansion of their ceramic package is much lower than that of commonly used PCB materials. Excessive failures of the solder bond between the LCC and the PCB occur under temperature-cycled conditions. Power cycling (power On/off) also has the potential, at least in consumer applications, to be a considerably larger reliability problem than simply that of temperature cycling. The LCC must be mounted to a substrate with a compatible temperature coefficient of expansion.

To overcome this reliability problem, LCC packages are sometimes reflow soldered or vapor-phase soldered to a ceramic motherboard, which is then attached to a PCB. Vapor-phase soldering is an extremely consistent soldering process. It also leaves the item relatively clean and dry after the soldering operation, and it is unique in that LCCs may be attached to both sides of the substrate in a single operation.

Since microcircuit reliability is directly related to junction temperature, effective thermal mounting is also important. There are two different types of LCCs, those with the "cavity up" and those with the "cavity down". With the "cavity up" design the major heat-radiating surface is facing the board. With the "cavity down" design the major heat radiating surface is away from the board. Depending upon the method of cooling, a significant junction temperature difference could occur between the two different package configurations. Thus, the thermal properties of the package must be addressed during the equipment design.

Epoxy-Encapsulated Packages

The reliability of an epoxy-encapsulated device is somewhat dependent upon the type of technology contained in the package. For example, chips which dissipate more power (for example, TTL) will tend to drive out any moisture which migrates into the package, thus reducing the probability of internal metallization corrosion.

In addition to their lack of hermeticity, there is also a second major reliability concern with the use of epoxy-encapsulated microcircuit packages. The thermal coefficient of expansion of the epoxy is usually greater than that of the gold bond wires. This can result in fracture of the wire under conditions of thermal cycling. Advanced epoxy molding compounds have been developed which match the thermal coefficients of expansion of the gold bond wires over a limited temperature range. However, if the epoxy expansion coefficient matches that of the gold bond wire then it necessarily does not match the expansion coefficient of the chip itself. Intermittent connections can also occur as a result of broken internal wire or lifted bonds caused by improper or uneven curing of the epoxy. These broken wires are frequently held in place by the epoxy and make a pressure contact at room temperature. Thus they are often not detected during normal electrical testing. However, as the operating temperature of the device increases the pressure contact becomes intermittently open.

Epoxy-encapsulated packages do offer some unique advantages. They are unsurpassed in their ability to withstand extremely high levels of shock and vibration. In addition, the existence of the encapsulant in contact with the chip eliminates the possibility of particulate contamination. This is a failure mode which is difficult and expensive to screen out in high-reliability hermetic packages. These advantages do qualify epoxy-encapsulated packages for some unique applications.

There is growing evidence that some epoxy-encapsulated devices can offer reliable performance under conditions of controlled temperature and humidity in office-type environments for equipment with a relatively short life (less than five years). However, for more severe environmental conditions and long-life equipment the hermetically sealed devices are preferred for high reliability applications.

Microcircuit Package Summary

The relative vulnerability of different microcircuit packages to degradation from mechanical environmental causes is reflected in MIL-HDBK-217 reliability prediction procedure. It is addressed in the form of a complexity factor C3, contained in a table relating that portion of the failure rate that is caused by mechanical effects to the package construction and to the number of pins used in the package.

2.4 SCREENING OF MICROCIRCUITS AND OTHER DISCRETE PARTS

The majority of part failures occur very early in the equipment's life during the infant mortality period. Therefore, these are the ones that should be attacked first to assure that they are eliminated before the equipment is delivered to the customer. A simple receiving inspection test of incoming parts will serve to eliminate dead-on-arrival parts, but it does not help to eliminate those still-functioning but marginally weak parts. These are the parts which are likely to fail during the early stages of use at the customer's facility.

Eliminating Infant Mortality Failures

Finding and eliminating these weak parts usually involves a two-step process. First, the part must be stressed to activate a failure mechanism, and then subsequently tested to detect a failure mode. Having now a detectable failure mode, these parts can be efficiently removed from the total part population. This requires the establishment of some type of screening program. The screening test, however, while finding and eliminating weak parts, must not damage or reduce the life of good parts.

A screening test by definition is a 100% test. Each and every item of the population must be subjected to the stress in order for it to be effective. Sample testing is not an acceptable substitute for a 100% screen. To establish an effective screening program we must first identify the nature of the part weakness that we are searching for, and then identify the specific type of stress that is most likely to bring out that specific weakness. Unfortunately, each type of part and each different part technology has its own unique weakness. Moreover, these weaknesses are often related to the choice of materials and the specific manufacturing process used to fabricate the part. This means that a thorough understanding of the part and its unique weaknesses are required to establish an effective screening program.

Such a screening program for microelectronic devices has been established and documented in MIL-STD-883. The effectiveness of this general approach is witnessed by the general acceptance of and frequent reference to this military specification by even commercial and industrial microcircuit users. A brief outline of MIL-STD-883 is shown in Table 2.3.

Method 5004 in MIL-STD-883 defines an effective screening program for monolithic microcircuits while Method 5005 defines both the initial qualification procedures and the periodic quality conformance procedures for monolithic microcircuits. Method 5008 addresses all three areas; screening, initial qualification, and periodic quality conformance requirements for hybrid and multichip devices. The standard defines the specific procedures and other details for performing each of the screening test steps, initial qualification test steps, and the periodic quality conformance tests required in the other numbered methods within the standard. A summary of the screens required by MIL-STD-883, Method 5004 for class S and class B microcircuits is shown in Table 2.4.

Table 2.3 MIL-STD-883 Outline

Method 1XXX	Environmental tests
Method 2XXX	Mechanical tests
Method 3XXX	Digital electrical tests
Method 4XXX	Linear electrical tests
Method 5XXX	Miscellaneous tests (including)
Method 5004	Screening procedures
Method 5005	Qualification procedures and quality conformance procedures
Method 5008	Hybrid and multichip test procedures

Table 2.4 Screening Requirements MIL-STD-883, Method 5004

Screen	Class S	Class B
Internal visual	Condition A	Condition B
Stabilization bake	24 hr	24 hr
Temperature cycling	10 cycles	10 cycles
Constant acceleration	30,000 g	30,000 g
Particle impact noise detection	Yes	No
Hermeticity (fine and gross)	Yes	Yes
Interim electrical parameters	Yes	No
Burn-in	168 + 72 hr	168 hr
Final electrical parameters at 25°C	Yes	Yes
Final electrical parameters at max. and min. temperature	Yes	Yes
X-ray radiograph	Yes	No
External visual	Yes	Yes

Source: *Test Methods and Procedures for Microcircuits* (MIL-STD-883), Naval Publications and Forms Center, Philadelphia.

In order to understand the nature of screening it is first necessary to understand the failure mechanisms associated with the parts which we wish to screen. Temperature is the primary contributor or accelerating factor for the majority of the part failure mechanisms.

Thermally Induced Failure Mechanisms in Microcircuits

The mismatch of thermal expansion coefficients of interfacing materials is a common failure mechanism attributable to thermal stress. The mechanical stress or strain at the interface will be a function of the temperature and the difference in the thermal expansion coefficients of the materials involved. When the force due to thermal mismatch exceeds the strength of the bonding material at the interface, a failure will occur. Microcircuit examples of such failures are die bond failures and package-seal failures in ceramic-dual-inline packages (CER-DIPs).

Dynamic thermal stress or temperature cycling may also produce failures. These failures are dependent on the number of temperature cycles and the range over which the temperature is cycled. When the rate of change of temperature is relatively slow, temperature-cycling-induced defects are usually attributed either to work-hardening of the material (due to repeated expansion and contraction cycles) or to the cumulative effects of mass transport caused by repeated occurrence of cycles with temperatures high enough to result in plastic flow. Work-hardening effects have been observed in microcircuit wirebonds utilizing aluminum wires. Work-hardening tends to make such bonds brittle and thus more susceptible to cracking.

Voiding of the die-to-subtrate bonds in microcircuits is an example of the net transport of mass due to temperature cycling. This occurs when the temperature during the cycle stresses the bonding material in excess of its elastic limit (the stress results from the thermal expansion mismatch between the die and the substrate). The temperature need not be high enough to cause a failure, but only just high enough to result in plastic flow of the bonding material. The net mass transport occurring during one temperature cycle is usually insignificant. However, the effects are cumulative. In microcircuits the primary means of conducting heat away from the die itself is through the die bond. Thus the progressive voiding of the bonding material increases the thermal resistance, resulting in still higher temperatures, again resulting in further voiding, eventually resulting in a failure.

When the rate of change of temperature is very fast as in thermal shock tests, failures are usually catastrophic and appear as cracks or breaks in the material. Thermal shock places severe stress on any material having a relatively large thermal coefficient of expansion. Thermal shock is not normally used as a screen. It is considered to be a destructive test and is typically used in component test labs as an accelerated test intended to simulate the long-term effects of temperature cycling. The extrapolation of thermal shock test results back to thermal cycling conditions is at best a controversial procedure.

At the part level, thermal cycling is usually performed by moving the part from a chamber with the air at a given temperature to another chamber with the air at a different temperature. In thermal shock, however, the part is moved from a liquid at a given temperature to another liquid at a different temperature.

Temperature-Dependent Failure Mechanisms

Solid-state diffusion and most chemical reactions are examples of part-level temperature-dependent failure mechanisms. Specific examples include the formation of intermetallics (such as purple plague) and corrosion of metals, respectively. The effects of thermal energy are cumulative with time for such failure mechanisms. In theory, such mechanisms occur at any temperature except at zero degrees Kelvin. Temperature merely influences the rate at which the mechanisms proceed. For the majority of these mechanisms, the thermal dependence has been shown to be adequately predicted by the previously described Arrhenius model.

Activation energies of typical microcircuit failure mechanisms range from less than 0.1 eV to more than 1.1 eV. Many of the common mechanisms exhibit activation energies of about 0.7 eV. Quantitatively, at 50°C a mechanism having an activation energy of 0.7 eV will proceed at more than eight times the rate that it would at 25°C. Thus, the probability of survival of a component having a failure mechanism of activation energy 0.7 eV is eight times better at 25°C than it is at 50°C.

Screening Programs

Screening of parts can be performed in a number of different ways. Screening can be performed by the part manufacturer, by the user in his own facility, or by an independent testing laboratory. No matter which agency is employed to do the screen tests, the user should first acquaint himself with the efficacy of the screening tests used by the vendor in his normal production. If such screens exist and they are effective, then additional screens may not be necessary, or if necessary, they can be designed to supplement the vendor's tests. If existing tests are unsatisfactory, then a more comprehensive screening program may be required.

It should be recognized, however, that not all of the screens are feasible at all of these locations. The precap visual inspection detailed in MIL-STD-883, Method 2010 can be performed only at a vendor's facility during the part manufacturing process itself, that is, before the lid of the integrated circuit is installed.

Table 2.5 Screening Tests for Die-Related Failure Mechanisms

Die failure mechanisms	Applicable screen test						
	(1)	(2)	(3)	(4)	(5)	(6)	(7)
Surface contamination/ leakage	X			X	X		
Foreign materials/ particles			X			X	X
Inversion/channeling	X			X			
Crystal imperfections	X	X			X		
Cracked die		X	X		X		
Oxide pinholes		X		X	X		
Oxide faults		X			X		

Oxide short/breakdown	X	X	X
Passivation defects	X		X
Diffusion anomalies		X	X
Diffusion spikes		X	X
Mask faults			X
Open metallization	X	X	X
Metallization shorts	X	X	X
Electromigration			X

Note: (1) stabilization bake; (2) temperature cycle/thermal shock; (3) constant acceleration/mechanical shock; (4) reverse bias and temperature; (5) dynamic operation and temperature; (b) X-ray; (7) PIND.

Source: *Microcircuit Screening Effectiveness* (TRS-1), copyright 1978 by IIT Research Institute, RADC/RAC, Griffiss Air Force Base, N.Y.

Table 2.6 Screen Tests for Package-Related Failure Mechanisms

Package failure mechanism	Applicable screen test								
	(1)	(2)	(3)	(4)	(5)	(6)	(7)	(8)	(9)
Broken wirebond		X			X	X			
Lifted wirebond		X			X	X			
Overbonded		X				X			
Misplaced bond					X	X			
Multiple bond			X						
Intermetallic formation	X	X				X			
Die attach defect		X	X			X	X		X
Broken wire		X			X	X			

	(1)	(2)	(3)	(4)	(5)	(6)	(7)	(8)	(9)
Shorted wire				X	X				
Poor lead dress	X	X							
Corroded wire	X			X	X				
Nonhermetic seal	X	X	X						
Excessive seal material								X	X
External lead defect								X	

Note: (1) stabilization bake; (2) temperature cycle/thermal shock; (3) constant/acceleration/mechanical shock; (4) hermeticity; (5) reverse bias and temperature; (6) dynamic operation and temperature; (7) X-ray; (8) external visual; (9) PIND.
Source: *Microcircuit Screening Effectiveness* (TRS-1), copyright 1978 by IIT Research Institute, RADC/RAC, Griffiss Air Force Base, N.Y.

When particular failure modes or mechanisms are known or suspected to be present, a specific screen should be selected to detect and eliminate these unreliable elements. Table 2.5 identifies a number of the failure mechanisms commonly associated with integrated circuit die together with the screen tests that are effective in their activation. Table 2.6 relates back to Chapter 2.3 and identifies a number of the failure mechanisms commonly associated with integrated circuit packages together with the screen tests that are effective in their activation. Examples of some of the ranges of rejects from key microcircuit screens are shown in Table 2.7. Table 2.8 compares the estimated cost of each of the individual screens for a class B part. The screening costs provided are for comparison purposes only. The intent is to illustrate the relative cost differences for screening tests on devices of varying complexity. For a simple SSI/MSI integrated circuit, screening test costs will be lower. For LSI and VLSI devices, the cost will approach the maximum indicated.

It should be noted that the cost of dynamic burn-in and the associated electrical tests are the most significant element in the total cost of the screen, and that this factor is determined primarily by the complexity of the part itself. From Table 2.5 it is evident that dynamic burn-in is also the most effective screen test.

The cost of precap visual inspection is also directly related to device complexity. Unfortunately, however, its effectiveness is inversely related to device complexity. Precap visual inspection consists of both a low-power magnification inspection and a high-power magnification inspection. The low-power magnification is intended to inspect the die bonds, the lead dress, particulate contamination, and so on. This inspection is effective and necessary, regardless of the

Table 2.7 Fallout from MIL-STD-883 Screens

Screen	Average fallout (%)	Range (%)
Precap visual	15	2.0–24
Hermeticity	5	0.1–10
Burn-in	3	0.1–20
Electrical testing	5	1.3–12
External visual	4	0.1–8

Source: MIL-HDBK-338, *Electronic Reliability Design Handbook*, Naval Publications and Forms Center, Philadelphia.

Table 2.8 Screening Costs for Class B Devices

	Cost in dollars ($)		
MIL-STD-883 method	Min.	Typical	Max.
1. Precap visual inspection	0.15	0.25	3.00
2. High-temperature storage	0.01	0.05	0.10
3. Temperature cycling	0.05	0.10	0.10
4. Constant acceleration	0.05	0.10	0.25
5. Fine leak	0.05	0.10	0.25
6. Gross leak	0.05	0.10	0.20
7. Dynamic burn-in	0.25	0.50	5.00
8. Final electrical test	0.25	0.50	2.00
9. External visual	0.05	0.10	0.25
Total (class B)	0.91	1.00	11.15

device complexity. The high-power inspection, in contrast, minutely inspects the entire surface of the die. This is a laborious exercise of questionable value with highly complex devices. In fact, MIL-STD-883 allows the substitution of a more rigorous electrical test in lieu of high-power magnification in some instances. Precap visual inspection is the only screen that must be performed at the device manufacturer's facility; it cannot be performed after delivery of the part as is possible with the remaining screens.

In addition to the screens shown in Table 2.8 two more screens are required as noted in Table 2.4 for class S (space) devices: These are the particle impact noise detect test (PIND) and the x-ray test. Both of these tests are intended to detect loose contamination inside of the package. This is a much more serious problem in space than on earth because of the propensity of the contaminants to move about more freely in a zero-specific-of-gravity environment; thus the added requirement. The precap visual inspection of class S devices is also more rigorous than for class B devices.

3

Part Application and Derating

3.0 INTRODUCTION

The preceding chapter on part selection, specification, and control assumed that the parts have an inherent reliable level. However, in order to achieve this high reliability these parts must also be properly applied and be capable of withstanding all of the stresses to which they may be subjected in the equipment.

Proper derating of the electronic parts incorporated within the equipment is one of the most powerful methods available to enhance the reliability of our equipment. Derating is defined as follows: limiting the stress applied to the parts to levels that are well within their specified or proven capabilities in order to enhance their reliability.

All part derating is done with reference to the *absolute maximum ratings*. These ratings are defined by the manufacturer in his specification or data sheet as those values which should not be exceeded under any service or test condition. There are usually several different absolute maximum ratings for a given part, such as voltage, current, and power. Each of these absolute maximum ratings is unique; each must be applied individually and not in combination with any other absolute maximum rating. The absolute maximum ratings usually state a maximum operating and/or storage temperature (junction or hot-spot temperature) and various electrical values based upon DC power conditions measured in free air at 25°C.

Derating can be accomplished by either reducing the stresses on the part or by increasing the strength of the part, that is, by selecting a part having greater strength.

Reliability Improvement with Stress Reduction

Derating is effective because the failure rates of most parts decrease as the applied stress levels are decreased below their rated value. The reverse is also true. Failure rates increase when a part is subjected to higher stresses and higher temperatures. The failure rate models of most electronic parts are stress and temperature dependent.

Electronic parts are prone to premature failure as a function of thermal stress. MIL-HDBK-217 failure rate data shows that part failure rates vary significantly with temperature. Certain parts are more sensitive to temperature than others. Reduction in failure rate can be achieved in these cases by reduction of temperature, that is, by improving the thermal design. Typical examples of failure rate reduction resulting from lower temperatures are shown in Table 3.1.

Derating of electronic parts and materials should be accomplished as necessary to assure that the equipment reliability is within specification. Actual derating procedures vary with different types of parts and with their application: resistors are derated by decreasing the ratio of operating power to rated power, capacitors are derated by reducing the applied voltage to a value lower than that for which the part is rated; semiconductors are derated by keeping their power dissipation and hence their junction temperature below the rated level.

Derating curves for electronic parts relate derating levels to critical environmental or physical factors using mathematical models which quantify a base failure rate in terms of stress ratio, temperature, and other part related parameters.

Solid-state device manufacturers usually provide curves of operating parameters vs. temperature, maximum and minimum storage temperatures, maximum operating junction temperature, and package thermal resistance. Unless specially selected premium parts are specified, parameter deviation from the mean reported values is significant for most parts. Maximum operating junction temperature must be derated by reference to failure rate vs. temperature data so that desired part reliability can be achieved.

It is not sufficient simply to compute the worst-case semiconductor junction temperature and then assume that the thermal design is adequate as long as the manufacturer's maximum junction temperature is not exceeded. The device may function under such conditions, but its reliability will generally be unacceptable. Maximum allowable semiconductor junction temperatures are of little use unless they are related to the required equipment reliability. This will normally require derating from the published ratings.

In addition to the equipment reliability requirement considerations, maximum junction temperature derating is also advisable to provide

Table 3.1 Reliability Improvement at Reduced Temperatures

Part description	Base failure rates[a] (per 10^6 hrs)		Δt (°C)	Failure-rate improvement
	Reduced temp. (°C)	High temp. (°C)		
PNP Silicon transistors	0.0012 at 40°	0.0092 at 160°	120	8:1
NPN Silicon transistors	0.0008 at 40°	0.0048 at 160°	120	6:1
Glass capacitor	0.0024 at 40°	0.0073 at 125°	85	30:1
Transformers and coils	0.0025 at 40°	0.0666 at 85°	45	27:1
Resistors	0.0003 at 40°	0.0054 at 120°	80	17:1

[a] From MIL-HDBK-217D (Notice 1), assuming a 10% stress level.

some margin for error. Derating provides allowance for system electrical transients and also allows for possible nonuniform part heating without catastrophic failure.

Since semiconductors as well as most electronic parts are sensitive to temperature, the thermal analysis of any design should accurately reflect the ambient temperatures needed for proper application of the part. Therefore, thermal analysis must be an integral part of the design process. It should be included in all tradeoff studies covering equipment performance, reliability, weight, volume, environmental control requirements, and cost.

Derating must be cost-effective. It should not be conservative to the point where costs rise excessively, for example, where lower than necessary part stresses are applied. Optimum derating occurs where a rapid increase in failure rate is noted for a small increase in temperature or stress. There is usually a practical minimum to derating. At some minimum stress level, circuit complexity increases unnecessarily, thus offsetting the reliability gain accomplished by further derating.

Data on failure rate vs. stress is available in the MIL-HDBK-217 for most types of electronic parts, and this data is used to determine reliability improvement through derating. Unfortunately, this type of data is not readily available for mechanical parts.

Additional Benefits of Derating

Derating can help to compensate for many of the variables inherent in any design. All electronic parts produced on an assembly line are not identical. Subtle differences and variations exist from one part to the next. Proper part derating will help to compensate for these part-to-part variations and minimize their impact upon the equipment reliability. Electronic parts with identical manufacturer's part numbers may be purchased from a variety of different suppliers. While these items are electrically interchangeable there may significant design, material, and manufacturing differences between them. Derating will help to compensate for these differences.

In the course of design, the designer will try to anticipate the various electrical and environmental extremes to which the equipment may be subjected. If he fails to anticipate properly the impact of all of these variations, derating can provide an additional margin of safety. It is also apparent that parts and their associated critical parameters are not completely stable over their entire life. Proper derating will help to assure that the circuit itself will continue to function properly in spite of these part parameter changes.

3.1 DETAIL STRESS RELIABILITY PREDICTION

During the useful life of an electronic part its reliability is a function of both the electrical and the thermal stresses to which the part is subjected. Increased thermal stress is reflected as increased junction temperature. This results in increased chemical activity within the part as described by the Arrhenius reaction rate model (as was addressed in Chapter 2.2), and thus results in an increased failure rate. Increasing the electrical stresses also increases the failure rate. If both thermal stress (that is, the junction temperature) and the electrical stress are simultaneously increased, the two factors are compounded, thus greatly increasing the failure rate. This is the basic theory behind the MIL-HDBK-217 failure rate prediction methodology. The major factors, other than temperature, included in the MIL-HDBK-217 that affect the reliability of piece parts are shown in Table 3.2 for some of the more common electronic parts.

In microcircuits and discrete semiconductors the thermal resistance junction-to-case θ_{jc} characteristics of the device package have a significant impact upon the device junction temperature and hence upon the device reliability. For a given power dissipation, the device's thermal resistance is a function of the size of the die, the method by which the die is attached to the substrate and the substrate material, its geometry, and the package materials employed.

Table 3.2 Major Factors (Other Than Temperature) Which Determine Device Reliability

Device	Principal factor
Digital ICs	Complexity—number of gates
Linear ICs	Complexity—number of transistors
Memory ICs	Type of memory (ROM, RAM, etc.) and complexity—number of bits stored
Discrete semiconductors	Material—silicon or germanium Type—PNP or NPN and voltage, power, and current derating
Resistors	Type (carbon, metal film, etc.) and power derating
Capacitors	Type (ceramic, plastic, etc.) and voltage derating
Inductive devices	Type, insulation grade
Connectors	Type and number of pins
Switches and relays	Type, application, and current derating

Thermal resistance is inversely proportional to die size. Thus, for a given power dissipation a small size die will yield a higher junction temperature than a large die.

The die itself may be mounted to the substrate with a eutectic (solder) bond, an epoxy bond, or a glass bond. Each of these die mounting methods have significantly different thermal characteristics and thus could yield significantly different junction temperatures in the same application. The eutectic bond would exhibit a lower thermal resistance than either of the other two mounting methods. Various different geometries and materials might be used for the substrate itself. Each of these would have its own thermal characteristics and hence produce different junction temperatures and device reliabilities.

Thermal resistance is also inversely proportional to the package size. A given die and a given power dissipation will yield a higher junction temperature in a small package than in a large package.

Typical package materials include metal, ceramic, or epoxy encapsulant. Of these three materials the epoxy would normally have the poorest thermal characteristics, that is, the greatest thermal resistance, and hence (other factors being the same) it would produce the highest junction temperature.

Thus the same microcircuit device (that is, the chip) could exhibit a very different junction temperature and very different reliability depending upon the materials and construction characteristics of the package.

Temperature-Stress Factors

The temperature-stress effect upon reliability can easily be observed by referring to the MIL-HDBK-217 failure rate models. For example, the part failure rate model for discrete semiconductors is expressed as:

$$\lambda_p = \lambda_b(\pi_E \times \pi_A \times \pi_{S_2} \times \pi_C \times \pi_Q \times \pi_R)$$

where:

λ_p is the final part failure rate.

λ_b is the base failure rate.

π_E is the environmental factor—it accounts for the influence of environmental factors other than temperature.

π_A is the application factor—it accounts for the effect of application in terms of circuit function.

π_{S_2} is the voltage stress factor—it adjusts the model for a second electrical stress (application voltage) in addition to power dissipation.

π_C is the complexity factor—it accounts for the effect of multiple devices in a single package.

π_Q is the quality factor—it accounts for the effects of different part-quality levels.

π_R is the rating factor—it accounts for effect of maximum power and current ratings.

The equation for the base failure rate, λ_b, is based upon the Arrhenius reaction rate model as was discussed in Chapter 2.2. Figure 3.1 shows how the base failure rate (λ_b) varies with temperature

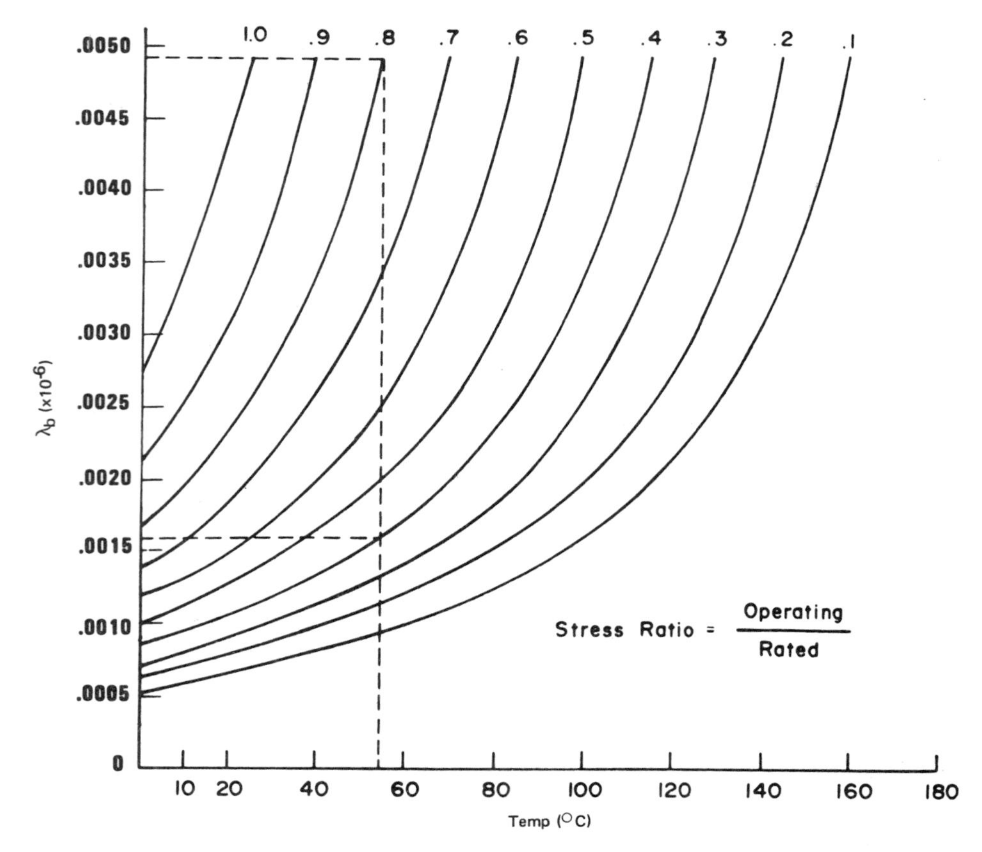

Figure 3.1 Stress/temperature plot. [From *Reliability Design Handbook* (RDH-376), copyright 1976 by IIT Research Institute, Reliability Analysis Center, RADC/RAC, Griffiss Air Force Base, N.Y.]

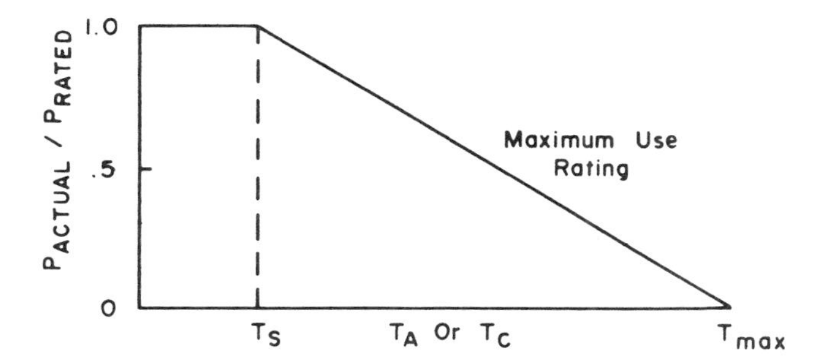

Figure 3.2 Typical derating curve. [From *Reliability Design Handbook* (RDH-376), copyright 1976 by IIT Research Institute, Reliability Analysis Center, RADC/RAC, Griffiss Air Force Base, N.Y.]

and stress for a typical silicon, NPN transistor. The data, taken from MIL-HDBK-217E, is based on the maximum junction temperature of 175°C (fully derated) and on the maximum temperature of 25°C at which full rated power/current is permitted. The data show that at higher temperatures (greater than 100°C) and at electrical stress higher than 40% (even at lower temperatures) the slope of the curves, and thus the failure rate, increase dramatically.

Power/Current Derating for Temperature

Transistor and diode power/current ratings are usually based upon operation in still air at 25°C. Thus the first step in derating an electronic part involves the determination of the device's true rating at its actual operating temperature using the applicable derating curve from the part manufacturer's specification. The objective of the power derating is to hold the worst-case junction temperature to a value below the normal permissible rating. The major failure modes associated with excessive temperature are degradation of the gain and increased leakage with prolonged use at elevated temperatures. A typical semiconductor derating curve is shown in Fig. 3.2 where:

T_s is the temperature derating point (usually 25°C).

T_m is the maximum junction temperature (usually 175°C, or 200°C for high power devices).

T_a is the ambient temperature.

T_c is the case temperature.

This conventional derating approach assumes that the thermal resistance, from junction-to-ambient θ_{ja} or junction-to-case θ_{jc} is a constant and that the junction temperature is equal to:

$$T_j = T_a + \theta_{ja} P_j$$

or

$$T_j = T_c + \theta_{jc} P_j$$

where:

T_j is the junction temperature.

T_a is the ambient temperature.

T_c is the case temperature.

θ_{ja} is the thermal resistance (junction-to-ambient, in °C per watt).

θ_{jc} is the thermal resistance (junction-to-case, in °C per watt).

P_j is the power dissipation (in watts).

These equations indicate that operation anywhere along the derating line between T_s and T_{max} will result in a junction temperature equal to T_{max} and that the thermal resistance, θ, is a constant equal to:

$$\theta = \frac{T_{max} - T_s}{P_{rating}} \text{ °C/watt}$$

where P_{rating} is the power rating (in watts at temperature T_s).

The assumption of constant thermal resistance is approximate. For microcircuits and low power transistors the assumption is valid because their actual thermal resistance has only a slightly negative droop as a function of the temperature of the bulk semiconductor material.

If the curvature of $T_j - T_{max}$ curve is large, this straight line assumption can result in significant errors for high-power devices. Therefore, suppliers frequently specify a multipoint derating curve to approximate more closely the curvature in the constant junction temperature $T_s - T_{max}$ curve. Take for example the derating curve for the 1N3263 power diode. In Figure 3.3 the three rating points are 160 amps at 125°C, 120 amps at 150°C, 0 amps at 175°C. Figure 3.3 shows that the two-point linear derating assumption from 160 amps at 125°C would result in an 80 amps rating at 150°C instead of the more accurate rating of 120 amps. This would cause a third or more of the device capability to be wasted at 150°C with a straight linear derating.

The curvature of $T_j = T_{max}$ can be somewhat different for θ_{ja} or θ_{jc}, therefore, slightly different results may be obtained using each of the two different ratings in reliability prediction computations.

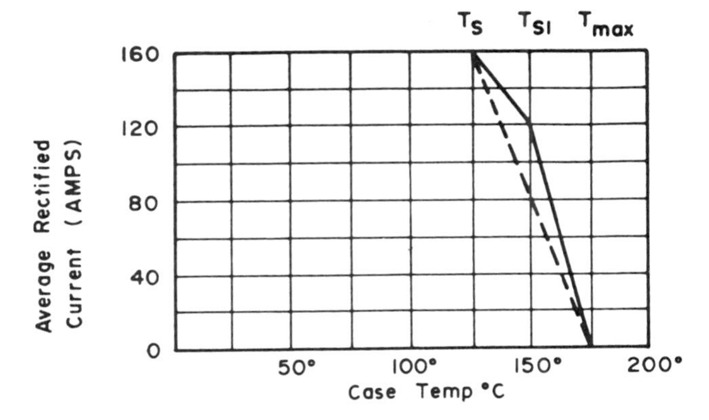

Figure 3.3 Multipoint derating curve. [From *Reliability Design Handbook* (RDH-376), copyright 1976 by IIT Research Institute, Reliability Analysis Center, RADC/RAC, Griffiss Air Force Base, N.Y.]

3.2 DERATING GUIDELINES

Specific guidelines for derating component parts used in electronic equipment are presented in Tables 3.3 through 3.17. These guidelines represent a composite summary of derating policies employed by various firms within the electronics industry. Specific derating guidelines together with application notes are addressed as follows:

Silicon transistors, Table 3.3
Silicon diodes, Table 3.4
Digital microcircuits, Table 3.5
Linear microcircuits, Table 3.6
Fixed capacitors, Table 3.7
Variable capacitors, Table 3.8
Fixed resistors, Table 3.9
Variable resistors, Table 3.10
Transformers and inductors, Table 3.11
Mechanical relays and switches, Table 3.12
Lamps and optical sources, Table 3.13
Optical detectors, Table 3.14
Fiber optic cables, Table 3.15
Connectors, Table 3.16
Rotating devices, Table 3.17

Transistor Application Notes (Table 3.3)

Generally avoid the use of faster semiconductor devices than the design requires, since faster devices frequently compromise other parameters and are thus more susceptible to secondary breakdown.

Power transistors may be subject to thermal fatigue failure when exposed to many temperature cycles. This includes normal on-off cycling. For maximum reliability, observe the case temperature change limits shown in Figure 3.4. For example, a device which is

Table 3.3 Silicon Transistor Derating

Transistor type	Parameter	Derating factor: Nominal	Derating factor: Worst-case
Bipolar	Power dissipation	50%	80%
	Current (I_c)	75%	85%
	Voltage (V_{ce})	60%	80%[a]
	Safe operating area (high power devices only)		
	V_{ce}	70%	90%
	I_c	60%	80%
	Junction temperature[b]	95°C	135°C
Field-effect	Power dissipation	50%	75%
	Breakdown voltage	60%	80%[a]
	Junction temperature	95°C	135°C
Thyristors, SCRs, and TRIACs	On-state current	50%	75%
	Off-state voltage	60%	80%[a]
	Junction temperature[b]	95°C	135°C

[a]The sum of the anticipated transient voltage peaks and the operating voltage peaks should not exceed the worst-case recommended derating factor.

[b]Assuming that the device is rated at 175°C. For devices rated at other that 175°C, use the formulas:

$$T_j(\text{nominal derated}) = 0.5(T_{j(max)} - 25°C) + 25°C$$

$$T_j(\text{worst-case derated} - 0.7(T_{j(max)} - 25°C) + 25°C$$

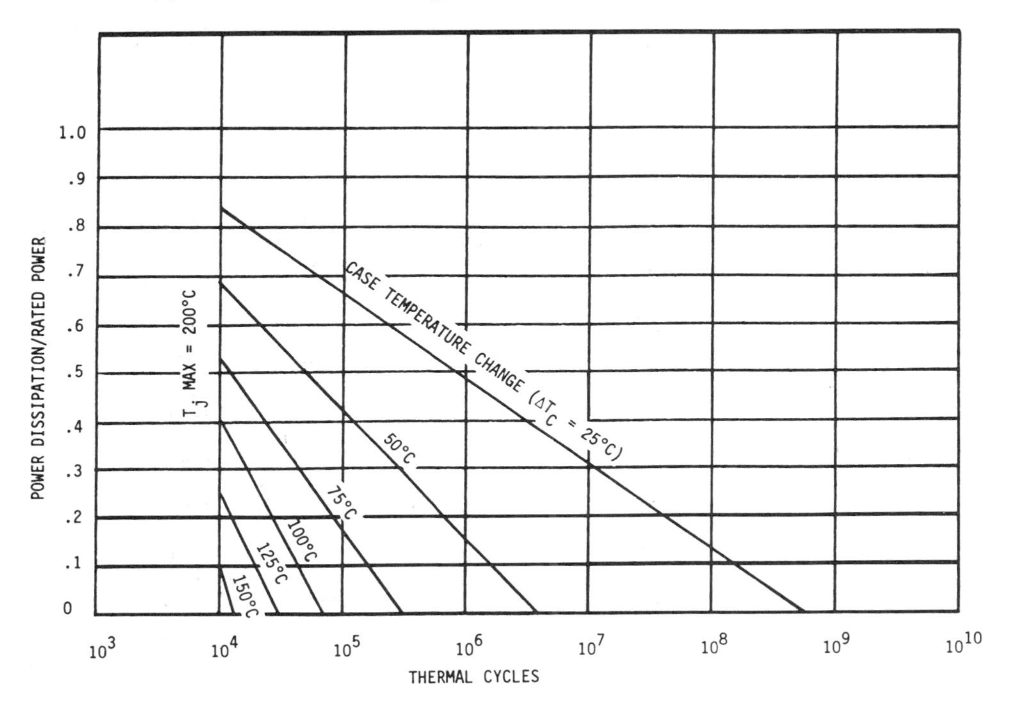

Figure 3.4 Thermal cycling ratings for power transistors. [From *Reliability Parts Derating Guidelines* (RADC TR-82-177), Rome Air Development Center, Griffiss Air Force Base, N.Y.]

operated at a 50% derated power dissipation and is expected to have 50,000 on-off cycles during its useful life should not exceed at 50°C case temperature change.

Design thyristor circuits for a "hard" turn-on (gate voltage and current well above minimum levels). Marginal or slow gate drive can cause device failure.

Diode Application Notes (Table 3.4)

Axial leaded diodes with a metallurgical bond exhibit better reliability than nonmetallurgically bonded devices and are preferred when they are available.

Do not overspecify both speed and forward voltage, as other characteristics may suffer and thus degrade reliability.

Microwave diode design frequently involves distributed constants, therefore, the diode, package, parasitics, interconnections, and other components must be considered as a single unit. This often prevents consideration of derating as a separate variable.

Capacitor Application Notes (Tables 3.5–3.7)

The reliability of capacitors normally varies as the third to the fifth power of the applied voltage, depending on the type of dielectric utilized.

The life of capacitors decreases with increased temperature. A simple rule of thumb is that the life is halved for each 10°C increase in temperature. At the low temperature end many dielectrics also exhibit a large decrease in capacitance with a small decrease in temperature.

Excessive derating of metallized film capacitors is not recommended. The applied voltage should be sufficient to promote self-healing of microshorts within these capacitors.

Aluminum electrolytic capacitors must be used with discretion in high-reliability applications. They are not normally hermetically sealed and can be damaged by the use of halogenated cleaning solvents unless special epoxy end seals are specified. They are not recommended where they may be subjected to low barometric pressures. They have definite wear-out mechanisms associated with use and also have limited shelf life. This storage deterioration can be overcome, however, by a simple dielectric reforming process documented in MIL-STD-1131 "Storage Shelf Life and Reforming Procedures for Aluminum Electrolyte Fixed Capacitors."

The storage life of aluminum electrolytic capacitors have improved significantly over the past few years. Nevertheless this failure mode has not been entirely eliminated.

Various quality grades of aluminum electrolytic capacitors are available commercially with different temperature specifications and storage lives.

Internal capacitor construction must be considered in selecting a capacitor for highly dynamic environments. Many types of aluminum electrolytic capacitors, for example, are easily damaged by vibration.

Solid-slug tantalum electrolytic capacitors tend to develop potentially destructive internal microshorts and thus require the use of current-limiting impedances in the circuit. For maximum reliability,

Table 3.4 Silicon Diode Derating

Diode type	Parameter	Derating factor	
		Nominal	Worst-case
Power rectifer	Power dissipation (Schottky only)	50%	80%
Power rectifer	Average forward current	50%	80%
	Reverse voltage	70%	80%[a]
	Junction temperature[b]	100°C	135°C
Small signal	Power dissipation (Schottky and PIN)	50%	75%
	Average forward current	50%	80%
	Reverse voltage	70%	80%[a]
	Junction temperature	100°C	135°C
Transient suppressor	Power dissipation	50%	70%
	Average current	50%	75%
	Junction temperature	95°C	125°C
Voltage reference/regulator	Power dissipation	50%	75%
	Current (I_Z)	$I_Z = I_{z(nom)} + 0.5(I_{z(max)} + I_{z(nom)})$	
	Junction temperature	95°C	125°C
Microwave	Power dissipation	50%	75%
	Reverse voltage	70%	80%[a]
	Junction temperature	95°C	125°C
LEDs	Average forward current	50%	75%
	Junction temperature	95°C	115°C

[a]The sum of anticipated transient voltage peaks and operating voltage peaks should not exceed worst-case recommended derating factor.
[b]Assuming that the device is rated at 175°C. For devices rated at other than 175°C, use the formulas:

$$T_j\text{(nominal derated} = 0.5(T_{j(max)} - 25°C) + 25°C$$

$$T_j\text{(worst-case derated)} = 0.7(T_{j(max)} - 25°C) + 25°C$$

Table 3.5 Digital Microcircuit Derating

Type	Parameter	Derating factor	
		Nominal	Worst-case
Bipolar and MOS	Fixed supply voltage	(Manufacturer's recommended) ±3%	±5%
Bipolar and MOS	Dynamic supply voltage	70%	85%
Bipolar	Frequency	80%	95%
MOS (and CMOS)	Frequency	80%	90%
Bipolar and MOS[a]	Output current	80%	90%
Bipolar[b]	Junction temperature	110°C	125°C
MOS[b]	Junction temperature	85°C	110°C

[a]Reducing fan-out may increase the part count and thus reduce equipment reliability. This situation should be avoided.
[b]Assuming full-temperature-range devices, that is −55°C to +125°C. For limited-temperature-range devices the case temperature should not exceed the manufacturer's maximum rated case temperature.

Table 3.6 Linear Microcircuit Derating

Type	Parameter	Derating factor	
		Nominal	Worst-case
Bipolar and MOS	Fixed supply voltage	(Manufacturer's recommended)	
	Dynamic supply voltage	70%	80%
	Input voltage	60%	70%
	Output current	70%	80%
	Junction temperature[a]	80°C	105°C

[a]Assuming full-temperature-range devices, that is −55° to +125°C. For limited-temperature-range devices the case temperature should not exceed the manufacturer's maximum rated case temperature.

Table 3.7 Fixed Capacitor Derating

Capacitor type	Parameter	Derating factor	
		Nominal	Worst-case
Paper/plastic	DC voltage[a]	50%	70%
	Temperature	(Max −10°C)	
Mica	DC voltage[a]	50%	70%
	Temperature	(Max −10°C)	
Glass	DC voltage[a]	50%	70%
	Temperature	(Max −10°C)	
Ceramic	DC voltage[a]	50%	70%
	Temperature	(Max −10°C)	
Tantalum electrolytic	DC voltage[a]	50%	70%
	Temperature	(Max −20°C)	
Aluminum electrolytic	DC voltage[b]	—	—
	Temperature	(Max −20°C)	

[a]The sum of the anticipated transient voltage peaks and the operating voltage peaks should not exceed the worst-case recommended derating factor.
[b]Derating does not normally enhance reliability. However, the maximum applied voltage including the sum of the operating, ripple and transient voltages should not exceed 90% of the capacitor voltage rating.

the recommended minimum circuit impedance is 3 ohms/applied volt. Polarized solid-slug tantalum electrolytic capacitors must not be exposed to reverse voltages in excess of 2% of their rated voltage.

Wet-slug tantalum electrolytic capacitors can be seriously damaged by even small amounts of reverse current, resulting in leakage of the corrosive electrolyte.

Fixed Resistor Application Notes (Tables 3.8 and 3.9)

The maximum voltage applied to a fixed resistor should be limited to:

$$V(\text{derated}) = 0.8 \times \sqrt{\text{applied power} \times \text{total resistance}}$$

Table 3.8 Variable Capacitor Derating

Capacitor type	Parameter	Derating factor	
		Nominal	Worst-case
Ceramic	DC voltage[a]	30%	50%
	Temperature	(Max −10°C)	
Piston, tubular, trimmer	DC voltage[a]	40%	50%
	Temperature	(Max −10°C)	

[a]The sum of the anticipated transient voltage peaks and the operating voltage peaks should not exceed the worst-case recommended derating factor.

Variable Resistor Application Notes (Table 3.10)

The maximum voltage applied to a variable resistor should be limited to:

$$V \text{ (derated)} = 0.8 \times \sqrt{\text{rated power/total resistance}}$$

The use of variable resistors is not recommended in high-reliability applications. They are not hermetically sealed and are susceptible

Table 3.9 Fixed Resistor Derating

Resistor type	Parameter	Derating factor	
		Nominal	Worst-case
Composition	Power	50%	70%
	Temperature	(Max −30°C)	
Film	Power	50%	70%
	Temperature	(Max −20°C)	
Wirewound (power)	Power	50%	70%
	Temperature	200°C	200°C
Wirewound (precision)	Power	30%	60%
	Temperature	125°C	125°C

Table 3.10 Variable Resistor Derating

Resistor type	Parameter	Derating factor	
		Nominal	Worst-case
Wirewound	Power	30%	60%
	Temperature	(Max −20°C)	
Nonwirewound	Power	30%	60%
	Temperature	(Max −20°C)	
Thermistor	Power	50%	60%
	Temperature	(Max −20°C)	

to degraded performance due to the ingestion of soldering flux, cleaning solvents, and conformal coatings during equipment fabrication. Also, variable resistors contain moving parts that are subject to wear.

Operation of thermistors above the maximum hot-spot temperature produces permanent resistance changes. Use current-limiting resistors to prevent negative-coefficient thermistors from going into thermal runaway.

Transformer and Inductor Application Notes (Table 3.11)

Winding voltages are fixed and are not derated to improve reliability. Operation at less than the rated frequency will result in overheating

Table 3.11 Transformer and Inductor Derating

Parameter	Derating factor	
	Nominal	Worst-case
Operating current[a]	60%	80%
Transient voltage	90%	90%
Dielectric withstanding voltage	50%	60%
Hot-spot temperature	(Max −30°C)	(Max − 15°C)

[a]For each winding.

and possible core saturation. Frequency cannot be derated for transformers and inductors to enhance reliability.

Transformers and inductors are among the most temperature-sensitive electronic devices. Increased temperature degrades the insulation, reducting the reliability and the life of the devices.

In miniature designs, transformers and inductors may be significantly contributors to the overall temperature rise.

Relay and Switch Application Notes (Table 3.12)

The life of a mechanical relay or switch is a function of the amount of current which it carries as shown in Figure 3.5.

Dry circuit applications, that is, those with less than 100 mw of power, should not be derated. Furthermore, these parts must always be used and tested under dry circuit conditions. Application of significant amounts of current to these devices will effectively destroy them for use in dry circuit applications.

Table 3.12 Mechanical Relay and Switch Derating

Load type	Parameter	Derating factor	
		Nominal	Worst-case
Loads rated in watts or volt-amperes	Contact power	40%	50%
Resistive	Switching current	75%	75%
Capacitive[a]	Surge current	75%	75%
Inductive[a]	Switching current	40%	40%
Motor[a]	Switching current	20%	20%
Lamp filament[a]	Switching current	10%	10%
N/A	Contact voltage	40%	50%
N/A	Coil operate voltage	Not less than 90%	
N/A	Coil dropout voltage	Not more than 110%	
N/A	Temperature	(Max −20°C)	

[a]If the relay or switch specification defines different inductive, motor, filament, or capacitive load ratings, then derate to 70% of the specified rating.

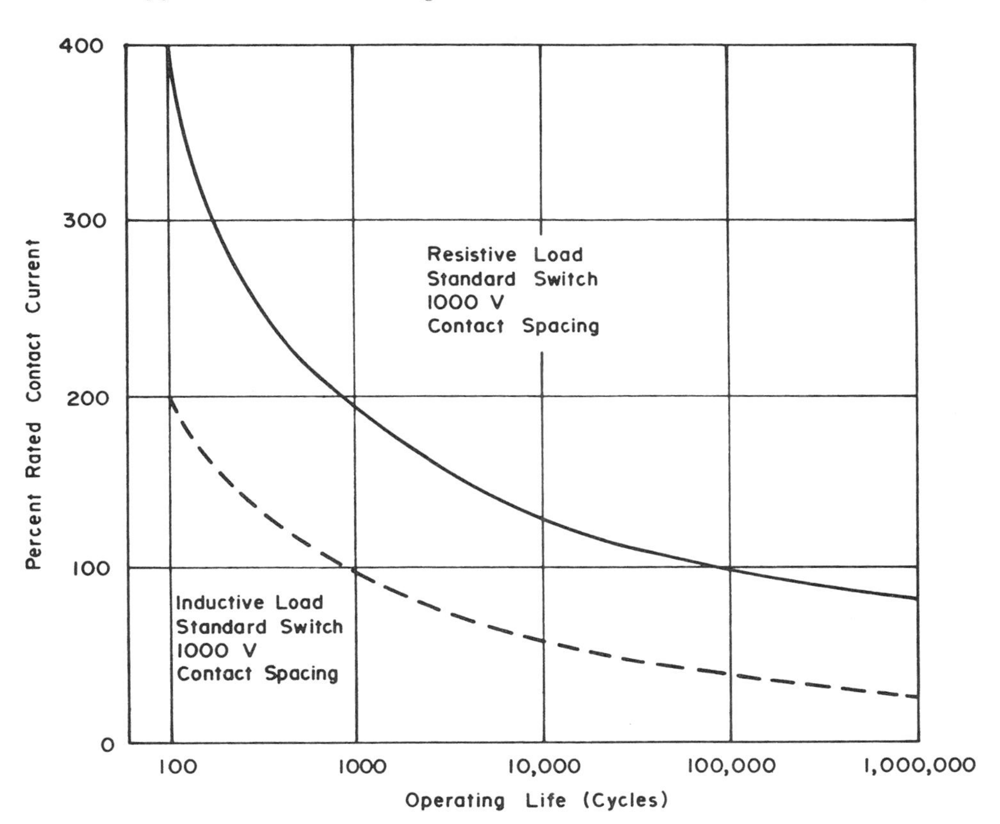

Figure 3.5 Effect of current upon operating life. [From *Reliability Design Handbook* (RDH-376), copyright 1976 by IIT Research Institute, Reliability Analysis Center, RADC/RAC, Griffiss Air Force Base, N.Y.]

Pure solid-state relays without mechanical contacts are also sensitive to nonresistive loads, surge currents can create high junction temperatures that degrade the device, causing it to fail.

If transient suppression clamping diodes are used with relay coils, their voltage rating should be not less than 1.5 times the normal coil voltage.

Lamp and Optical Source Application Notes (Table 3.13)

The recommended derating for both incandescent and gaseous lamps will double lamp life with only a 16% reduction in light output.

Table 3.13 Lamps and Optical Sources Derating

Type	Parameter	Derating factor	
		Nominal	Worst-case
Incandescent	Voltage	94%	94%
Gaseous lamp	Current	94%	94%
LED	Average forward current	50%	75%
ILD[a]	Power output	50%	70%
LED and ILD[a]	Junction temperature	95°C	110°C

[a]ILD stands for injection laser diode.

For gaseous lamps the starting and sustaining voltage levels are lamp characteristics and cannot be derated. Starting voltage can be increased for quicker response time, but life will be decreased.

Temperature cycling is especially damaging to incandescent lamps. The application of standby power (50% voltage applied) can significantly enhance reliability under these conditions.

Opto-isolators utilize both solid-state sources and detectors; therefore, the requirements of both Tables 3.13 and 3.14 apply.

Optical Detection Application Notes (Table 3.14)

For APD and PIN photodiodes, power derating is not necessary because internal dissipation is not significant. Reverse voltage cannot be derated for APD devices as the voltage is used to set or adjust device gain, and is typically set slightly below the breakdown voltage.

Table 3.14 Optical Detectors Derating

Type	Parameter	Derating factor	
		Nominal	Worst-case
PIN	Reverse voltage	70%	70%
APD[a] and PIN	Junction temperature	95°C	125°C

[a]APD stands for avalanche photodiode.

Table 3.15 Fiber Optics Cable Derating

Parameter	Derating factor
Temperature	20°C inside both the upper and lower limtis
Fiber tension	20% of proof test
Cable tension	50% of rated tensile strength
Bend Radius	200% of minimum

Table 3.16 Connector Derating

	Derating factor	
Parameter	Nominal	Worst-case
AC/DC voltage	50%	80%
Current	50%	85%
Insert temperature	(Max −50°C)	(Max −20°C)

Table 3.17 Rotating Devices Derating

	Derating factor	
Parameter	Nominal	Worst-case
Temperature	(Max −40°C)	(Max −15°C)
Bearing load	75%	90%

Connector Application Notes (Tables 3.15 and 3.16)

Additional voltage derating may be required for equipment subjected to low barometric pressure to prevent flashover.

When pins are connected in parallel at the connector to increase current capacity, allowance should be made for a 25% surplus of pins over that required to meet the current derating for each pin.

Rotating Device Application Notes (Table 3.17)

The voltage must be maintained at the nominal level to achieve maximum efficiency and life.

Excessive loads or low speed can create high winding temperature and excessive bearing loads. Moisture should be minimized to prevent corrosion, insulation degradation, and low resistance to electrical leakage.

3.3 PLACEMENT AND MOUNTING OF PARTS

Placement of Parts

Part operating temperatures are effected by the arrangement of the parts within the equipment. Equipment reliability can be enhanced by optimum part placement. Some guidelines for part placement to enhance reliability are as follows:

1. Provide the maximum separation possible between power-dissipating parts.
   (a) Avoid placing temperature-sensitive parts near high-power-dissipating parts, close to each other, or next to hot spots.
   (b) In free-convection-cooled equipment, avoid placing other parts directly above high-power-dissipating parts. Instead, stagger them horizontally.
2. Position temperature-sensitive parts in the coolest regions in order to maximize the reliability of the assembly.
   (a) With free-convection-cooled equipment, locate temperature-sensitive parts at the bottom and the others above them.
   (b) With forced-convection-cooled equipment, place temperature-sensitive parts nearest the coolant inlet side and less-sensitive parts on the outlet side.
   (c) With coldwall-cooled circuit cards, place sensitive parts close to card edge.

Mounting Parts

The thermal design objective in mounting parts is to minimize the thermal resistance between the case and the sink. Specific guidelines are as follows:

1. Use short paths in order to minimize the thermal resistance to conduction. Minimize the thickness of adhesive bonds used to attach parts to a module or cold plate.
2. Use large mounting areas, that is, maximize the areas of all conduction paths and interfaces between the parts and the heat sink, in order to minimize the thermal resistance to conduction.
   (a) High-power parts cooled by free convection and radiation or by impingement of cooling air should be mounted on heat-transfer fins to increase the area of the conduction path.
   (b) Avoid mounting heat-dissipating parts so that the leads are the only conduction path to the heat sink.
   (c) With free-convection cooling avoid the use of tightly spaced fins. Do not use more than four fins per inch, and avoid using fins higher than one inch.
3. Use materials having high thermal conductivity in order to minimize the thermal resistance to conduction.
   (a) Consider the use of metals such as copper and aluminum for heat-conduction paths and mounting brackets.
   (b) In spacecraft and high-altitude avionic equipment where free convection is very small or nonexistent, fill all gaps along the heat-flow path with thermally conductive compounds.
   (c) Multilayer printed circuit boards are often recommended to provide both a better ground plane and a good thermal path for heat dissipation. When using multilayer printed wiring boards, plated through-holes may be used to reduce the thermal resistance to conduction through the board (these copper-plated holes are thermal paths called thermal vias).
   (d) Minimize the use of dry interfaces between contacting surfaces as thermal paths. Instead, consider the use of zinc-oxide-filled silicon compound (thermal paste) between the surfaces to decrease thermal resistances.
4. When contact interfaces must be used, the following practices will help to minimize the contact thermal resistance:
   (a) Use as large a contact area as possible.
   (b) Ensure that the contacting surfaces are flat and smooth.
   (c) Use soft contacting materials.
   (d) Torque all bolts to achieve high contact pressure.
   (e) Use enough fasteners to assure uniform contact pressure.
   (f) When using coldwall-cooled cards, avoid the use of spring-loaded card guides to provide contact pressure between the card guide and the card edge. Instead, use positive restraints (for example, wedge clamps or cam-operated guides).

Each system design should have a formal thermal performance analysis based on its specific characteristics. However, there are some

Table 3.18 Guidelines to Reduce Component Temperatures

Semiconductor devices

1. Minimize thermal contact resistance between the device and its mounting. Use large area, smooth contacting surfaces, and specify the use of thermal compounds where required.
2. Heat transfer may be enhanced by mounting high-power hybrid microcircuit chips on molybdenum tabs having an area larger than the chip. The use of heat sinks on the top of dual-in-line packages is not particularly effective since the die is usually located near the bottom surface rather than the top surface.
3. Locate such devices remote from high temperature parts.
4. Use heat sinks with fins positioned vertically and in direction of the air flow. Painted or coated surfaces may be used to improve radiation characteristics.

Capacitors

1. Locate capacitors remote from heat sources.
2. Thermally insulate capacitors from other heat sources.

Resistors

1. Locate resistors for favorable convection cooling.
2. Provide mechanical clamping or encapsulating material to improve heat transfer to heat sinks.
3. Use short leads whenever possible.

Transformers and inductors

1. Provide conduction paths for the transfer of heat from these devices.
2. Locate them for favorable convection cooling.
3. Provide cooling fins where appropriate.

Printed wiring boards

1. Specify larger area conductors where practical.
2. Use an intermediate metal core in multilayer boards. Provide good conduction paths from these layers to support members and intermediate heat sinks.
3. Use protective coatings and encapsulants to improve heat transfer to lower temperature supports and heat sinks.

Source: *Reliability Design Handbook* (RDH-376), copyright 1976 by IIT Research Institute, Reliability Analysis Center, RADC/RAC, Griffiss Air Force Base, N.Y.

general rule-of-thumb approaches associated with specific components to obtain suitable thermal performance. Some guidelines to achieve reliable design through temperature reduction of specific components are itemized in Table 3.18.

4

Electrostatic Discharge

4.0 INTRODUCTION

The Nature of Static Electricity

When any two substances, solid, liquid, or gas make contact and are separated, an electrostatic charge is developed. This static charge results from an interaction of two materials involving the removal of electrons from the surface atoms of one of the materials. This transfer of electrons is related to the amount of energy (the work function) required to move the electrons over some separation distance. The basic factors which determine the magnitude of the static charge (Q) are the intimacy of contact, the rate of separation of materials, and the electrical conductivity of the materials. The polarity of the charge is a function of the placement of the materials on the triboelectric series as shown in Table 4.1.

The triboelectric series is a ranking of materials in the order of their ability to become positively charged (that is, their deficiency of electrons) during separation. Another way of visualizing static electrification via a triboelectric series chart is that it is a listing of materials by their relative electron densities. As long as there is a difference in electron density, charges will be transferred at each encounter. The cleanliness and the smoothness of the material surfaces in contact with one another also enhance the amount of surface electrification, or charge transfer. The apparent voltage (V) present at any specific instant is a function of the capacitance between the two objects since $Q = CV$.

Nonconductive materials such as common plastics can easily build up static voltages in excess of 20,000 volts. Electrostatic charges can also be induced on other objects by virtue of their proximity to the charged object. This is due to the electrostatic force field

Table 4.1 Triboelectric Series

Positive	Air
(+)	Asbestos
	Rabbit fur
	Glass
	Mica
	Human hair
	Nylon
	Wool
	Fur
	Lead
	Silk
	Aluminum
	Paper
	Cotton
	Steel
	Wood
	Amber
	Sealing wax
	Hard rubber
	Nickel, copper
	Brass, silver
	Gold, platinum
	Sulfur
	Acetate, rayon
	Polyester
	Celluloid
	Orlon
	Saran
	Polyurethane
	Polyethylene
	Polypropylene
	PVC (vinyl)
	Kel-F (CTFE)
Negative	Silicon
(−)	Teflon

emanating from a charged body, nonconductive materials have few free electrons available, and charges are not uniformly distributed on their surface. Conversely, conductors may have large amounts of surface charges but these are uniformly distributed. These charges can easily be drained to ground if such a path exists. High-moisture conditions or even a monolayer of moisture on a nonconductive surface will allow charge redistribution. This increases the apparent conductivity of the material and allows bleed-off of the charge to ground.

Electrostatic charges may be brought into the work areas by people or may be generated by people during their normal movements. Clothing and articles of common plastic such as cigarette and candy wrappers, styrofoam cups, parts trays and bins, tool handles, packing containers, highly finished surfaces, waxes floors, work surfaces, chairs, processing machinery, and numerous other articles are prime sources of static charges. These electrostatic charges can be great enough to damage or cause the malfunction of electronic parts, assemblies, and equipment during their discharge to ground or to an object at a lower potential.

ESD-Related Damage

Electrostatic discharge (ESD) can cause intermittent upset failures as well as hard catastrophic failures of electronic equipment. Intermittent or upset failure of digital equipment is usually characterized by a loss of information or temporary disruption of a function or functions. In this case no apparent hardward damage occurs and proper operation may be resumed after the ESD exposure. Upset transients are frequently the result of an ESD spark in the near vicinity of the equipment.

While upset failures occur when the equipment is in operation, catastrophic or hard failures can occur at any time. Most ESD hard damage to semiconductor devices, assemblies, and electronic components occurs below the human sensitivity level of approximately 4,000 volts.

In microcircuits, two primary hard failure mechanisms exist, both of which can cause permanent damage. First voltage punch-through can occur with very thin oxide dielectric layers. This is typical of the damage which occurs in CMOS and MOS devices. The second failure mechanism P-N junction degradation, is caused by excessive power dissipation and is typical of the damage inflicted upon bipolar circuits. In both cases reduction of circuit dimensions increases the probability of part damage due to electrostatic discharge.

Other failures may not be catastrophic, but may result in a slight degradation of key electrical parameters such as (1) increased leakage current, (2) lower reverse breakdown voltages of P-N junctions, or (3) softening of the knee of the V-I curve of P-N junctions in the forward direction. Some ESD part damage is more subtle. It can remain latent until additional operating stresses cause further degradation and ultimately catastrophic failure. For example, an ESD overstress can product a dielectric breakdown of a self-healing nature. When this occurs the part can retest well but it may contain a weakened area or a hole in the gate oxide. With use, metal may eventually migrate through the puncture, resulting in a direct short through the oxide layer.

Specific ESD Failure Modes and Mechanisms

There are a number of ESD-related failure mechanisms. Typically they include:

Thermal Secondary Breakdown (Avalanche Degradation)

For very small active area junctions (for example, emitter-base junction of a microcircuit transistor), the thermal time constant of the semiconductor material is generally large compared to the transient time of an ESD pulse. There is little diffusion of heat from the area of power dissipation, and a large temperature gradient can form in the part, resulting in a localized junction melting. The smaller the part, the less power the device can dissipate.

Parts which may be damaged by avalanche breakdown include:

1. Discrete MOS field-effect transistors (FETs)
2. Diodes (PN, PIN, Schottky)
3. Bipolar transistors
4. Junction field-effect transistors (JFETs)
5. Thyristors
6. Bipolar microcircuits (both digital and linear)
7. Input protective circuits on MOS and CMOS microcircuits

Some common failure modes for these parts include: (1) high leakage current between gate to source and gate to drain for JFETs, (2) degradation of beta (gain) and soft reverse characteristics for bipolar transistors, and (3) no output or latch-up for digital and analog microcircuits.

Metallization Melt

Failure can also occur when ESD pulses increase part temperature sufficiently to melt metal or fuse bond wires. This can occur, for example, in microcircuits where the metal strips have reduced cross-sections as they cross oxide steps or in nonuniform areas resulting in localized high current density with resultant hot spots in the metallizations. Parts sensitive to metallization failure include:

1. Hybrid microcircuits
2. Monolithic microcircuits
3. Multiple finger overlay switching and highfrequency transistors

Open circuit of a metallization strip is the normal failure mode for metallization melt type failures.

Dielectric Breakdown

When a potential difference is applied across a dielectric region in excess of the region's inherent breakdown characteristics, a puncture (or punch-through) of the dielectric occurs. Depending on pulse energy, this can result in either total or limited degradation of the part. The thinner the dielectric the lower its breakdown voltage. Parts utilizing MOS structures are most susceptible to dielectric breakdown. These include:

1. MOS FETs (discrete)
2. MOS and CMOS microcircuits
3. Semiconductors with metallization crossovers (in both digital and linear microcircuits)
4. MOS capacitors (in both linear and hybrid microcircuits)

The typical failure mode for these part types is an electrical short (high leakage) between gate and drain or gate and source.

Gaseous Arc Discharge

For parts with closely spaced, unpassivated, thin electrodes, gaseous arc discharge can cause degraded performance. The arc discharge condition causes melting and fusing of electrode metal. Part types affected by this failure mechanism include:

1. Surface acoustic wave (SAW) devices
2. Thin metal unpassivated, unprotected semiconductors and microcircuits

Surface Breakdown

For perpendicular junctions the surface breakdown is explained as a localized avalanche multiplication process caused by narrowing of the junction space charge layer at the surface. The destruction mechanism of surface breakdown results in a high leakage path around the junction, effectively nullifying the junction. Another mode of surface failure is the occurrence of an arc around the insulating material. This phenomenon resembles metallization-to-metallization gaseous discharge, except in this case the discharge is between metallization and semiconductor. Parts utilizing shallow junctions can be damaged as a result of surface breakdown.

Bulk Breakdown

Bulk breakdown is the result of changes in junction parameters caused by excessive local temperatures within the junction area. This effect is usually preceded by thermal secondary breakdown. Part types that commonly fail from thermal secondary breakdown are also candidates for bulk breakdown failure.

Other ESD Failure Mechanisms

Film resistors (for example, thick and thin film resistors in hybrid microcircuits, monolithic microcircuit thin film resistors, and encapsulated discrete film resistors) may fail as a result of dielectric breakdown, metallization melt or both. The failure is indicated by a shift in resistance, increase in degree of instability, and a change in the temperature coefficient.

A different type of failure mechanism is noticed for crystal oscillators. Crystal fracture results from mechanical forces when excessive voltage is applied, as occurs during a high voltage ESD pulse.

4.1 DEVICE AND EQUIPMENT SUSCEPTIBILITY TO ESD

Device or equipment susceptibility to ESD or both are usually determined by "V-ZAP" testing utilizing the simulated Human Body Test Model. This model is the most widely used and standardized ESD-susceptibility testing model. It is the model specified in both DOD-STD-1686/DOD-HDBK-263 "Electrostatic Discharge Control Program for Protection of Electrical and Electronic Parts, Assemblies, and Equipment" and MIL-STD-883 Method 3015. A schematic of the test circuit is shown in Figure 4.1. IEC Specification 801-2 "Electrostatic Discharge (for Industrial Process Control)" utilizes essentially the same schematic but specifies a capacitance value of 150 pF and a resistance value of 150 ohms.

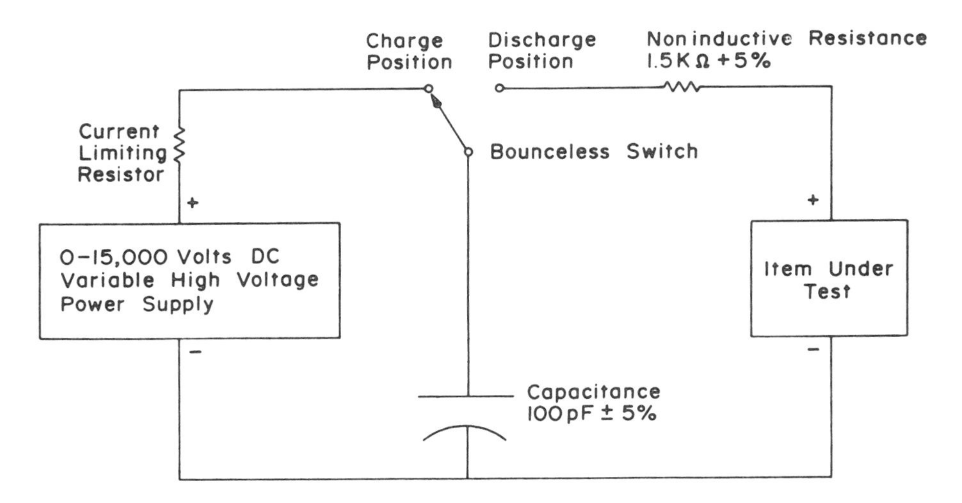

Figure 4.1 ESD test circuit. [From *Electrostatic Discharge Control Handbook for Protection of Electrical and Electronic Parts, Assemblies and Equipment (Excluding Electrically Initiated Explosive Devices)*, DOD-HDBK-263, Naval Publication and Forms Center, Philadelphia.]

This model simulates the situation in which a charged person or object comes into contact with a device that has one or more pins tied directly or resistively to ground. The current, voltage, power, and energies are easily calculated from the source capacitance and source resistance. The waveform, as in all discharge models, is theoretically a decaying exponential (as shown in Figure 4.2) with the pulse width depending upon the resistance and capacitance. Experimental evidence has shown that the range of capacitance of a human body is normally between 50 pF and 200 pF and the range of resistance is 1000 to 5000 ohms.

The reason for the requirement for a bounceless switch is to control the exact number of pulses which the device receives and thus assure reproducable test results. In a real-life situation the subject device may indeed experience an indeterminate number of discharge pulses. However, this is a nonpredictable function of the ionization path of the air between the charged object and the subject device. Multiple discharges will effect the degree of damage to the device.

The human body test model is also an effective method for testing the input protection capability of a device. This is the model which is specified in the individual MIL-M-38510 slash sheet test procedures and to date has been the most widely used and understood model. The majority of V-ZAP test data recorded at present utilizes this test model.

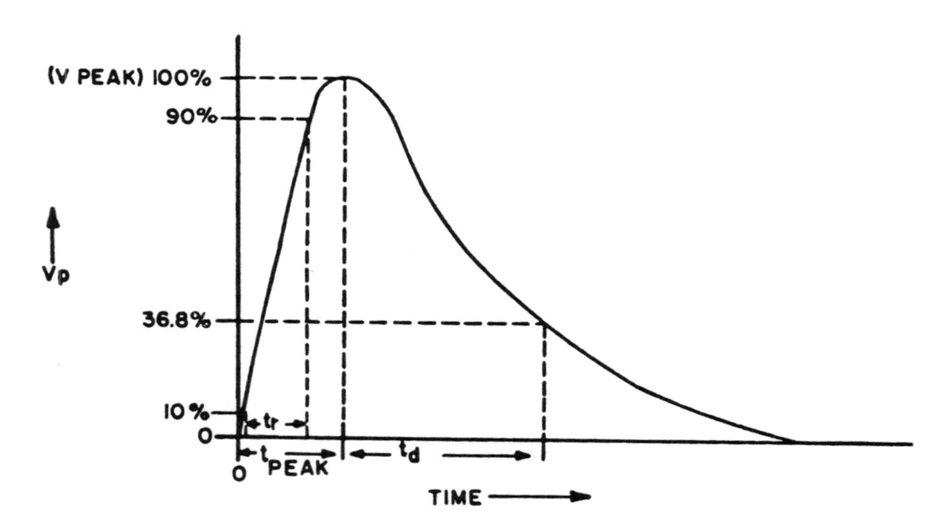

Figure 4.2 ESD test pulse shape. [From *Test Methods and Procedures for Microcircuits* (MIL-STD-883), Naval Publications and Forms Center, Philadelphia.]

Sensitivity Classification Schemes

DOD-STD-1686 and DOD-HDBK-263 define three ESD-sensitivity classifications. These classifications, applicable to both the equipment level and the device level, are defined as:

Class 1: Items which can be damaged by 0–1000 volts
Class 2: Items which can be damaged by 1000–4000 volts
Class 3: Items which can be damaged by 4000–15000 volts

Many industrial and commercial equipment manufacturers as well as military equipment manufacturers are actually performing qualification tests of their final products at these static voltage levels to assure that they will not be damaged by ESD under normal-use conditions.

Unfortunately MIL-STD-883 Method 3015 establishes a different two ESD-sensitivity classification systems for devices. These classifications are defined as:

Class A: Devices which can be damaged by 20–2000 volts
Class B: Devices which can withstand 2000 volts without damage

JAN microcircuits are actually marked to indicate their ESD sensitivity. As part of the qualification process documented in MIL-M-38510 they must pass a 2000 volt ESD V-ZAP test in accordance with MIL-STD-883 Method 3015 utilizing the Human Body Test Model. All JAN microcircuits which do not pass this test must be marked as ESD sensitive.

Recent studies, however, have developed additional theories of discharge which simulate situations quite different than these portrayed by the human body model. It has been shown by several investigators that the human body model discharge does not necessarily duplicate all or even the worst-case failure modes.*,†

Charged Device Model

The charged device model assumes that a charge can be placed on the lead frame of a device and then discharged to ground through one or more of its pins through a very low resistance. One obvious way in which this charge could be placed on the device is by sliding

---

*P. R. Bossard, R. G. Chemelli, and B. A. Unger (Bell Laboratories). "ESD Damage from Triboelectrically Charged IC Pins," EOS/ESD Symposium Proceedings, EOS-3, Sept. 1980, pp. 17–22.
†D. E. Fisch (Mosteck Cort.), "A New Technique for Input Protection Testing," IEEE 19th Annual Reliability Physics Proceedings, April 1981, pp. 212–217.

the device out of a shipping tube onto a grounded surface. The sliding motion of the device on the plastic tube material generates a triboelectric charging of the microcircuit pins and lead frame.

Since the lead frame is conductive, a constant charge density results (an equipotential surface) due to the capacitance of the device. This capacitance coupled with the charge stored on the de-is the cause of the potential ESD hazard. Damage from the charged device model assumes a very low resistance to ground upon device discharge. This results in a high-amplitude, short-duration pulse. In the situation where a device slides out of a tube and contacts a conducting surface, the surface itself does not have to be grounded for device damage to occur. Since the capacitance of the surface is usually much higher than that of the device, it is capable of accepting most or all of the charge on the device.

The capacitance of a device is relatively small compared to that normally used to model a human being. This capacitance, along with the low resistance to ground discharge, causes a much shorter and potentially more damaging transient pulse than is the case with the human body model.

Other Models

In addition to the human body model and the charged device model other significant ESD-damage models also exist. These include the floating device model and the field-induced model. These models, however, are beyond the scope of the text. An excellent reference source for additional information regarding ESD is the annual EOS/ESD Symposium Proceedings (see Chapter 15.4).

4.2 ESD DESIGN PRECAUTIONS

The first consideration in the design of equipment which is not susceptible to ESD is the selection of parts which are the least susceptible to ESD damage. This is not an easy decision for a number of reasons. Each new family of microcircuits is usually more sensitive to ESD than that which preceded it, and the manufacturers seldom give data regarding to ESD susceptibility of their devices.

It should be noted that JAN microcircuit parts are the only parts that are specifically marked as ESD sensitive. Commercial microcircuits, "883" parts, and vendor equivalent parts are never marked as ESD sensitive even though they are just as sensitive and possibly even more sensitive than the JAN parts.

Also, the ESD sensitivity of parts can increase with time as the manufacturer scales down the die size to increase his yield (i.e., the number of dies per water). This factor is not controlled in anything but the JAN product line where scaling of the die is not allowed without specific government approval and subsequent requalification.

The Reliability Analysis Center (see Chapter 15.4) has a publication which specifically deals with this problem. V-ZAP-1 "Electrostatic Susceptibility Data" contains tabulated data on the ESD sensitivity of microcircuits and discrete semiconductors to assist the designer in selecting those devices which are the least sensitive to ESD.

Built-in Protective Networks

Various protection networks have been developed to protect sensitive MOS (but not bipolar) devices. These circuit-protective networks provide limited protection against ESD. Many of the protective networks designed into MOS devices reduce the susceptibility to ESD to a maximum of about 800 volts. Examples of some of the commonly used protective circuits are shown in Figure 4.3.

The protection afforded by specific protective circuitry is limited to a maximum voltage and a minimum pulse width. ESD pulses beyond these limits can subject the part's constituent elements to damage or can damage the protective circuit elements themselves. The protective circuits themselves are often made of elements which are moderately or marginally sensitive to ESD. Damage to the protective circuit constituents can result in degraded part performance or can make the part more susceptible to subsequent ESD damage. This degradation, for example, could be a change in the speed characteristics of the ESD-sensitive part.

Multiple ESD pulses at voltages below the single ESD pulse sensitivity level can also weaken the part, cause failure of the part, or degrade performance of protection-circuitry constituents resulting in degradation and subsequent failure of the ESD-sensitive part. Furthermore, loss of the protective circuitry is not usually apparent after an ESD event.

While the present ESD protective circuits may be somewhat effective against Human body model pulses they are totally ineffective against charged device model pulses. This is because of the latter's much faster rise time and higher instantaneous currents.

Protection networks thus reduce but to not eliminate the susceptibility of a part to ESD. Fortunately this reduction in ESD sensitivity does result in a significant reduction in the incidence of ESD part failures.

The sensitivity of the same type of ESD-sensitive part can vary from manufacturer to manufacturer and within the same manufacturer's product as the die design is scaled down to increase the number of chips from a given-size wafer. Similarly, the design and effectiveness of protective circuitry also varies from manufacturer to manufacturer.

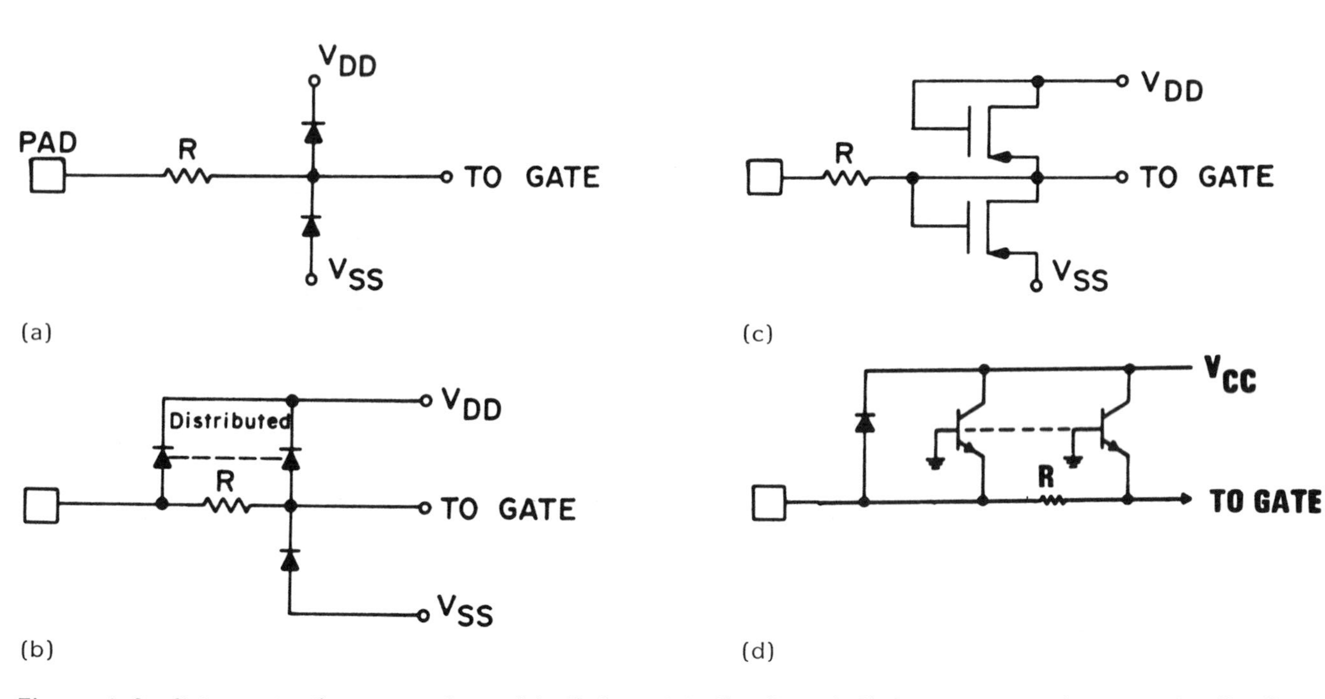

Figure 4.3 Gate protection networks: (a) diodes; (b) distributed diodes; (c) transistors; (d) distributed transistors.

ESD Design Precautions

Various design techniques have been employed in reducing the susceptibility of parts and assemblies to ESD. Diffused resistors and limiting resistors provide some protection, but are limited in the amount of voltage they can handle. Zener diodes usually require more than 5 nanoseconds to switch and thus may not be fast enough to protect an MOS gate. Furthermore, Zener schemes, diffused resistors, and limiting resistors often tend to reduce the performance characteristics of the part, where, in many instances, these characteristics are the primary considerations for which that part was originally selected.

Considerations for Assembly Protection

Leads to sensitive devices mounted on printed circuit boards (PCBs) should not be routed directly to connector terminals without series resistance, shunt paths, voltage clamps, or other protective means. Assemblies containing ESD-sensitive items should be reviewed for the possible incorporation of ESD protective circuitry.

Systems incorporating keyboards, control panels, or key locks should be designed to dissipate personnel static charges directly to chassis ground, thus bypassing ESD-sensitive parts. Antistatic security locks may be required in some installations.

Additional protection can sometimes be obtained for MOS devices by adding external series resistors to each input.

Equipment containing parts susceptible to transient upset from ESD sparks should be designed using radio frequency interference (RFI) and electromagnetic interference (EMI) protection schemes. These schemes include the following design practices.

RFI/EMI Design Zoning

Integral to a transient-upset-proof system are shielding, single-point grounds, and isolation of EMI-sensitive parts from EMI-generating assemblies. This concept can be used to isolate ESD-sensitive parts from the environment. Figure 4.4 represents the zoning concept as applied to a circuit board with ESD-sensitive parts. ESD-sensitive parts are moved to the center of the PCB where accidental contact is unlikely. The power supply leads of all the parts should be capacitively bypassed, and all the leads of ESD-sensitive parts that go to external locations should be buffered for maximum protection. There should be no signal leads nearby that can couple radio frequency (RF) energy to the ESD-sensitive parts. Although zoning is simple in concept, violations can be subtle since there are numerous possible paths for allowing RF energy into highly sensitive areas. Some coupling can be as obvious as ESD breakdown of panel indicators,

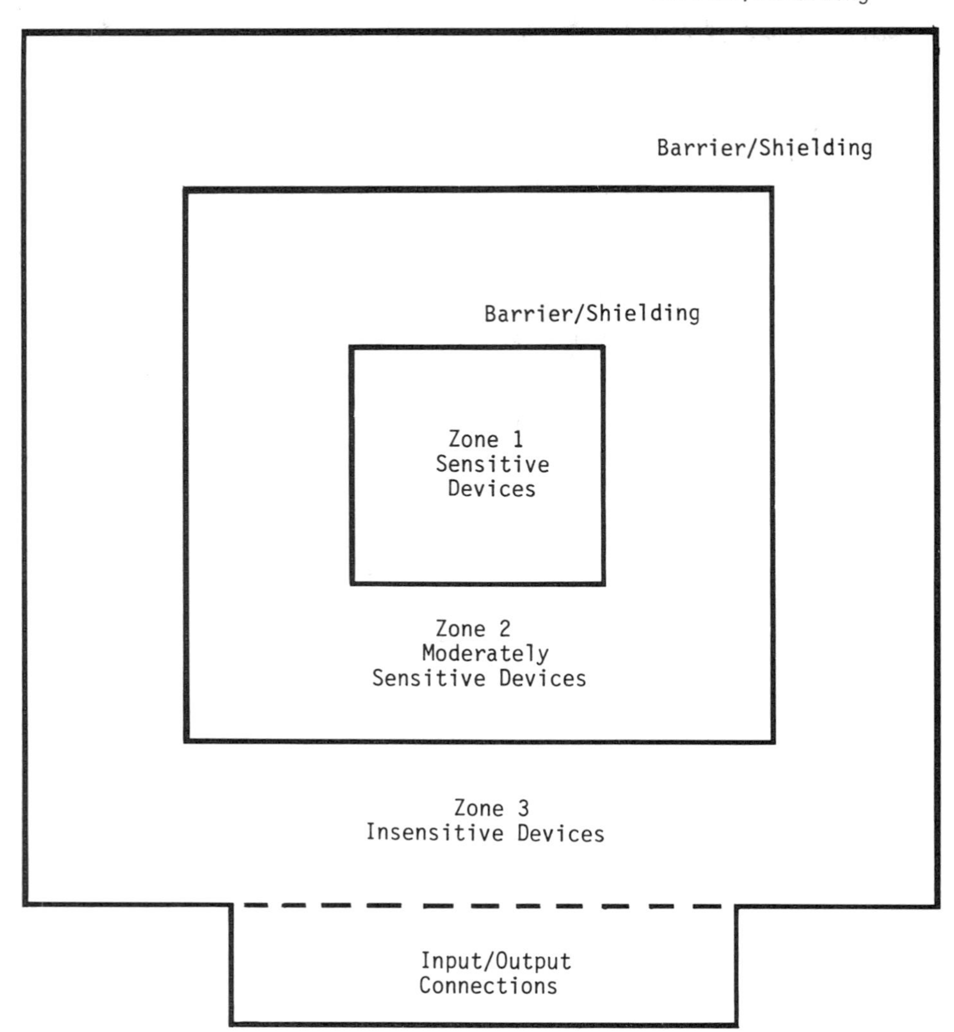

Figure 4.4 Zoning for ESD protection. [From *Requirements for the Electrostatic Discharge Protection of Electronic Components and Assemblies* (NAVSEA OD 46363), Naval Sea Systems Command.]

or as subtle as RF coupling between the ESD source and sensitive areas. This loop coupling is an especially insidious problem for operating frequency bands over 500 MHz. This wide-susceptibility bandwidth makes good ground planes a necessity.

Faraday Shielding and Masking

For many applications, simple Faraday shielding using conductive films or foils can protect very sensitive parts. Care must be exercised that these shields are not used where higher performance RF shields are required. A simplified shielding concept for localized protection of an ESD-sensitive part is shown in Figure 4.5. One may also reduce the sensitivity of an assembly by increasing the amount of ground conductors on a PCB. These extra conductors mask the sensitive-part leads so that the probability of an ESD spark coupling to a signal line is reduced. A similar process recognizes that an ESD spark is more likely to strike a conductor edge with its microscopic sharp edges than strike the smooth surface of the conductor. Thus, by etching a pattern on the ground plane, the likelihood of an ESD pulse striking a ground rather than a signal line is increased.

Grounding

Grounding schemes include the multipoint uncontrolled ground, the fishbone type ground, and the single-point ground. Each of these

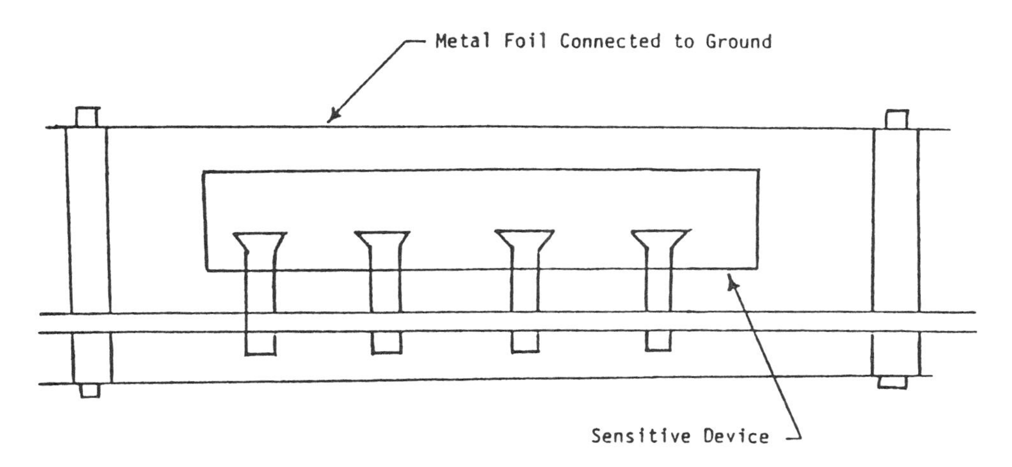

Figure 4.5 Device Faraday shield. [From *Requirements for the Electrostatic Discharge Protection of Electronic Components and Assemblies* (NAVSEA OD 45363), Naval Sea Systems Command.]

schemes leads to difficulties for complex systems. A hybrid grounding technique is shown in Figure 4.6. This technique is based upon implementation of the circuit zoning and shielding concepts previously described, and it exhibits some aspects of the multipoint, fishbone, and single-point grounding systems. At each level of zoning, a ground plane is provided.

This ground plane is connected to the parts within the zone using a multipoint grounding system (care must be taken to ensure that each component finds only one path to the ground plane). The ground plane is also bonded to the shield of the zone. Each zone

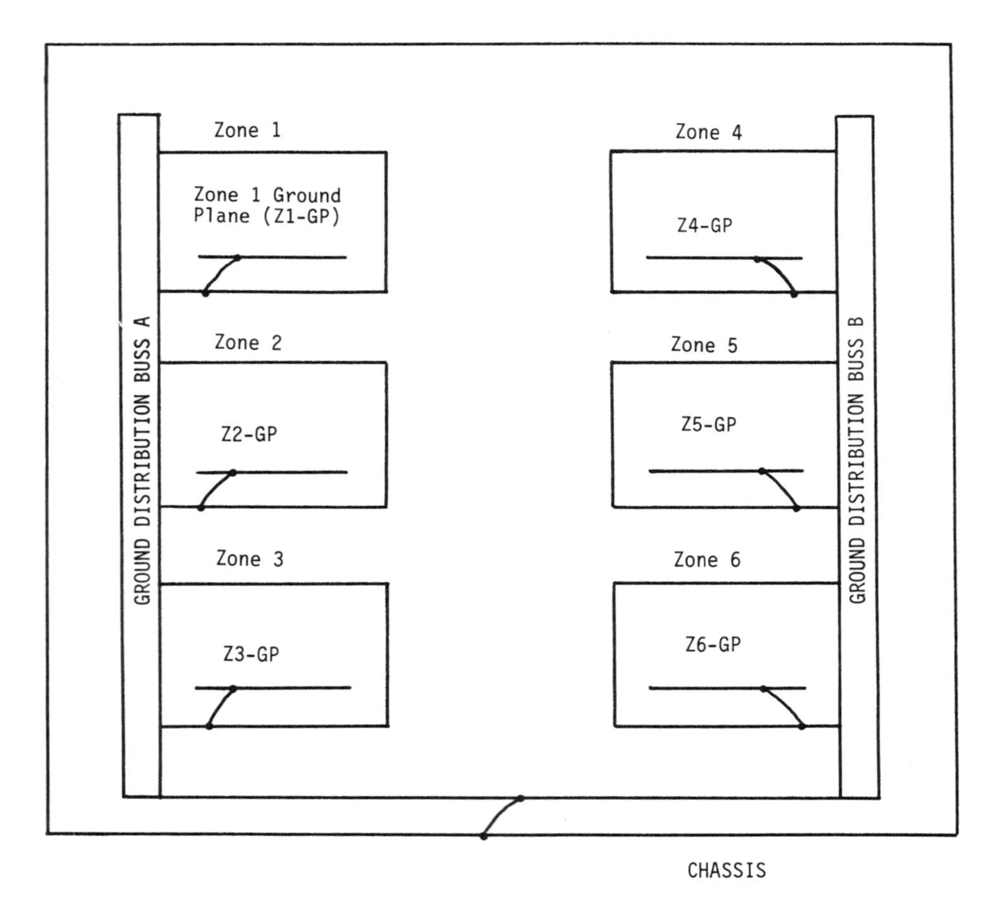

Figure 4.6 Grounding technique. [From *Requirements for the Electrostatic Discharge Protection of Electronic Components and Assemblies* (NAVSEA OD 46363), Naval Sea Systems Command.]

within an overall chassis is then connected to the chassis ground plane. Depending on the complexity of the subsystem being designed, it may be necessary to provide more than one chassis ground plane. For example, in a rack, separate ground planes may be required for each drawer. In any case, the chassis ground planes are then bonded to the chassis.

The general solution for many of the common impedance problems lies in adequate bonding techniques. If a low impedance path is available throughout the ground plane, then common impedance problems will usually be significantly reduced.

Cabling Practices

Besides direct contact interjection of an ESD pulse, cable coupling of ESD transients can also be a problem. Such ESD-coupled transients can occur in several ways:

By electromagnetic energy penetrating through the cable shielding and coupling into the inner conductors
By electromagnetic-energy transfer from one conductor to another within the cable
By electromagnetic-induced surface currents on the cable sheath which inject energy into the system at cable terminals

Wiring to ESD-susceptible items should be shielded where possible to prevent transient coupling. The best shields are solid materials like conduit, but metal-braid shields can also be used effectively. It is imperative that the continuity of shields be maintained at the back shells of inerface connectors. In particular, the overall shield must have circumferentially complete termination to the connectors. Pigtail terminations do not provide acceptable continuity. The use of special EMC/RFI protective connector back shells is recommended where possible. The order to preference for wiring to ESD-sensitive items is: (1) twisted shielded pair, (2) shielded pair, and (3) plain twisted pair.

4.3 ESD PROTECTIVE MATERIALS AND EQUIPMENT

The protection of ESD-sensitive electronic parts, assemblies, and equipment can be enhanced through the implementation of low-cost ESD controls, many of which have been used in the ordinance industry for decades. ESD-sensitive parts are susceptible to damage during processing, assembly, inspection, handling, packaging, shipping, storage, testing, installation, and maintenance throughout the equipment's life-cycle both at the manufacturer's and the user's facility. An effective ESD control program will normally utilize a variety of protective materials and equipment to prevent damage to ESD-sensitive items.

ESD-Protected Work Station

All activities involving ESD-sensitive items should be performed at an ESD-protected work station. Such a work station is shown in Figure 4.7. The ESD-protected work station should be clearly identified by prominently placed signs, and access to such areas should be limited to only properly trained and suitably equipped personnel.

Wrist Straps

Wrist straps are the first line of defense in the battle against ESD. The purpose of the wrist strap is to provide a permanent path to ground for the individual operator in order to prevent unsafe static charge levels from being generated during ordinary work-related movements. The wrist strap provides prompt and effective removal of these charges. Essentially all ESD prevention programs should utilize wrist straps especially if no other precautions are taken. The writs strap must contain a current-limiting resistor (nominally 1 megohm) to protect the operator in case of accidental contact with line voltage.

Protective Work Surfaces

The purpose of an ESD protective work surface is to drain static charges out of an operator's general work area. After wrist straps,

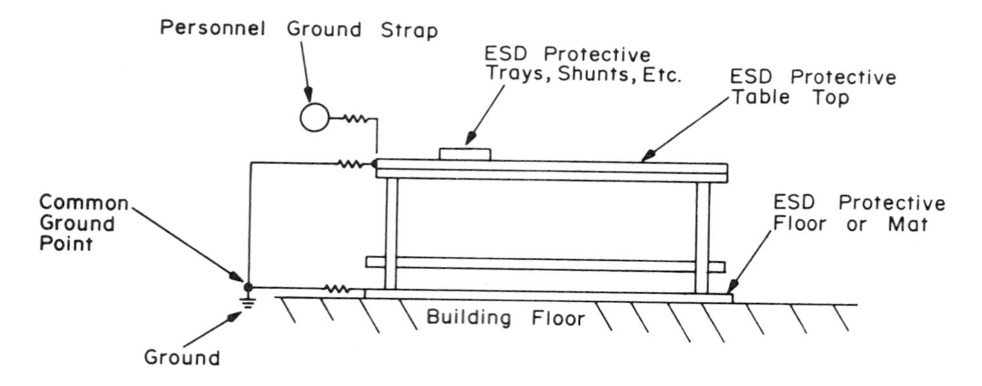

Figure 4.7 ESD protected work station.

protective work surfaces are the most important defense against ESD damage. There are a considerable number of different generic types of ESD protective work surfaces presently available with new ones coming on the market continually. In selecting the most effective work surface for a given application a number of factors, in addition to cost, must be considered. These factors include:

1. Electrical conductivity
2. Life, durability, or wear resistance
3. Chemical compatibility
4. Surface hardness
5. Operator comfort
6. Grounding method

ESD protective materials are frequently grouped into three surface-resistivity ranges:

Conductive material: 10^5 ohms per square maximum
Static dissipative materials: greater than 10^5 but less than 10^9 ohms per square
Antistatic material: greater than 10^9 but less than 10^{14} ohms per square

Static dissipative material is the optimum material for a protective work surface because it presents a compromise between being too conductive and not conductive enough.

Protective Packaging

ESD protective packaging encompasses an array of different items. The various forms of protective packaging include:

1. Protective bags
2. Antistatic cushioned packaging materials
3. Conductive foam
4. Conductive shunts or shorting plugs
5. DIP sticks or magazines and microcircuit carriers
6. Tote boxes, bins, and trays

An effective ESD control program will utilize a number of these items in the battle against ESD.

To be completely effective, electrostatic protective packaging requires the application of three separate and distinct principles: (1) equipotential bonding, (2) the prevention of charge generation caused by triboelectric contact and separation, and (3) protection from strong electrostatic fields by Faraday cage shielding.

Equipotential Bonding

Bonding is the process of connecting two or more conductive objects together by means of a conductor. There is practically no potential difference between two metallic objects that are connected by a bond wire because the current through a bond wire is generally quite small. However, the situation may be different with an object connected to ground. Grounding (earthing), is the process of connecting one or more objects to ground in a specific form of bonding. An object that is connected to ground may, under heavy current flow, develop a high potential difference with respect to ground since $E = I \times R$. Thus, when all the terminals of an electronic device are bonded together (but not grounded), this device is virtually immune to ESD. However, a bonded device that is charged might be subject to damage if suddenly grounded.

ESD-sensitive devices may be equipotentially bonded, for example, by inserting them into conductive foam. Equipotential bonding of modules and PCBs containing ESD-sensitive parts may be accomplished by the use of shorting plugs or conductive shunts placed on PCB edge connectors. This will help ensure that during shipping and handling all of the circuitry on the card will be maintained at the same potential.

Triboelectric Charging

As previously stated, when any two materials are placed in intimate contact and then separated, a triboelectric charge is generated. The triboelectric series (Table 4.1) identifies which material will become negatively charged. Triboelectric charging can be reduced in three ways. The first way is by proper selection of packing material. The second way, by the use of packing materials with sufficient conductivity to allow any such charges to be quickly bled away and dissipated before they can build up to damaging levels. The third way is by the introduction of a lubricant or other surface contamination between the materials to reduce the intimacy of their contact. In particular, antistats, when deposited on the surface of a material, act in such a way as to prevent charge generation.

Faraday Cage Shielding

Charges placed on an insulated hollow conductive object reside entirely on its outer surface. Because of this, no charge is apparent inside the conductive object. Michel Faraday performed experiments to prove this. He built a large metal-covered sphere which he mounted on insulating supports and charged with a powerful electrostatic generator. Faraday wrote of his experiment: "I went into the sphere and lived in it, and using lighted candles, electrometers, and other tests of electrical states, I could not find the least influence upon them . . . and though all the time the outside of the sphere was very powerfully charged, and large sparks and brushes were darting off from every part of its outer surface."

It is this behavior that enables conductive bags to protect ESD-sensitive devices from external charges and electrostatic fields. More importantly, it illustrates the need to have ESD-sensitive devices contained within a conductive enclosure to be protected.

Higher Assembly Considerations

Each of the three principles equipotential bonding, prevention of triboelectric charging, and Faraday cage shielding, must be applied in all stages of shipping and handling; from the receiving of the ESD-sensitive devices through the completion of the finished product containing these devices. Assembly of an ESD-sensitive device into a higher level assembly does not render it insensitive. Because of their greater electrical capacitance, PCBs can store much more charge than the device itself. In some cases, then, the device is more vulnerable to ESD when it is installed on a PCB than the device would be by itself. In addition, the circuit paths themselves on the PCB can act as antennas and can intensify the potential to which the device is subjected when the PCB is exposed to an electrostatic field. Thus, the same or comparable precautions must be taken with higher level assemblies. For example, PCBs containing ESD-sensitive parts must be handled as ESD-sensitive assemblies, and finished products containing ESD-sensitive parts may require shipment with attached shorting plugs to protect the sensitive circuits therein.

Each of these three principles, equipotential bonding, prevention of triboelectric charging, and Faraday cage shielding, apply individually and cannot be interchanged. Unfortunately, however, no single package scheme incorporated all three principles. Therefore the proper ESD packaging will usually be a two- or three-step process so that all three types of protection can be incorporated.

Air Ionizers

Wrist straps and protective work surfaces are effective in eliminating electrostatic charges on conductive items. Wrist straps and protective work surfaces are not, however, effective in eliminating electrostatic charges on nonconductive items such as clothing and nonconductive tote trays. Air ionizers can neutralize static charges on nonconductive items by supplying them with a constant stream of both positive and negative air ions. The charged object attracts the oppositely charged ions, thereby neutralizing itself in situ. The unused ions or those having the same polarity as the charged surface are repelled and eventually recombine with other ions or are themselves neutralized by contact with grounded conductive surfaces.

Electrostatic Detectors, Voltmeters, and Monitors

Electrostatic detectors are used to determine the presence or absence of electrostatic charges in the work area. They may be used to determine the polarity and the relative magnitude of the charge. They are not, however, designed to make accurate measurements of electrostatic charges, or measure the rate of decay of electrostatic charges. These types of measurements require the use of more elaborate laboratory-grade electrostatic voltmeters.

A third class of instruments, electrostatic monitors, may be used for the continuous automatic surveillance of excessive electrostatic potentials in the work area. They may be used to record the event, notify the operator, and/or actuate air ionizers.

Conductive Floors, Floor Mats, and Footwear

Conductive floors, floor mats, and footwear are rather specialized forms of ESD protection. Although they may not be required in every instance, they do fulfill a definite role in the arsenal of ESD protective weapons. Their primary role is in those areas where, for whatever reason, it is not possible to employ all the previously discussed ESD protective tools. For example, wrist straps may not be used near moving conveyor belts and in wave soldering operations. In this type of application conductive floors or floor mats and conductive footwear should be utilized.

Conductive floors or floor mats alone, without conductive footwear, are of limited use in providing ESD protection. The conductivity of normal footwear varies greatly depending upon materials and construction. High-quality leather-soled shoes may well provide adequate conductivity with the floor whereas man-made shoe materials seldom provide adequate conductivity. Therefore, it is essential that heel straps or similar means be used to provide conductivity between the operator and the conductive floor or floor mat.

Chairs and stools used in conjunction with conductive floors should also be conductive. Both the chair or stool legs which make contact with the floor and the seat surface itself should be conductive to assure that the operator remains properly grounded even with his or her feet off the floor.

Garments and Clothing

Usually specific garments or clothing are not required for ESD control; however, some exceptions do exist. Shop or lab coats and smocks are probably the most important examples. Where these types of garments are required to be worn it is essential that ESD protection also be considered.

"Clean rooms" probably represent the biggest challenge in this area. Synthetic outer garments must be worn in these areas

(a)

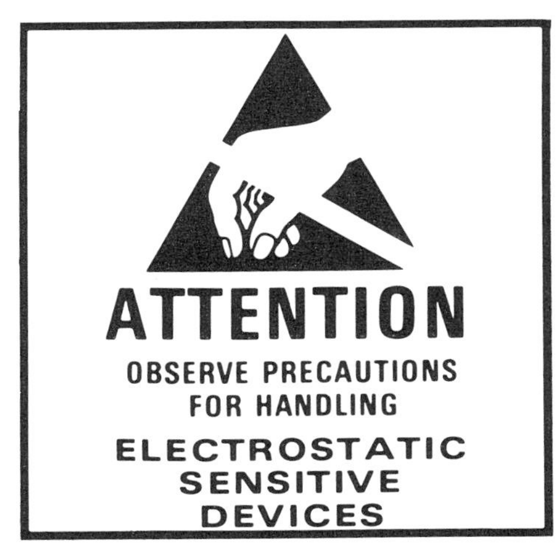

(b)

(c)

Figure 4.8 Common ESD symbols: (a) military standard; (b) EIA/-JEDEC approved; (c) also commonly used.

because of the tendency for natural fibers to shed lint and other contaminate particles. These synthetic materials tend to be capable of generating significant static charges. Therefore either highly specialized fabrics must be utilized or the garments themselves must be treated with a topical antistat after each laundry or dry-cleaning to assure their safe use in a clean room.

Topical Antistats

Topical antistats may also be an important weapon in the battle against ESD. They function in two different ways. First, they reduce the materials' coefficient of friction by increasing surface lubricity. This tends to reduce the maximum potential charge that can be generated in a frictional or triboelectric situation. Second, they increase surface conductivity, thus allowing any charges to be bled off and dissipated more rapidly.

Topical antistats are typically used in applications such as the cleaning of work surfaces and floors and surface treatment of items which are not amenable to other ESD control techniques, such as the exposed common plastic surfaces of CRT displays, computer terminals, and other equipment found in the work area.

Identification of ESD-Sensitive Items

To ensure that all personnel who handle ESD-sensitive items are aware of their static sensitivity, the items and their protective packaging must be suitably identified. All packaging containing ESD-sensitive items should be prominently marked with one of the commonly used ESD-sensitive symbols. These three symbols, shown in Figure 4.8, are the MIL-STD-129 symbol of three inward-pointing arrows, the EIA-approved hand symbol, and the slash-circle symbol.

ESD Summary

The expression *caveat emptor*, let the buyer beware, is especially germane to the purchase of ESD protective materials. For maximum quality assurance, substantial purchases should be made only after a thorough review of the market, the implementation of a formal product qualification program, and lot sample testing to assure consistent product quality.

Part III
Circuit Design and Analysis

5

Circuit Analysis Methods and Techniques

5.0 INTRODUCTION

Part parameters (for example, operating characteristics and values) change with time due to aging effects and stress. These part parameter changes can have a harmful effect on circuit performance and they must be recognized as a significant cause of both system and circuit failure.

Failure rates calculated in accordance with MIL-HDBK-217 do not normally address circuit failures caused by part changes due to aging. Parts such as resistors and capacitors change with age and with stress, and this aging degradation can result in out-of-tolerance failures of the circuit or the system. As a result of gradual aging parameters change, and eventually a point is reached where the collective effect of parameter changes causes circuit performance to be unacceptable.

In typical quantity manufacture, critical parts characteristics have statistical distributions. These characteristics (for example, resistance) have a nominal or mean value, and a variance above and below it. The extreme values of the variance are designated as the "tolerances." The part distributions are generally related to manufacturing lots and to the techniques used for the selection of close-tolerance parts out of wide-tolerance lots.

In addition to the manufacturing distributions, there are also distributions of these characteristics resulting from environmental effects (temperature, humidity, etc.), stresses (pressure, power dissipation, voltage, etc.), and time (cold flow, drift, aging, etc.). These distributions or tolerances must be combined with the manufacturing distributions or tolerances in order to determine the part's operational parameter distribution.

Typical examples of part parameter changes are shown in Figures 5.1 for resistors and 5.2 for capacitors. These figures show the average change from initial value versus time and the standard deviation of change from initial value versus time for the resistance of a particular resistor type. They also show the change in capacitance of a particular capacitor type. The resistor data is plotted for two stress levels, while the capacitor data is plotted at rated voltage. Another type of presentation (Figure 5.3) shows the initial tolerance and nominal value for a parameter, and then plots the change in these parameters under one specified stress and temperature condition for a period of time.

Compensating for Part Degradation

The design process must ensure that all of the various part parameter distributions or tolerances cannot combine in such a way as to interfere with the intended circuit function. In a circuit it is necessary to consider the overall effect of the expected range of manufacturing tolerance, operational environment, stresses, and the effect of time. Ideally the circuit should be designed to operate satisfactorily at the parameter extremes of all of its associated parts.

Two fundamentally different design approaches may be utilized to overcome these part degradation problems. They are:

1. Parameter change control: The device and material parameter changes may be controlled and held within specified limits for a specified time under specified conditions.
2. Tolerant circuit design: The circuits and systems may be designed to be tolerant of device and material parameter changes such that they can accommodate anticipated drifts and degradations with time.

Parameter Change Control

In this design approach, it is necessary to control not only the parameter value specified, but also to control its life history. Screening techniques, such as preconditioning or burn-in, can be used to eliminate or reduce part parameter changes. Thus screening produces more stability in the part parameter, so that there is a reduced probability of failure of the circuit due to part parameter changes.

An example of this concept is the use of the LTPD (lot tolerance percent defective) requirement in the microcircuit specification MIL-M-38510. This specification requires that critical part parameters be measured and recorded both before and after burn-in. The pre-burn-in and post-burn-in values are then compared and the entire lot may be rejected if the changes exceed a specified requirement thus indicating an unstable lot.

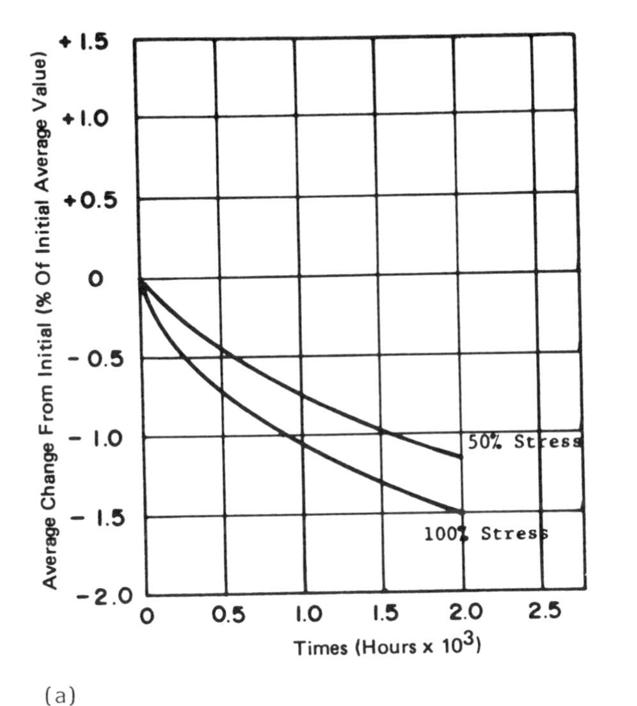

(a)

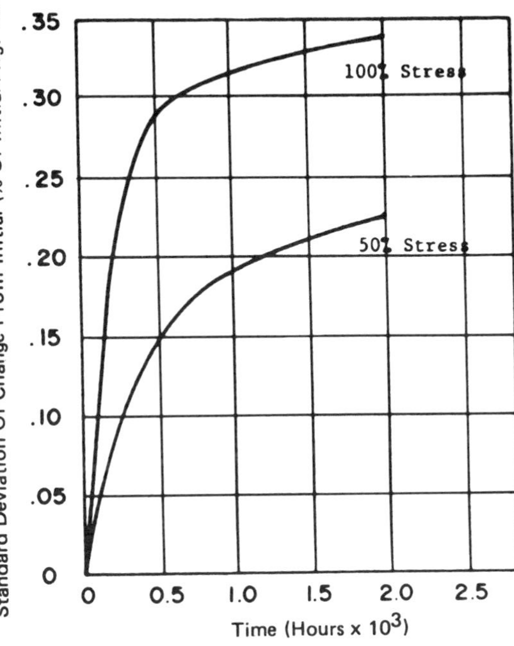

(b)

Figure 5.1 Resistor changes due to aging: (a) average change from initial resistance for MIL-R-11 carbon-composition resistors at 70°C ambient and various electrical stresses; (b) standard deviation of change from initial resistance for MIL-R-11 carbon-composition resistors at 70°C ambient and various electrical stresses. (From *Reliability Design Handbook* (RDH-376), copyright 1976 by IIT Research Institute, Reliability Analysis Center, RADC/RAC, Griffiss Air Force Base, N.Y.]

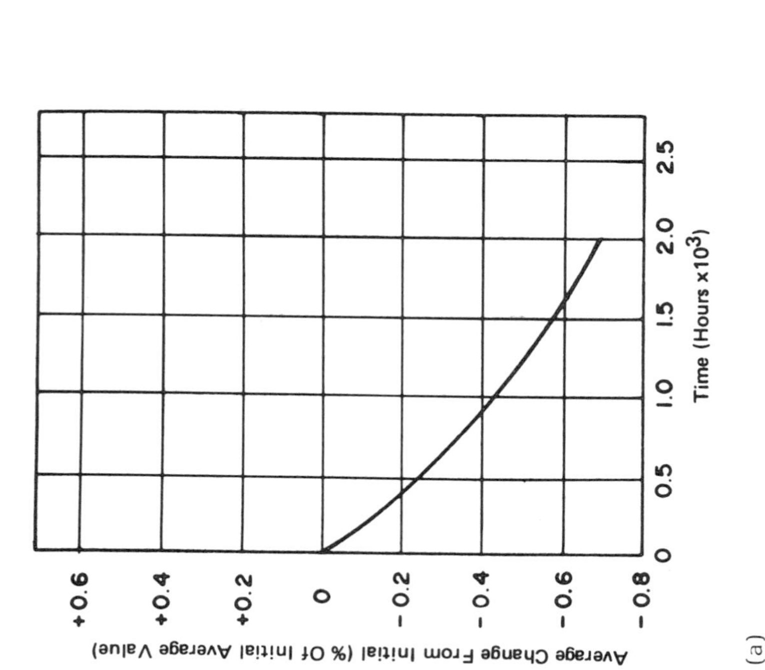

(a)

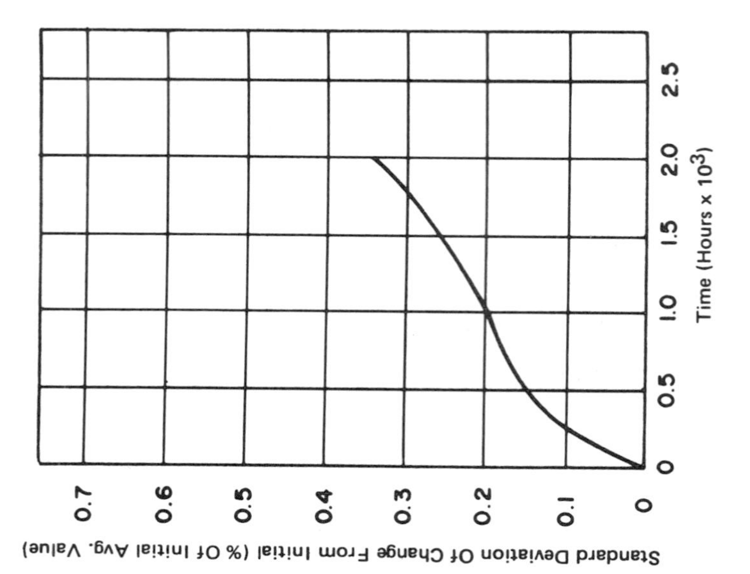

(b)

Figure 5.2 Capacitor changes due to aging: (a) average change from initial capacitance for MIL-C-20 temperature-compensating ceramic capacitors at rated voltage and 85°C ambient; (b) standard deviation of change from initial capacitance for MIL-C-20 temperature-compensating ceramic capacitors at rated voltage and 95°C ambient. [From *Reliability Design Handbook* (RHD-376), copyright 1976 by ITT Research Institute, Reliability Analysis Center, RADC/RAC, Griffiss Air Force Base, N.Y.]

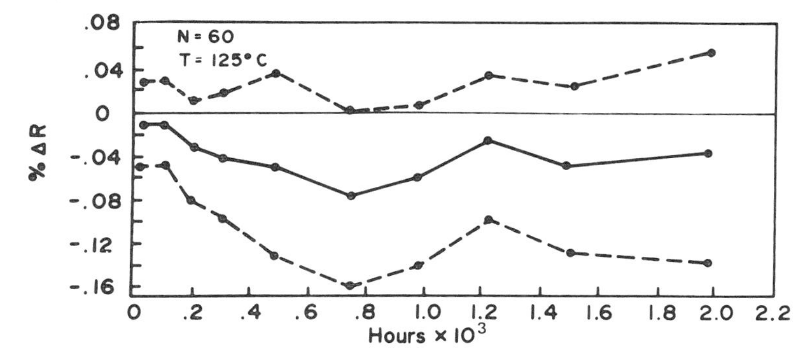

(a)

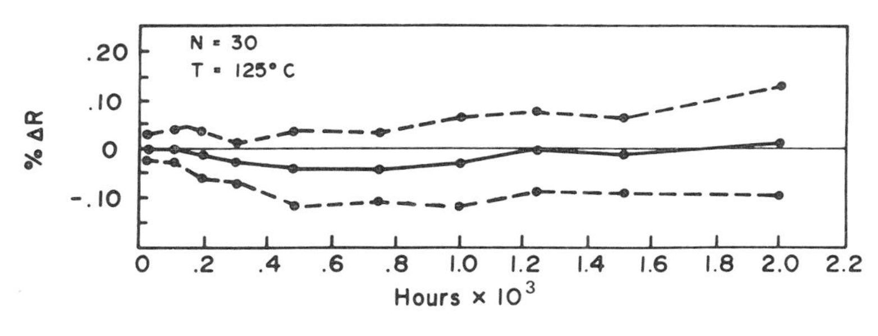

(b)

Figure 5.3 Resistance changes with life (1/8 watt metal film resistor): (a) 1/32 watt dissipation; (b) 1/16 watt dissipation. [From *Reliability Design Handbook* (RDH-376), copyright 1976 by IIT Research Institute, Reliability Analysis Center, RADC/RAC, Griffiss Air Force Base, N.Y.]

Parameter change control thus requires both detailed testing of the parts and control of the materials going into the parts. It requires strict process control, proven designs, and device testing to obtain valid parameter change data over the useful life of the parts. Both parameter value distribution for a single part and for a quantity of parts (population) related to changes in time and stress severity must be considered.

Tolerant Circuit Design

The second approach is to design circuits which are themselves tolerant of part parameter changes. Two different techniques are available to implement tolerant circuit design. The first is the use of negative feedback to compensate electrically for part parameter variation. This is typically used in analog circuits to provide performance stabilization.

The second technique is to prove by detailed mathematical analysis that the design of the circuitry is such that it provides the minimum required circuit performance, even though the part performance varies. This latter approach makes use of specific detailed analysis procedures such as:

1. Worst-case analysis
2. Parameter variation analysis
3. Moment method
4. Monte Carlo method

The objective of each of these analyses is to do one of the following:

1. Examine the circuit specification and then determine the allowable limits for each part parameter variation. The part is then selected considering the anticipated environments to which it will be subjected.

2. Examine the amount of parameter variation expected in each part and the expected range of inputs. The outputs are then determined under worst-case combinations, statistically expectant combinations, or other types of combinations. The circuit tolerance or resistance to degradation is then determined as the probability of satisfactory performance for a specific period of time.

We will examine and compare each of these detailed circuit analysis techniques in greater detail in Chapter 5.1.

Induced Part Damage

Parts may also be damaged or destroyed by transient overstress conditions. These transient conditions may result from a variety of causes (lighting strikes, power switching transients, noise spikes, and so on). Our circuits are expected to be both undamaged by these spikes and to continue to operate satisfactorily in spite of these transient conditions. Overstress and transient analysis will be addressed in Chapter 5.2.

Hidden Circuit Effects

After our design is complete it is still possible for hidden or sneak circuits to exist within our equipment, awaiting only the right combination of unique circumstances to play havoc with our design and reduce its reliability.

A unique type of analysis has been developed to address specifically these types of deficiencies. Sneak circuit analysis is the topic considered in Chapter 5.3.

This chapter will address a variety of circuit analysis techniques which have been developed to assist the designer in combating these various circuit-degrading influences and thus to prevent the reliability degradation of his circuit design.

5.1 CIRCUIT TOLERANCE ANALYSIS METHODS

An overview of the major attributes of the four best-known circuit tolerance analysis methods is shown in Table 5.1.

Circuit analysis employs mathematics and modeling to reduce design schematics from pictorial form to descriptive equations, so that the circuit's full range of behavior may be established numerically. Accurate mathematical evaluation of electronic circuits is very important.

The basic approach in each method involves the systematic manipulation of a suitable circuit model (or models) to give the desired information. They are all dependent upon the speed and accuracy of a digital computer to manipulate the mathematical models and to process the data.

Worst-Case Analysis

A worst-case analysis is a rigorous critique of the ability of a design to meet its operational requirements under the worst possible combination of circumstances. It can be used to determine whether it is possible, with given part parameter tolerance limits, for the circuit performance characteristics to fall outside of the performance specifications. This is done using circuit models in which parameters are set at either their upper or lower tolerance limits. Parameter values are then chosen to cause each performance characteristic to assume first its maximum and then its minimum expected value. If those performance characteristic values fall within specifications, the designer can be confident that the circuit is highly immune to drift. If specifications are exceeded, however, drift-type failures are possible, but the probability of their occurrence is not known.

A thorough worst-case analysis will usually encompass the following elements:

1. A narrative description of the specific design extending from the system block diagram down to the part level. It states the function of each part utilized in the circuit.

Table 5.1 Comparison of Circuit Tolerance Analysis Methods

Method of analysis	Type of model	Program output	Objectives
Worst-case	Nonstatistical	Worst-case values for outputs, all parameters at cumulative worst-case limits	Determine if failure is possible and under what conditions
Parameter variation	Nonstatistical	Range of variability data for Schmoo plots	Establish realistic tolerance limits for parameters
Moment	Statistical	Mean values of outputs, indices of variability, and redesign information	Reliability estimate, redesign if necessary
Monte Carlo	Statistical	Output histograms	Reliability estimates

2. A listing of the operational requirements placed on the circuit including the numerical data associated with its interfaces and the environments in which it must operate.
3. A presentation of the tolerance values applied to the electrical parameters of every piece part used in the circuit.
4. A listing of the performance attributes assigned to each circuit and the scenario(s) considered to be the worst case for each attribute.
5. Appropriate logical and/or mathematical proof of the ability of the circuit to achieve its performance attributes in the chosen worst-case scenario(s).
6. A concise statement of the analysis results, comparing the required circuit attributes with those analytically derived.
7. Conclusions regarding the successful, marginal, or deficient performance of the circuit and any recommendations for improving or eliminating noncompliant areas.

The exact format of a worst-case analysis is flexible. However, the analyst must have a thorough knowledge of electronic parts and an understanding of the way they react to various electrical and environmental stresses in order to determine the ability of the circuit to meet given performance requirements. This includes knowledge of the important part characteristics, for example, their normal performance parameters plus any stress-related and application limitations. Design problem areas are often dominated by subtle aspects of part behavior. The extent to which these subtle aspects are addressed and evaluated is often a clue to the integrity of a circuit analysis.

Computer-aided analyses are readily available and popular today. While the analyst may shift the details of numerical analysis to the computer, he must still have an ability to visualize the circuit design as a network. The network model, whether manually or computer generated, must be an accurate and thorough representation. The analyst must also be throughly familiar with the techniques employed in reducing the network to equations and numerical equivalents.

Network analysis involves various mathematical operations. Initial equations are formulated to describe the loop currents and node voltages associated with the network. These equations utilize network reduction techniques such as circuit isolation, Thevenin and Norton equivalent circuits, superposition theorem, Matrix manipulation, and time and frequency domain conversions. Additional mathematical techniques such as gain and phase evaluations (Bode plots, and so on), piecewise linearization of otherwise nonlinear elements, sensitivity analysis, impulse response, and additional statistical methods are then used to derive useful results from the initial equations. Occasionally complex variables, Fourier and other series analysis techniques, and numerical methods may also be employed for adequate performance evaluation.

Computer analysis programs can eliminate many tedious operations. However, the analyst cannot blindly accept these labor-saving tools. All computer methods are subject to limitations somewhere in their modeling and analytical process. While these programs may replace the mathematical chores, the analyst must still be certain that the results are not only plausible, but are also accurate. This requires a full and accurate knowledge of the methods used to arrive at the results and a sensitivity to the intimate details of those results. Otherwise the analyst will risk erroneous conclusions by total dependence on computer-derived results.

Some aspects of circuit design analysis are application oriented rather than mathematical. Thus the analyst must be able to associate the present parts and circuit design with those of other successful (or unsuccessful) designs.

Parameter Variation Method

The parameter variation analysis method provides a means for determining the maximum and minimum values for the input parameters of a circuit which will result in satisfactory circuit operation. Input parameters, either one at a time or two at a time, are varied in steps from their maximum to their minimum limits or vice versa, while all other input parameters are held at their nominal value. Data are thus generated to develop safe operating envelopes known as *Schmoo plots* for the various parameters. Schmoo plots were so named because of the observed unique operating characteristics of early computer memory cores and the similarity of these characteristics to cartoonist Al Capp's fictional animal in the "Little Abner" comic strip.)

If the values of these parameters are maintained within the limits determined from the Schmoo plots, the circuit will function successfully. Schmoo plots are a powerful tool used by the part specialist in characterizing a new part being considered for use in a design. They may also help to identify potential circuit-related weaknesses in the part which may not be obvious from a study of the part data sheet itself.

Figure 5.4 is an example of a Schmoo plot of the interaction between the Vdd and Vbb power supply voltages of a random access memory (RAM). The manufacturer's data sheet implies that the device should function properly with any power supply voltage between Vdd = 10.8 and Vdd = 13.2 volts, and Vbb = −4.5 and Vbb = −5.5 volts. However, as seen in Figure 5.4, with this specific device, if the two power supply voltages simultaneously reach either their upper or their lower specification limits the device fails to function properly.

Vdd (volts)									
14.4	0	0	0	0	0	0	0	0	0
	0	*	*	*	0	0	0	0	0
13.2	*	*	*	*	*	*	0	0	0
	*	*	*	*	*	*	*	0	0
12.0	0	*	*	*	*	*	*	*	0
	0	0	*	*	*	*	*	*	*
10.8	0	0	0	*	*	*	*	*	*
	0	0	0	0	*	*	*	*	0
9.6	0	0	0	0	0	*	*	0	0
	−6.0		−5.5		−5.0		−4.5		−4.0

Vbb (volts)

Figure 5.4 RAM "Schmoo plot." 0 = unsatisfactory operation; * = satisfactory operation. [From *Characterization of Random Access Memorie* (ECR-93), Electronikcentralen, Horsholm, Denmark, 1979.]

For a RAM the combined upper voltage corner (Vdd maximum and Vbb maximum) is the noise-sensitive corner (an increase in temperature will also increase the noise effects, primarily because the increase in supply current). The opposite corner, the speed corner (Vdd minimum and Vbb minimum), is also a potential problem area for this device. This is a result of device timing sensitivities.

Thus if the circuit designer decides to utilize this device and the device is expected to function at close to its upper frequency limits or in an environment with excessive electrical noise, he should ensure that the two power supplies are not boot-strapped together so that their voltages track together.

Moment Method

Statistics are combined with circuit analysis techniques in the moment method to estimate the probability that performance will remain within specified limits. This method applies the propagation-of-variance formula to the first two moments of component-part frequency distributions to obtain the moments of performance-characteristic frequency distributions. Using this information, the probability of specific circuit parameters drifting out of their acceptable range can then be computed.

Monte Carlo Method

In the Monte Carlo method a large number of alternate replicas of a circuit are simulated by mathematical models. Component values are randomly selected and the performance of each replica is determined for this particular set of components. The performance of the replicas is then compared with the specification limits to yield an accurate estimate of the circuit performance reliability.

Method Selection Tradeoffs

The nonstatistical, worst-case circuit analysis method gives basic information concerning the sensitivity of a configuration to variability in the parameters of its component parts. This information is useful to the designer in selecting economical, but adequately stable, components for the circuit, and in modifying the configuration to reduce the critical effects on certain parameters.

In contrast, the statistically based moment method and the Monte Carlo method use actual parameter variability data to simulate real-life situations and predict the probability that the circuit performance will be within the tolerance specifications. The moment method prediction of performance variability is usually less accurate than the Monte Carlo method, but it is satisfactory for most purposes. The moment method provides information that is extremely useful to the designer in pinpointing sensitive circuit areas and reducing this sensitivity because of parameter variability.

In addition to providing data on drift-type failures, these techniques are all capable of giving stress level information of the type needed for estimating catastrophic failure rates. Thus they are useful powerful tools for predicting overall circuit reliability.

However, because of the variety of variations and the large number of parameters for even a small number of parts, the task quickly grows out of proportion for manual computation, thus, computer programs have been generated for most if not all of these analysis methods.

Available Circuit Analysis Programs

Over the years a number of computerized circuit design analysis programs have been developed. Many were developed specifically to analyze circuits for transient radiation effects. Although they are not necessarily reliability oriented, they are an important aid to the electronic circuit designer. Some of these programs include: ASTAP, BELAC, CIRC, CIRCUS 2, ECAP, ECAP II, LISA, MARTHA, NET-2, PREDICT, SPICE, IG-SPICE, SCEPTRE, SUPERSCEPTRE, SYSCAP, TRAC.

Table 5.2 Circuit Analysis Programs for Personal Computers

Name	Description	Source
PC-USPICE	150 Elements, AC, DC, & transient analysis, IBM PC, AT	Electronic Software Products 18013 Sky Park Circle Irvine, CA 92714 (714) 261-1777
MICRO-CAP	40–60 Nodes, AC, DC, & transient analysis, IBM PC, APPLE, HP-150, & CP/M	Spectrum Software 1021 S. Wolf Rd. Dept. E Sunnyvale, CA 94087 (408) 738-4387
MICRO-CAP II	100 Nodes, AC, DC, & Fourier analysis, IBM PC, HP-150, & Apple-Macintosh	Spectrum Software 1021 S. Wolf Rd. Dept. E Sunnyvale, CA 94087 (408) 738-4387
ECA	64 Nodes, 127 Branches, AC & DC analysis, IBM PC, TRSDOS, & CP/M	Tatum Labs P. O. Box 722 Hawleyville, CT 06440 (203) 426-2184
AC NAP2 DC NAP	21 Nodes, 60 elements, AC, DC, & transient analysis, IBM PC, TRSDOS, & CP/M	BV Engineering Suite 207 2200 Business Way Riverside, CA 92519 (714) 781-0252
P SPICE	Up to 120 transistors per circuit, AC, DC, & transient analysis	Micro Sim Corp. 14101 Yorba St. Tustin, CA 92680 (714) 731-8091

These programs use a variety of different methods to perform the circuit analysis. Some use a nodal analysis method based upon Kirchhoff's current law, some a state variable analysis method based upon the formulation of a set of first-order differential equations. Others use a topological analysis method based on Mason's formula, which constructs the transfer function of a signal flow, or a tableau analysis method based upon Kirchhoff's voltage law, Kirchhoff's current law, and the functional relationships of the components.

A major weakness of earlier circuit analysis programs was the fact that they required access to large main-frame computers. With the miniaturization of computers this is no longer true. A number of circuit analysis programs are now available on large minicomputers, and, in addition, some circuit analysis programs are also available which utilize personal computers. Some examples of these personal computer circuit analysis programs are shown in Table 5.2.

5.2 OVERSTRESS AND TRANSIENT PROTECTION

Electronic circuits are very susceptible to disturbance from electrical transients. Electronic components are often overstressed and permanently damaged by short duration voltage/current transients. These transients may be caused by lightning, switching of electrical loads, capacitive and inductive effects, power supply fluctuations, testing errors, and so on. Some transients are unique to a given application. For example, in an automotive environment very severe transients may be caused by abrupt reduction in the alternator load (load dump) or by alternator field decay when the ignition switch is turned to "OFF".

Semiconductor circuit malfunctions arise from two general sources: transient circuit disturbances, and component burnout. Transient upsets are generally the more common because they can occur at much lower energy levels.

Transients in circuits can prove troublesome in many ways. Flip-flops and Schmitt triggers can be inadvertently triggered, counters can change count, data in memory cells can be altered one-shot multivibrators can pulse, switches can change state, CMOS devices can "latch-up" (requiring reset), the transient itself can be amplified and interpreted as a control signal, and so on. All exposed circuit nodes can be susceptible to transients including the input node, the output node, the power supply node, or any combination of these nodes.

Characteristics of Transients

Circuit transients can be generally characterized as follows:

1. Transients are remarkably independent of exact wave shape, their seriousness depends largely on the peak value of the transient and the time duration over which the transient exceeds the DC threshold. This waveform independence allows relatively easy experimental determination of circuit behavior with simple square wave pulses.

2. All circuit nodes are susceptible to transients, although, the input leads (or signal reference leads) are generally the most susceptible to transient upset. However, for ultimate reliability, all circuit nodes should be protected from electrical transients.

3. Circuit threshold regions for upset are generally very narrow, that is, there is a very small voltage amplitude difference between signals which have no probability of causing upset and signals which will certainly cause upset.

4. The DC threshold for response to a very slow input swing can be calculated from the basic circuit schematic. This can establish an accurate bound for transients that exceed the DC threshold for times longer than the circuit propagation delay (as defined in the manufacturer's specification).

Microcircuits and small semiconductor devices are particularly vulnerable to overstress, because of the limited power dissipation capability of the chips and the very low thermal inertia of their wire bonds.

The Wunsch-Bell Damage Model

It is primarily the high-amplitude, short-duration transient pulse which causes the destruction of electronic components. The extent of the damage suffered by a semiconductor device due to a transient pulse can be predicted to some degree using the Wunsch-Bell* electrical overstress model. This model is shown graphically in Figure 5.5. Along with the peak voltage and current incurred in a device, it has been shown that the pulse duration (t) is of prime importance in determining the amount of power required to cause the failure of a bipolar junction. The amount of power required to cause failure (P_f) is of the following general form (see table on p. 120).

The $Pf = Kt^{-1}$ region corresponds to the situation in which the pulse is short enough to exhibit an adiabatic thermal characteristic. That is, the pulse duration is shorter than the time it takes the heat dissipated near the junction to spread to surrounding areas.

---

*D. C. Wunsch and R. R. Bell (BDM Corp.), "Determination of Threshold Failure Levels of Semiconductor Diodes and Transistors Due to Pulse Voltages," IEEE – Transactions on Nuclear Sciences, NS-15, No. 6, December 1968, pp. 244–258.

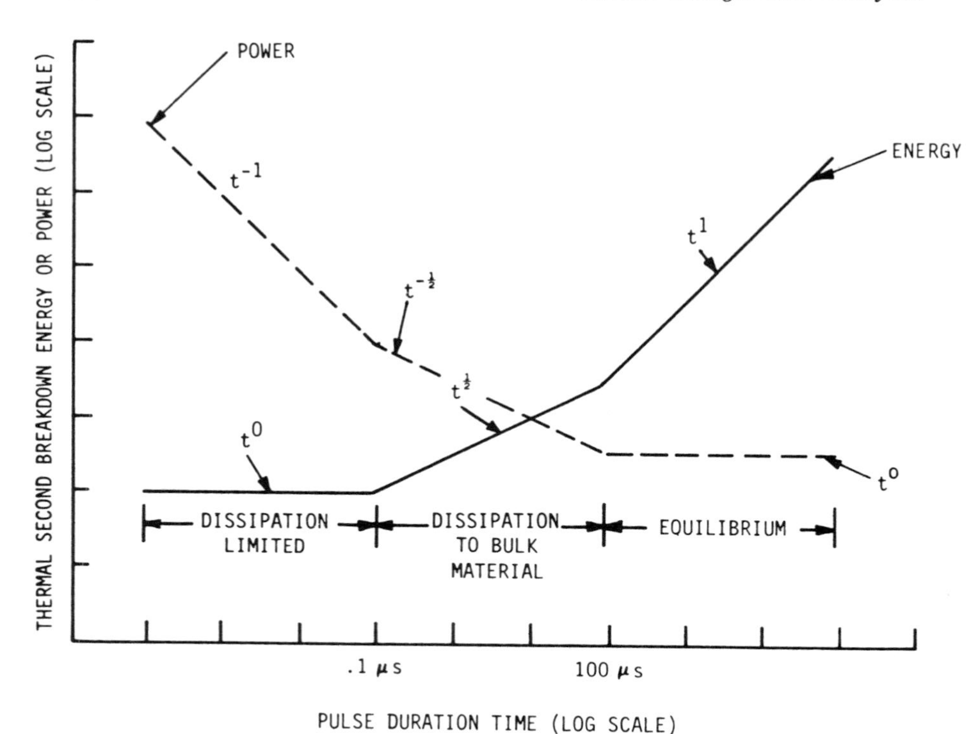

Figure 5.5 Failure power and energy vs. pulse width. [From *Electrostatic Discharge Susceptibility of Electronic Devices* (VZAP-1), copyright 1983 by IIT Research Institute, Reliability Analysis Center, RADC/RAC, Griffiss Air Force Base, N.Y.]

Pulse width	Power to fail
$t < 0.1$ μs	$P_f = Kt^{-1}$
$0.1 < t < 100$ μs	$P_f = Kt^{-0.5}$
$t > 100$ μs	$P_f = Kt^{0} = K$

The $P_f = Kt^{-0.5}$ region corresponds to a semiadiabatic thermal characteristic. The pulse width is long enough to allow some of the heat to diffuse into the surrounding areas. In the $P_f = Kt^0$ region, the pulse width is sufficiently long to allow the heat to dissipate into the surrounding areas at a constant rate while the pulse is occurring. In the equation, k is a thermal constant dependent on specific device parameters including: thermal conductivity, density, specific heat, and the melting point of the semiconductor material.

The plot of relative energy required to cause device failure as a function of pulse width is also depicted in Figure 5.5. Since the energy required to cause device failure (E_f) is the integral of the power to failure, the pulse width and associated energy required to cause device failure are:

Pulse width	Energy to fail
$t < 0.1\ \mu s$	$E_f = Kt^0 = K$
$0.1\ \mu s < t < 100\ \mu s$	$E_f = Kt^{0.5}$
$t > 100\ \mu s$	$E_f = Kt^1$

The failure mechanisms modeled by this kind of pulse-width dependency are normally power or energy-related failure mechanisms, such as thermal secondary breakdown.

Protective Measures

Four major techniques are available for limiting the damage to semiconductor junctions due to transient. There are as follows:

1. By shunting the transient to ground before it can cause problems. Examples: shielding and power supply decoupling capacitors which act as a short circuit path to ground for high-frequency transient pulses.
2. By limiting the peak value of the transient. Example: clamping diodes.
3. By slowing down the transient. Example: simple integrating circuits to limit high-frequency components in the waveform, and thus round off the leading edge of the pulse.
4. By using a less susceptible device, that is, one with a larger size die (capable of dissipating more power).

Specific Protection Examples

Passive devices can also be damaged by transient voltages. However, the energy levels required for such damage are generally much higher than for small semiconductor devices. Therefore, passive devices are frequently used along with less sensitive semiconductors to protect other more sensitive semiconductors.

Common forms of transient suppression are the use of gas tubes, metal-oxide varistors (MOVs), and silicon avalanche devices on incoming power lines, data lines, and telecommunication lines of electronic equipments to protect them from voltage spikes. Gas tubes crowbar, that is, they break down under surge conditions conducting the transient directly to ground. Metal-oxide varistors function by a rapid change in resistance when exposed to a transient overvoltage, thereby clamping the voltage at a specified value and diverting the transient to ground. Silicon avalanche devices also perform a clamping function.

Power supply decoupling capacitors are an example of another type of transient protection. Capacitors are added to the DC power bus to filter out transient riding on the power bus. Power supply decoupling is especially important, for example, with the newer 54/74HC devices, because both halves of the CMOS transistor are "on" briefly adding transients to the DC power bus which must then be filtered out.

Logic devices which interface with inductive or capacitive loads, or which are mated with test connectors, may also require additional transient voltage protection. This protection can be provided by techniques such as placing a capacitor between the voltage line to be protected and ground to absorb high-frequency transients, or using clamping diode protection to prevent voltages from rising beyond a fixed value and series resistances, to limit current values.

Many different techniques are available for providing transient suppression for semiconductor devices and their associated circuits. Some simple examples are shown in Figures 5.6 through 5.8. They illustrate diode protection (Figure 5.6), inductive load protection (Figure 5.7), and silicon-controlled rectifier (SCR) gate protection

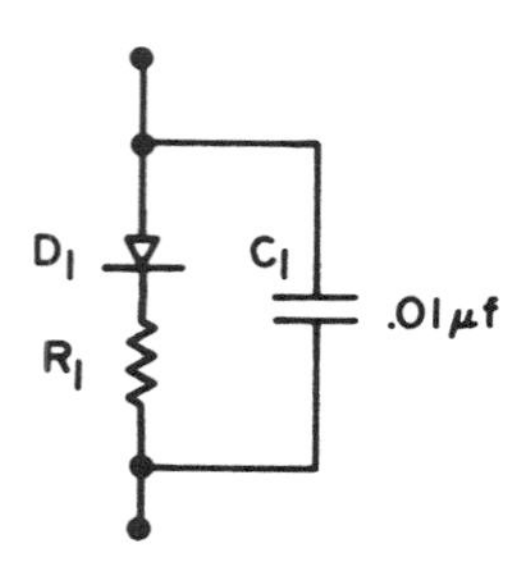

Surge Current Limit Resistor (R_1) and Transient Suppression Capacitor (C_1)

Figure 5.6 Diode protection. The best protection for a diode is sufficient overrating of the reverse breakdown voltage (PIV), forward surge current (I_S), and power dissipation capability (P). [From *Reliability Design Handbook* (RDH-376), copyright 1976 by IIT Research Institute, Reliability Analysis Center, RADC/RAC, Griffiss Air Force Base, N.Y.]

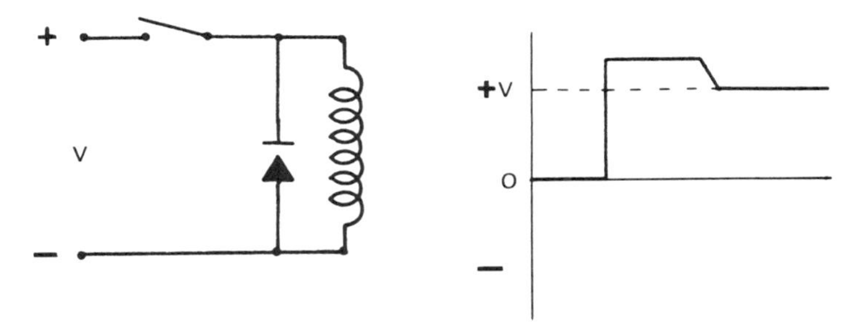

Figure 5.7 Inductive load protection.

(Figure 5.8). These simple techniques are representative of generally applicable methods and are not intended as an exhaustive list.

A number of the circuit analysis programs mentioned in Chapter 5.1 are also capable of calculating circuit performance under transient conditions. It is essential, however, that the designer thoroughly analyze the impact of any additional transient protection upon the normal operation of the circuit. Improperly implemented transient protection can inadvertently degrade one or more of the performance characteristics of the circuit itself, resulting in a worse situation than that which existed prior to the addition of the transient protection. For example, the inductive load protective circuit shown in Figure 5.7 will extend the "drop-out" time of the relay in question.

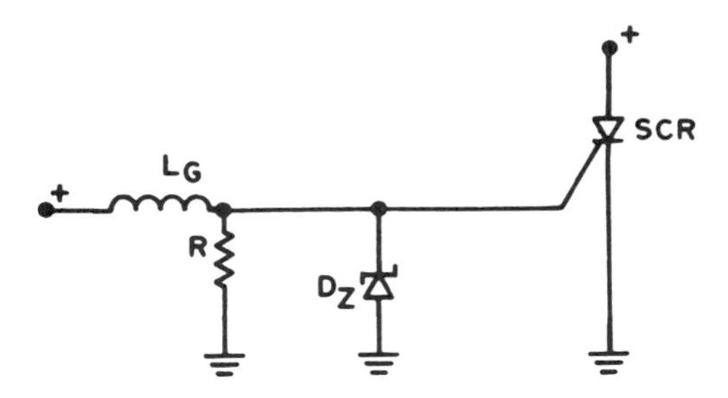

Figure 5.8 SCR gate protection. Integrator (L_G, R) serves to limit the initial surge current when the gate is turned on. Diode (D_Z) protects against voltage transients. The PIV of SCR should be chosen to provide sufficient anode-to-cathode protection. [From *Reliability Design Handbook* (RDH-376), copyright 1976 by IIT Research Institute, Reliability Analysis Center, RADC/RAC, Griffiss Air Force Base, N.Y.]

This could be a possible deterrent to its use if the drop-out time of the relay were a critical parameter in that application.

5.3 SNEAK CIRCUIT ANALYSIS

The sneak circuit analysis (SCA) is a relatively new analysis tool which has become increasingly popular during the past decade.

A sneak circuit is defined as an unexpected path or logic flow within a system which can, under certain donditions, initiate an undesired function or inhibit a desired function. The sneak path may consist of hardware, software, operator actions, or combinations of these elements. Sneak circuits are not the result of hardware failure but are latent conditions, inadvertently designed into the equipment or coded into the software, which can cause it to malfunction under specific operating conditions.

General Description of SCA

General categories of hardware sneak circuits are:

1. Sneak path: A sneak path allows current or energy to flow along an unsuspected path or in an unintended direction. There are two distinct subsets of this category. They are:
   (a) *Enable sneak path*: This occurs when the sneak path initiates an undesired function or result under certain conditions, but not under all conditions.
   (b) *Inhibit sneak path*: This occurs when the sneak path prevents a desired function or result under certain conditions, but not all conditions.
2. Sneak timing: A sneak timing condition causes a function to be inhibited or to occur to an unexpected or undesired time or in a conflicting sequence.
3. Sneak indications: These cause an ambiguous or false display of system operating conditions, and thus may result in an undesired action being taken by an operator.
4. Sneak labels: These are incorrectly or imprecisely labeled system functions, for example, system inputs, controls, displays, buses. They may mislead an operator into applying an incorrect stimulus to the system.

Figure 5.9 depicts a simple sneak circuit example. With the ignition off, the radio turned on, the brake pedal depressed, and the hazard switch engaged, the radio will turn on and off with the flashing of the brake lights.

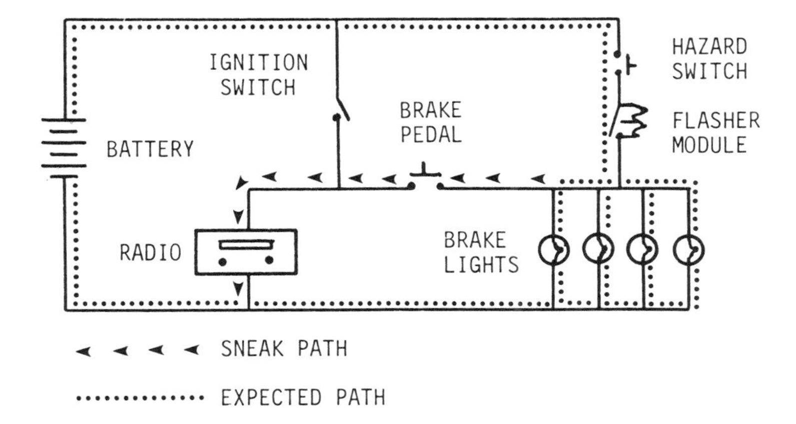

Figure 5.9 Automotive sneak circuit. [From *Contracting and Management Guide for Sneak Circuit Analysis (SCA)* (NAVSEA TE001-AA-GYD-010/SCA), Naval Sea Systems Command.]

Sneak Circuit Methodology

The first consideration in identifying sneak circuit conditions is to insure that the data being used for the analysis represent the actual "as-built" circuitry. Functional, integrated, and system-level schematics do not always properly represent the actual hardware construction. Detail manufacturing and installation schematics must be used since these drawings specify exactly what is to be built. Manufacturing and installation schematics rarely show complete circuits. These schematics are laid out to facilitate connection by technicians without regard to circuit or segment function. Thus analysis from detail schematics is extremely difficult. So many details and apparent continuities exist in these drawings that it is easy to become entangled and lost in the maze. Yet these schematics are the data that must be used if analytical results are to be based on true electrical continuity.

Network Tree Production

The initial task in the sneak analysis is to convert this accurate detailed information into a form usable for analytical work. The magnitude of data manipulation required for this conversion necessitates the use of computer automation.

Computer programs have been developed to encode simple continuities in discrete "from-to" segments extracted from the detail schematics and wire lists. Data encoding can be accomplished wthout detailed knowledge of the circuit function. The computer program connects associated points into paths and collects these paths into node sets. The node sets represent the interconnected nodes that make up each circuit. Output plots of node sets and other reports are generated by the program enabling the analyst to easily sketch accurate topological trees. These reports also provide complete indexing of every component and data point to its associated tree. This feature is especially useful in cross-indexing functionally related or interdependent trees, in incorporating changes, and in troubleshooting during operational support.

Topological Pattern Identification

Once the network trees have been produced, the analyst must identify the basic topological patterns that appear in each tree. Five basic patterns exist for hardware SCA: (1) single line (no-node) topograph, (2) ground dome, (3) power dome, (4) combination dome, and (5) the "H" pattern. These patterns are illustrated in Figure 5.10. One of these patterns or several in combination will characterize the circuitry shown in any given network tree.

Although a given circuit may appear more complex than these basic patterns, closer inspection reveals that the circuit is actually composed of these basic patterns in combination. As the sneak circuit analyst examines each node in the network tree, he must identify the topographical pattern or patterns incorporating the node and apply the basic clues that have been found to typify sneak circuits involving that particular pattern.

Clue Application

Associated with each topological pattern is a specific list of clues to help identify sneak circuit conditions. The clue list provides a guide to possible design flaws that can occur in a circuit containing that specific topological configuration. The list consists of a series of questions that the analyst must answer regarding the circuit to ensure that it is sneak-free. For example, the single line topograph (Figure 5.10) would have clues such as: (1) Is it possible for switch S to be opened when load L is desired? (2) Is it possible for switch S to be closed when load L is not desired? (3) Does the label for switch S clearly identify its relationship with load L?

Obviously, sneak circuits are rarely encountered in this simple topograph. This, of course, is an elementary example given primarily as the default case covering circuitry not included by the other topographs.

With each successive topograph, the clue list becomes longer and more complicated. The clue list for the "H" pattern includes over

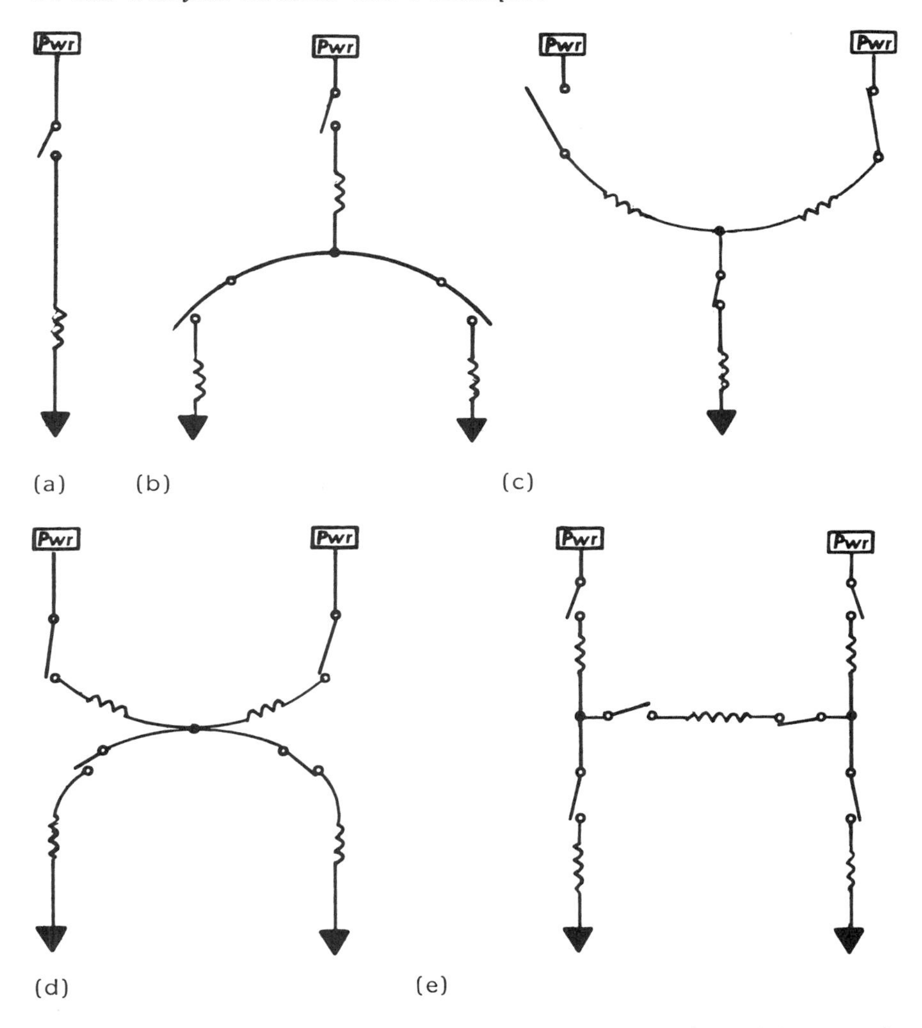

Figure 5.10 SCA topological patterns: (a) single line; (b) ground dome; (c) power dome; (d) combination dome; (e) "H" pattern. [From *Contracting and Management Guide for Sneak Circuit Analysis (SCA)* (NAVSEA TE001-AA-GYD-010/SCA), Naval Sea Systems Command.]

100 clues. This pattern, because of its complexity, is associated with more sneak circuits than any of the other patterns. Almost half of the critical sneak circuits identified have been attributed to the "H" pattern. Such a design configuration should be avoided if possible. The possibility of current reversal in the "H" crossbar is the most commonly used clue associated with "H" pattern sneak circuits.

Software Sneak Analysis

A software sneak is defined as a logic control path which causes an unwanted operation to occur or which bypasses a desired operation. Failures of the hardware system to respond as programmed are not included in this definition.

Software sneak analysis has been found to work equally well with various different software languages. Furthermore the analysis does not require execution of the software to detect problems.

The software sneak analysis technique has evolved along lines very similar to hardware sneak circuit analysis. Topological network trees are used with electrical symbology representing the software commands to allow easy cross analysis between hardware and software trees and to allow the use of a single standardized analysis procedure.

Topological pattern recognition is the keystone of both sneak circuit analysis and software sneak analysis, hence, the overall methodologies are quite similar. The software package to be analyzed is encoded, processed, and reduced to a standardized topographical format. Basic topological patterns are identified and the appropriate problem clues are applied to each pattern. For software, six basic patterns exist: the single line, the return dome, the iteration/loop circuit, the parallel line, the entry dome, and the trap circuit, as shown in Figure 5.11.

As with the hardware SCA, a given software tree may appear to be more complex than these basic patterns. However, closer inspection will reveal that the code is actually composed of these basic structures in combination. As each node in the tree is examined, the pattern or patterns which include that node must be identified. The basic clues found to typify the sneaks involved with that particular structure are then reviewed. The clues are in the form of questions that must be answered about the use and structure. These questions are designed to aid in the identification of the sneak conditions in the instruction set which could produce undesired program outputs.

Software sneaks are classified into four basic types:

1. Sneak output: The occurrence of an undesired output.
2. Sneak inhibit: The undesired inhibition of an output.

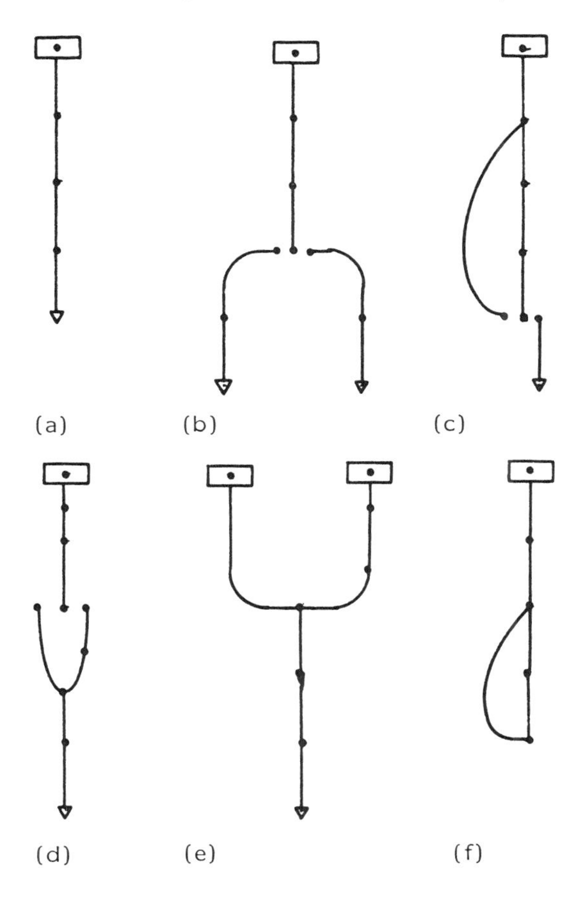

Figure 5.11 Software topographs: (a) single line; (b) return dome; (c) iteration loop; (d) parallel line; (e) entry dome; (f) trap circuit. [From *Contracting and Management Guide for Sneak Circuit Analysis (SCA)* (NAVSEA TE001-AA-GYD-010/SCA), Naval Sea Systems Command.]

3. Sneak timing: The occurrence of an undesired output by virtue of its timing or mismatched input timing.
4. Sneak message: The occurrence of a program message that does not adequately reflect the condition.

Integration of Hardware and Software Analysis

After a hardware sneak circuit analysis and a software sneak analysis have been performed on a system, the interactions of the hardware with the system software can readily be determined. Diagrammatic representations of these two elements of the system are now available in a single standardized format. Thus the effect of a control operation that is initiated by some hardware element can be traced through the hardware trees until it impacts the software. The logic flow can then be traced through the software trees to determine its ultimate impact on the system. Similarly, the logic sequence of a software-initiated action can be followed through the software and electrical network trees until its eventual total system impact can be assessed.

The joint analysis of a system's software and hardware circuitry is termed simply "sneak analysis." This is a powerful system safety tool which helps provide visibility of the interactions between a system's hardware and software. Hence, it can help reduce the difficulties involved in the proper integration of two diverse, complex systems designs. As hardware and software systems increase in complexity, the use of interface bridging analysis tools, such as sneak analysis, becomes imperative to help ensure the safety of the total system.

SCA Summary

It is helpful to contrast SCA with other analyses commonly performed on reliability programs. Sneak circuit analysis is expensive! It is usually performed late in the design cycle after the design documentation is complete. This makes subsequent design changes difficult and costly to implement. Therefore, SCA should be considered only for those components and circuits which are critical to safety and/or mission success.

Sneak circuit analysis generally concentrates on interconnections, interrelationships, and interactions of system components rather than on the components themselves. It addresses more what might go wrong in a system rather than verifying proper operation under some set of test conditions. The SCA technique originated from the study of other systems which have "gone wrong," not because of part failures, but because of design oversight or because a human operator made a mistake. Because of this different perspective, SCA tends to concentrate on and find problems which may be hidden from the perspectives of other analytical techniques.

Compare, for example, the failure mode, effects and criticality analysis (FMECA) and the SCA. The FMECA predicts and quantifies the response of a system to failures of individual parts or subsystems and then analyzes expected failure modes and their effect on system performance. Thus the FMECA results are often used in maintainability predictions, in the preparation of maintenance dependency charts, and in establishing sparing requirements. In contrast, the SCA considers possible human error in providing system inputs, while FMECA does not. In this regard the two types of analysis tend to complement one another.

Another comparison would be with the fault tree analysis (FTA). Fault tree analysis is a deductive method in which a catastrophic, hazardous end result is postulated, and the possible events, faults, and occurrences which might lead to that end event are determined. Thus, FTA and SCA overlap somewhat in their goals, since the FTA is concerned with all possible faults, including component failures as well as operator errors.

The original SCA computer programs were developed under government (NASA) contract with Johnson Spacecraft Center, Houston, Texas, on the Apollo program. Thus they are freely available to industry and government agencies.* These programs, however, are not current. Several companies have purchased these programs and spent development funds to update them. Thus, the updated programs and the accompanying analysis techniques are considered proprietary by most companies.

---

*They can be purchased from *Computer Software Management and Information Center (COSMIC), University of Georgia, 112 Barrow Hall, Athens, GA 30602.*

6

Redundancy

6.0 INTRODUCTION

Webster defines redundancy as needless repetition. In reliability engineering, however, redundancy is defined as the existence of more than one means for accomplishing a given task. Thus all of these means must fail before there is a system failure.

Under certain circumstances during system design, it may become necessary to consider the use of redundancy to reduce the probability of system failure—to enhance system reliability—by providing more than one functional path or operating element in areas that are critically important to system success. The use of redundancy is not a panacea to solve all reliability problems, nor is it a substitute for a good initial design. By its very nature, redundancy implies increased complexity, increased weight and space, increased power consumption, and usually a more complicated system checkout and monitoring procedure. On the other hand, redundancy may be the only solution to many of the problems confronting the designer of a complex electronic system.

Redundancy as a Design Technique

Given a simple system consisting of two parallel elements as shown in Figure 6.1 with A_1 having a probability of failure q_1 and A_2 a probability of failure q_2, the probability of total system failure is:

$$Q = q_1 q_2$$

Hence the reliability or probability of no failure is

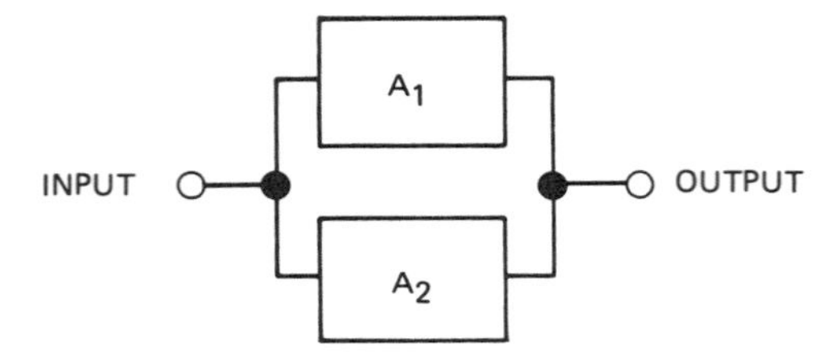

Figure 6.1 Parallel network.

$$R = 1 - Q = 1 - q_1q_2$$

Assume, for example, that A_1 has a reliability r_1 of 0.9 and A_2 a reliability r_2 of 0.8. Then their unreliabilities q_1 and q_2 would be:

$$q_1 = 1 - r_1 = 0.1$$

$$q_2 = 1 - r_2 = 0.2$$

The probability of system failure (that is, failure of both items) would then be:

$$Q = (0.1)(0.2) = 0.02$$

Hence the system reliability would be:

$$R = 1 - Q = 0.98$$

This is a significantly higher reliability than either of the component parts acting singly. Parallel redundancy may therefore be an effective design tool for increasing system reliability when all other approaches have failed.

Most examples of redundancy will consist of various groupings of series and parallel elements. It should be noted that while the use of redundancy reduces mission failures it increases logistics failures. In general, with m components in parallel, the overall probability of failure in time t is:

$$Q(t) = q_1(t)\, q_2(t) \cdots q_m(t) \tag{6.1}$$

The probability of operating without failure is:

$$R(t) = 1 - Q(t) = 1 - q_1(t)q_2(t) \cdots q_m(t) \quad (6.2)$$

Since $q_i(t) = 1 - r_i(t)$ for each component, this can also be given as:

$$R(t) = 1 - [1 - r_1(t)][1 - r_2(t)] \cdots [1 - r_m(t)] \quad (6.3)$$

Where each of the component reliabilities is equal, the above equations reduce to:

$$Q(t) = [q(t)]^m \quad (6.4)$$

$$R(t) = 1 - [q(t)]^m \quad (6.5)$$

$$= 1 - [1 - r(t)]^m \quad (6.6)$$

So far it has been assumed that the parallel components do not interact and that they may be activated when required by ideal failure sensing and switching devices. This latter assumption in particular is difficult to meet in practice. Therefore, the total potential benefits of redundancy cannot usually be realized fully.

Redundancy in Time-Dependent Situations

The previous discussion of reliability at a given point in time did not consider the time-dependent reliability function. As a rule, the previously given results can also be extended to the time-dependent situation. As in the previous example, assume that A_1 and A_2 have constant failure rates of q_1 and q_2 and that each have exponential time-to-failure distributions. Then the overall reliability is given by:

$$R(t) = 1 - q_1(t)q_2(t) \quad (6.7)$$

For each element $r(t) = e^{-\lambda t}$; hence $q(t) = 1 - e^{-\lambda t}$. Therefore:

$$R(t) = 1 - (1 - e^{-\lambda_1 t})(1 - e^{-\lambda_2 t})$$

$$= e^{-\lambda_1 t} + e^{-\lambda_2 t} - e^{-(\lambda_1 + \lambda_2)t}$$

The previously given basic redundancy formulas can then be used to solve any series-parallel combinations.

An important point to remember, however, is that the constant failure rates of the elements in a redundant configuration cannot be combined in the usual manner (addition) to obtain the system failure rate. In a redundant system configuration the system failure rate is no longer constant, that is, it does not follow an exponential distribution, but it increases with time, because the number of paths for successful operation decreases as each redundant path fails.

The system mean life, however, may be found from:

$$\Theta_s = \int_0^{\infty} R(t)dt \tag{6.8}$$

For the example given in Equation (6.7), the redundant system mean life would be:

$$\Theta_s = \int_0^{\infty} e^{-\lambda_1 t}\, dt + \int_0^{\infty} e^{-\lambda_2 t}\, dt - \int_0^{\infty} e^{-(\lambda_1 + \lambda_2)t}\, dt$$

$$\Theta_s = \frac{1}{\lambda_1} + \frac{1}{\lambda_2} - \frac{1}{\lambda_1 + \lambda_2} \qquad \text{for } \lambda_1 \neq \lambda_2 \tag{6.9}$$

$$\Theta_s = \frac{3}{2\lambda} \qquad \text{for } \lambda_1 = \lambda_2 = \lambda \tag{6.10}$$

Thus it can be seen that the mean life of a redundant system containing two parallel elements of equal reliability is 1.5 times the mean life of a single element. For n equal components in parallel:

$$\Theta_s = \frac{1}{\lambda} + \frac{1}{2\lambda} + \cdot \cdot \cdot + \frac{1}{n\lambda} \tag{6.11}$$

$$Rp(t) = 1 - (1 - e^{-\lambda t})^n \tag{6.12}$$

The reliability of redundant combinations is expressed in probabilistic terms of success or failure for a given mission period, a given number of operating cycles, or a given number of time-independent "events," as appropriate. The mean-time-between-failures (MTBF) measure of reliability is not readily usable for redundant combinations. The reliability of redundant combinations is "time dependent," therefore, reliability should be computed at discrete points in time, as a probability of success for a discrete time period.

Tradeoff Considerations

The decision to use redundant design techniques must be based on an analysis of the tradeoffs involved. Redundancy may prove to be the only available method, when other techniques of improving reliability, for example, derating, simplification, better components, and so on, have been exhausted, or when methods of item improvement are shown to be more costly than item duplication. Given adequate preventive maintenance, the use of redundant equipment can allow for repair with no system downtime. Occasionally, situations exist in which equipments cannot be maintained, for example, satellites. Then redundant elements may be the best approach to prolonging operating time significantly.

The application of redundancy is not without penalties. It will increase weight, space requirements, complexity, and cost. The increase in complexity results in an increase in unscheduled maintenance. Thus, safety and mission reliability are gained at the expense of adding items in the unscheduled maintenance chain.

The redundant systems described in this chapter represent only the tip of the iceberg as far as the variety and complexity of redundant system reliability models are concerned. Reading the IEEE Transactions on Reliability (see Chapter 15.4) will give further appreciation of the diversity of redundant reliability models.

For systems where very high safety or reliability is required, more complex redundancy is frequently applied. For example, redundancy is used in an electronic voting machine to insure the integrity of the data (that is, of the votes). The data is independently stored in four different media: EPROMs, E^2PROMs, static RAMs (with on-board battery back-up power), and on a paper tape printout.

One must be careful to ensure that single-point failures, which can greatly reduce the effectiveness of redundancy, are considered in assessing redundant systems. If, for example, redundant electronic circuits are included within a single microcircuit package, one failure such as a leaking hermetic seal could cause both circuits to fail. Such statistically dependent failures are also referred to as common-cause system failures. During design analysis of redundant system configurations, a great deal of emphasis should be devoted to the possibility of single-point, that is, common-cause failures.

Redundancy-with-Repair

It is sometimes more practical to design a system with built-in "on-line" maintenance features to overcome a serious reliability problem than it is to concentrate on improving the reliability of the components themselves. Redundancy-with-repair can yield substantial improvements in reliability over redundancy without repair. The exact amount of the gain is contingent on the rate with which the element failures can be detected and repaired or replaced. The system thus continues in an operational status while its redundant elements are being repair or replaced, as long as these repairs are completed before their respective redundant counterparts also fail.

Failure Detection

In general, two different types of monitoring may be used to detect a failure in systems employing redundant elements.

1. Continuous monitoring: With this approach element failures are recognized at the instant they occur and repair or replacement action begins immediately. Here it is assumed that repairs can be made at the rate of u per hour, where u is the mean of an exponential distribution of repair times.

2. Interval monitoring: In this case the system is checked for element failures every T hours. Any failed elements are then replaced with operable elements. Here it is assumed that the time required to monitor the elements and make the replacements is negligible.

6.1 MAJOR FORMS OF REDUNDANCY

Depending on the specific application, a number of different approaches is available to improve reliability with a redundant design. These approaches are normally classified on the basis of how the redundant elements are introduced into the circuit to provide an alternate signal path.

In general, there are two major classes of redundancy:

1. Active (or "fully on") redundancy: With this class of redundancy external components are not required to perform a detection, decision, and switching function when an element or path in the structure fails.
2. Standby redundancy: External elements are required with this class of redundancy to detect, make a decision, and then to switch to another element or path as a replacement for the failed element or path.

Techniques related to each of these two classes are depicted in the simplified tree structure shown in Figure 6.2. Figure 6.3 illustrates each of the five redundancy techniques identified in Figure 6.2.

Parallel Redundancy

Of the many different types of redundancy available for use in equipment design, parallel redundancy is the most commonly used. Parallel redundancy, however, does include a variety of different forms.

In general, the reliability gain for additional redundant elements decreases rapidly for additions beyond a few parallel elements. As illustrated in Figure 6.4, for simple parallel redundancy there is a diminishing gain in reliability as the number of redundant elements

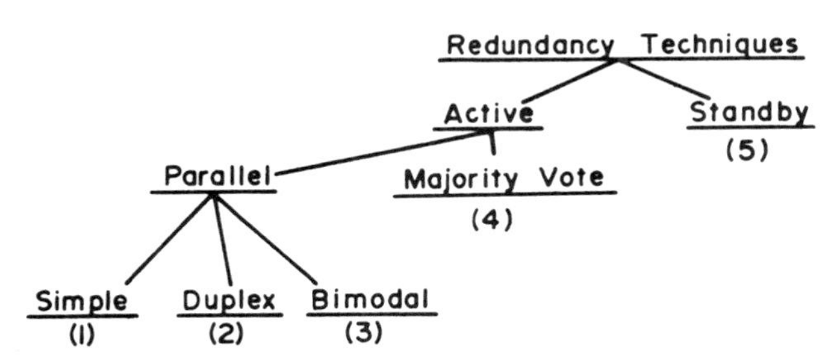

Figure 6.2 Redundancy techniques tree. [From *Reliability Design Handbook* (RDH-376), copyright 1976 by IIT Research Institute, Reliability Analysis Center, RADC/RAC, Griffiss Air Force Base, N.Y.]

is increased. For the simple parallel case, the greatest gain is achieved through addition of the first redundant element. As was previously shown, it is equivalent to a 50% increase in the system life.

In addition to increased maintenance cost due to repair of the additional elements, reliability of certain redundant configurations may actually be less than that of a single element. This is due to the serial reliability of the switching or other peripheral devices needed to implement the particular redundancy configuration. Care must be exercised to insure that reliability gains are not offset by increased failure rates due to switching devices, error detectors, and other peripheral devices needed to implement the redundancy configurations.

One case where the reliability of switching devices must be considered is that of switching redundancy. This occurs when redundant elements are energized but do not become part of the circuit until switched in after the primary element fails. Figure 6.5 is an example of redundancy with switching for two parallel elements. The mathematical model for this block diagram, written in terms of unreliability (R), considers two modes of failure associated with the switching mechanism. This equation indicates that the redundancy gain is limited by the failure mode or modes of the switching device, and the complexity increases due to switching.

The effectiveness of certain redundancy techniques (for example, standby redundancy) can be significantly enhanced by repair. Standby redundancy allows repair of the failed unit (while operation of the good unit continues uninterrupted) by virtue of the switching function built into the standby redundant configuration. Through continuous or interval monitoring, the switch-over function can provide an indication that failure has occurred and operation is

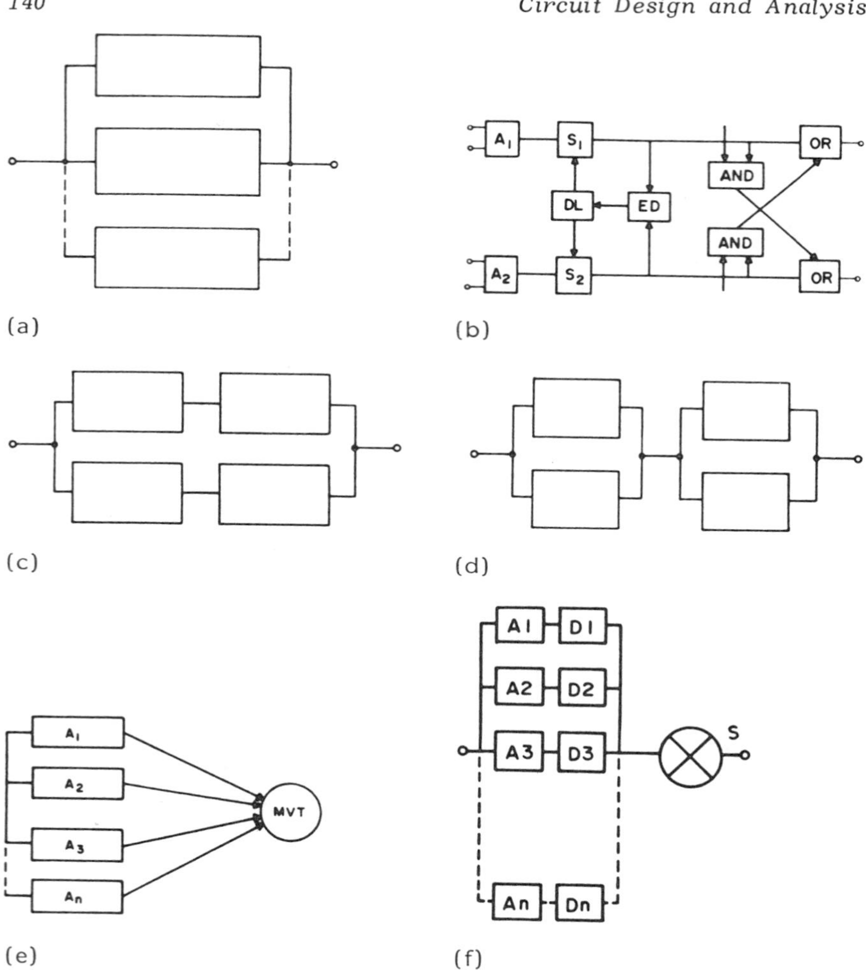

Figure 6.3 Redundancy techniques: (a) simple parallel; (b) duplex; (c) series-parallel; (d) parallel-series; (e) majority voting; (f) standby. [From *Reliability Design Handbook* (RDH-376), copyright 1976 by IIT Research Institute, Reliability Analysis Center, RADC/RAC, Griffiss Air Force Base, N.Y.]

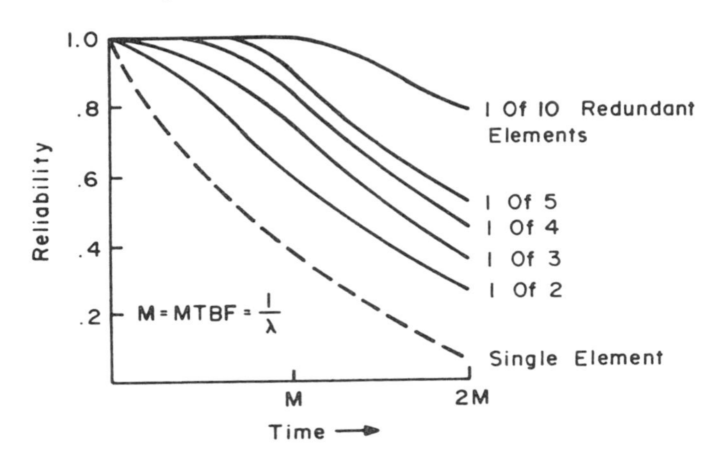

Figure 6.4 Decreasing gain in reliability. [From *Reliability Design Handbook* (RDH-376), copyright 1976 by ITT Research Institute, Reliability Analysis Center, RADC/RAC, Griffiss Air Force Base, N.Y.]

continuing on the alternate channel. With a positive failure indication, delays in repair are minimized. A further advantage of switching is related to built-in-test (BIT) objectives (see Chapter 11.3). Built-in-test can be readily incorporated into a sensing and switchover network for ease of maintenance purposes.

An illustration of the enhancement of redundancy with repair is shown in Figure 6.6. The achievement of increased reliability brought about by the incorporation of redundancy is dependent on effective isolation of redundant elements. Isolation is necessary to prevent the failure of one item from adversely affecting other parts of the redundant network. Figure 6.7 shows redundant power supplies; note the addition of the two isolation diodes to prevent a failed power supply from loading down the good unit.

In other cases, fuses or circuit breakers, overload relays, and so on, may be used to protect the redundant configuration. These items protect a configuration from secondary effects of an item's failure so that system operation continues after the element failure.

The susceptibility of a particular redundant design to failure propagation may be assessed by application of failure mode and effects analysis (FMEA) discussed in Chapter 7. The particular techniques addressed there offer an effective method of identifying likely fault propagation paths.

Redundancy may be incorporated into protective circuits as well as the functional circuit which it protects. Redundant protection

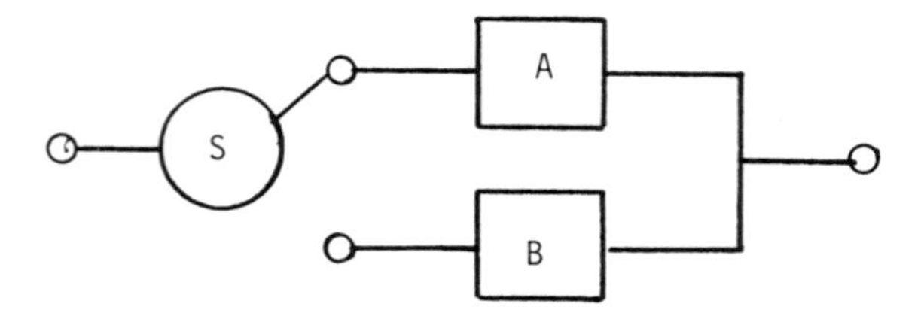

$R = p_a q_b q'_a + q_a p_b q_s + q_a q_b$

where:

q_s = probability of switch failing to operate when it is supposed to

q'_a = probability of switch operation without command (prematurely)

$q_{a,b}$ = probability of failure or unreliability of element A or B

$p_{a,b}$ = probability of success or reliability of element A or B

Figure 6.5 Redundancy with switching. [From *Electronic Reliability Design Handbook* (MIL-HDBK-338), Naval Publications and Forms Center, Philadelphia.]

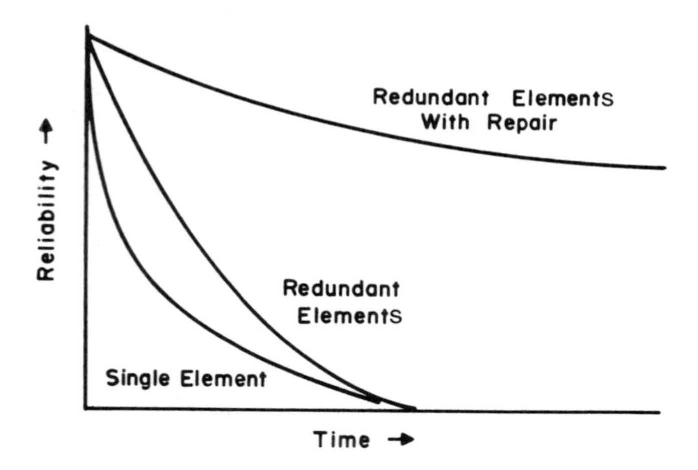

Figure 6.6 Reliability improvement: with and without repair. [From *Reliability Design Handbook* (RDH-376), copyright 1976 by ITT Research Institute, Reliability Analysis Center, RADC/RAC, Griffiss Air Force Base, N.Y.]

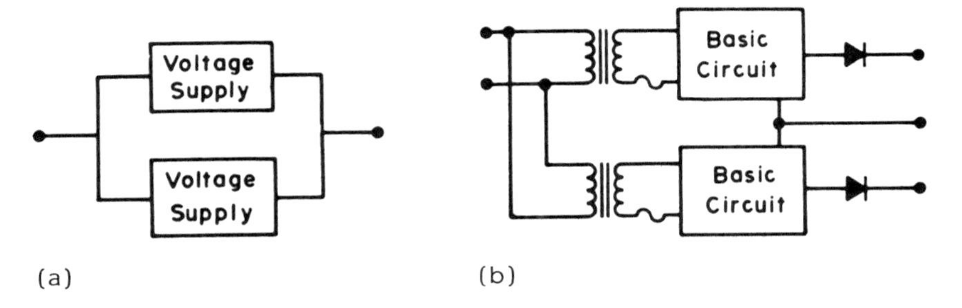

Figure 6.7 Redundant voltage supply protection: (a) reliability block diagram; (b) redundant circuit configuration. [From *Reliability Design Handbook* (RDH-376), copyright 1976 by IIT Research Institute, Reliability Analysis Center, RADC/RAC, Griffiss Air Force Base, N.Y.]

devices (for example, fuse, circuit breaker) can be used to reduce the possibility that the protected circuit is not completely disabled, should the protective circuit device open prematurely or fail to open due to overcurrent.

Duplex Redundancy

The objective of duplex redundancy (see Figure 6.3(b) is to prevent incorrect logic elements from upsetting other circuits. It is used primarily in computer applications such as the redundant logic sections A1 and A2 operating in parallel. A1 and A2 can be used in duplex or active redundant modes, or as separate elements. An error detector (ED) at the output of each logic section detects non-coincident outputs, and uses diagnostic logic (DL) to identify and disable the faulty element via switches S_1 or S_2.

Some of the advantages of duplex redundancy are:

It protects against both opens and shorts.
Faulty units can be repaired without disrupting computation.
It can maintain the function for n - 1 failures.

Some of the disadvantages of this form of redundancy are:

It may require an additional diagnostic program.
It increases complexity due to sensing and switching.
Storage capacity requirements may increase due to redundant data requirements.

Bimodal Redundancy

A combination of series and parallel redundant elements may be used to provide protection against both shorts and opens (see Figure 6.3(c) and 6.3(d). A direct short across the network due to a single element shorting is prevented by redundant elements in series. An open across the network is prevented by parallel elements. Network (c) of Figure 6.3 is useful when the primary element failure mode is short. Network (d) of Figure 6.3 is useful when the primary element failure mode is open.

Caution must be exercised in the selection and use of various redundant configurations in specific applications. Consider a parallel-series quad-redundant configuration, where the elements are identical. The reliability of this parallel-series configuration is:

$$R = 2e^{-2\lambda t} - e^{-4\lambda t}$$

Figure 6.8 shows a reliability plot of both single element reliability and the quad-redundant configuration reliability. The crossover point of the two curves illustrates that a reliability gain is provided by the parallel-series configuration up to the crossover point. Beyond the crossover point, there is a reliability loss associated with the parallel-series configuration compared with the use of the single nonredundant configuration. The normalized time, t/MTBF, for elements in this configuration determines where the reliability advantage is lost, that is, the point where reliability for the single element is the same as for the redundant configuration.

Figure 6.9 shows an example of parallel-series redundancy at the circuit level. The quadruple redundant configuration is centered about a transistor circuit with its biasing network. This configuration protects against both "fail open" and "fail short" failure modes. The mathematical model for this example assumes that opens and shorts are equally likely to occur. If they are not equally likely, the model would be more complicated. For a "no output" failure mode of the redundant configuration, either A and B, or A and D, or B and C, or C and D must fail open. This requires a double failure. For an "erroneous output" failure mode of the redundant configuration, either A and C, or B and D must fail short. Therefore, there are four combinations for a "no output" failure mode and two combinations for an erroneous output failure mode. It should be noted that an FMEA would consider these factors for analysis of this item. Double failures are required in each case to cause failure of the primary function.

Majority Voting

Decision can be built into the basic parallel redundant model by inputing signals from parallel elements into a voter to compare each

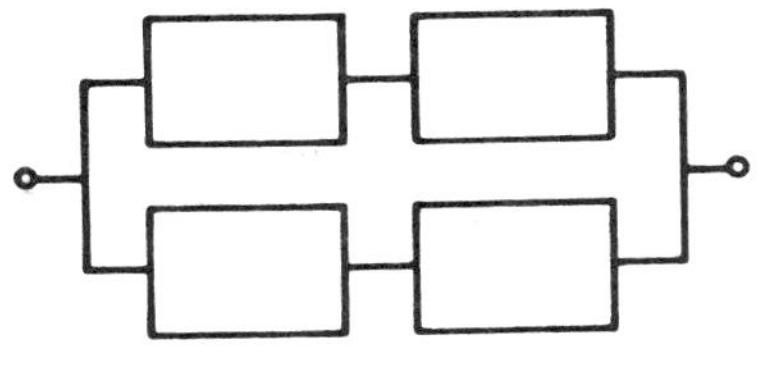

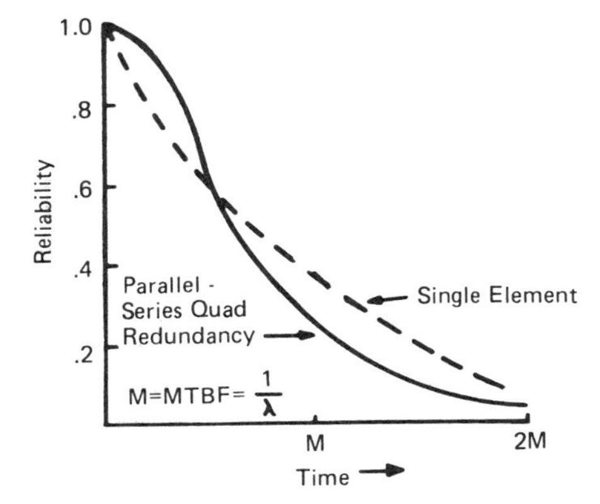

Figure 6.8 Single element vs. quad redundancy. [From *Reliability Design Handbook* (RDH-376), copyright 1976 by IIT Research Institute, Reliability Analysis Center, RADC/RAC, Griffiss Air Force Base, N.Y.]

signal with the remaining signals (see Figure 6.3(e). Valid decisions are made only if the number of useful elements exceeds the number of failed elements. This type of redundancy is used with logic circuitry for either continuous or intermittent operation. With minor modifications, this technique may be put in the form of adaptive majority logic or gate connector logic.

Some of the advantages of majority voting redundancy are:

It can be implemented to provide an indication of defective elements.
It can provide a significant gain in redundancy for a short mission time (less than one element MTBF).

Some of the disadvantages of this form of redundancy are:

It requires voter reliability that is significantly better than the element reliability.
Lower reliability can result for long mission time (that is, greater than one MTBF).

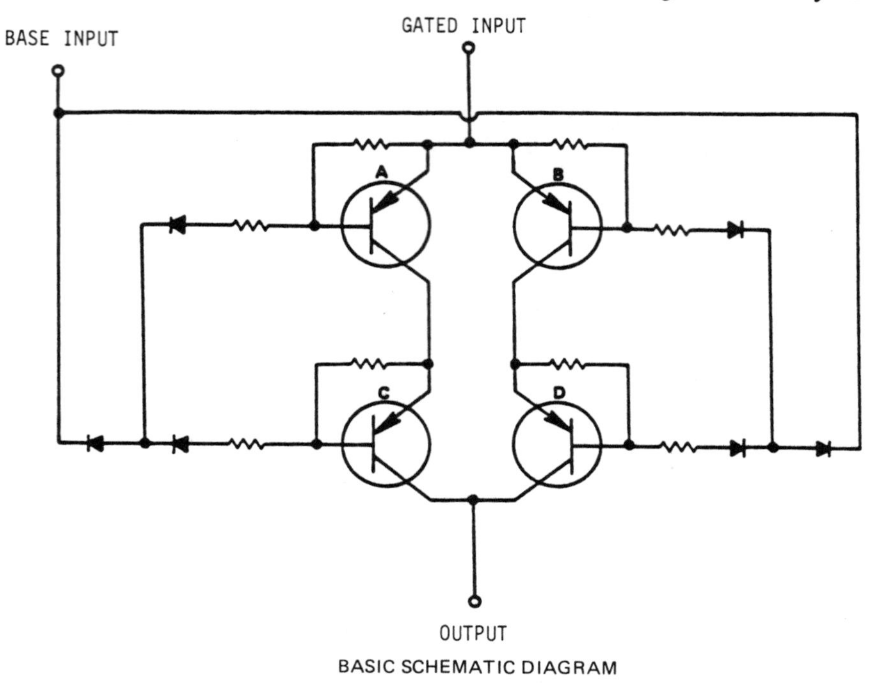

BASIC SCHEMATIC DIAGRAM

A C

B D

RELIABILITY BLOCK DIAGRAM PARALLEL - SERIES

Figure 6.9 Parallel-series redundant circuit. [From *Electronic Reliability Design Handbook* (MIL-HDBK-338), Naval Publications and Forms Center, Philadelphia.]

It requires more space, weight, and power.

Some specific examples of majority voting redundancy are:

1. Triple voting redundancy is used for the flight control computers on the automated landing system (category IIIb) for the Boeing 757 and 767 commercial aircraft.
2. The Concorde aircraft's electrical flight controls utilize triple voting active redundancy. A sensing system automatically switches off one system if its signals do not match those of the other two. There is also a manual backup system. (Thus a complete reliability evaluation of this system would include all three primary systems, the sensing system, and the manual standby system.)

Standby Redundancy

With this redundancy technique (see Figure 6.3(f) a particular redundant element of a parallel configuration can be brought into an active circuit by a switching element. Two different configurations are possible:

1. Cold standby: The element may be isolated by a switch until it is required. Power is not applied to the element until then.
2. Hot standby: All redundant elements are continuously powered up and are connected to the circuit, when needed, by a switching action.

Operating redundancy, a further refinement of standby redundancy, uses single-mode redundancy with a sensor (Dn) on each unit plus switching capability when a failure is detected. It is used when long starting time must be avoided and only a single output can be tolerated. This technique may be reconfigured to a standby technique by altering the switching arrangement to activate the elements as they are switched into the circuit.

An example of standby redundancy is the use, in both military and commercial aircraft, of dual or triple active redundant hydraulic power systems with a further emergency (standby) backup system in case of a failure of all the primary circuits.

6.2 ADDITIONAL REDUNDANCY EXAMPLES

K-out-of-N Configuration

A system consisting of n components or subsystems, of which only k need to be functioning for system success, is called a k-out-of-n configuration. For such a system, k is less than n. An example of such a system might be an air traffic control system with n displays of which k must operate to meet the system reliability requirement.

For the sake of simplicity, assume that the units are identical, that they are all operating simultaneously, and that failures are statistically independent. Then, R + Q = 1, where R is the reliability of one unit for a specified time period, and Q is the unreliability of one unit for a specified time period.

For n units:

$$(R + Q)^n = 1$$

$$(R + Q)^n = R^n + nR^{n-1}Q + \frac{n(n-1)}{2!} R^{n-2}Q^2 + \frac{n(n-1)(n-2)R^{n-3}Q^3}{3!} \cdots\cdots + Q^n = 1$$

This is the familiar binomial expansion of $(R + Q)^n$. Thus:

$$P\text{ (at least } n - 1 \text{ surviving)} = R^n + nR^{n-1}Q$$

$$P\text{ (at least } n - 2 \text{ surviving)} = R^n + nR^{n-1}Q + \frac{n(n-1)R^{n-2}Q^2}{2!}$$

$$P\text{ (at least 1 surviving)} = 1 - Q^n$$

For the case of four air-traffic-control display equipments which meet the previously mentioned assumptions the equation would be:

$$(R + Q)^4 = R^4 + 4R^3Q + 6R^2Q^2 + 4RQ^3 + Q^4 = 1$$

from which:

$$R^4 = P\text{ (all four will survive)}$$

$$4R^3Q = P\text{ (exactly 3 will survive)}$$

$$6R^2Q^2 = P\text{ (exactly 2 will survive)}$$

$$4RQ^3 = P\text{ (exactly 1 will survive)}$$

$$Q^4 = P\text{ (all will fail)}$$

We are usually interested in k out of n surviving, thus:

$$R^4 + 4R^3Q = 1 - 6R^2Q^2 - 4RQ^3 - Q^4 = P \text{ (at least 3 survive)}$$

$$R^4 + 4R^3Q + 6R^2Q^2 = 1 - 4RQ^3 - Q^4 = P \text{ (at least 2 survive)}$$

$$R^4 + 4R^3Q + 6R^2Q^2 + 4RQ^3 = 1 - Q^4 = P \text{ (at least 1 survive)}$$

If the reliability of each display for some time t is 0.9, what is the system reliability for time t if 3 our of 4 displays must be working?

$$R_S = R^4 + 4R^3Q = (0.9)^4 + 4(0.9)^3(0.01)$$

$$= 0.6561 + 0.029 = 0.685$$

If we were not interested in which particular unit fails, we can set up expressions for at least 1, 2, or 3 units surviving. For example:

$$P \text{ (at least 2 units surviving)} = R_1R_2R_3 + R_1 + R_2 + Q_3 + R_1Q_2R_3 + Q_1R_2R_3$$

This simple combinational reliability model is primarily for illustrative purposes to demonstrate the basic theory involved.

"Graceful Degradation"

In a complex electronic system containing many parallel redundant elements all capable of sharing the load, degraded but still tolerable performance may result from the failure of a limited number of the elements. With the failure of each successive element the output becomes progressively more degraded rather than suffering catastrophic failure.

In many applications where equipment operates on a continuous basis, somewhat degraded performance is much more tolerable than total loss of output or system shutdown. This is frequently referred to as "graceful degradation." This form of redundancy is typically employed in applications such as that of an electronically steered radar antenna. The antenna itself may be composed of a large number (possibly hundreds) of discrete transmitting and receiving elements in an array pattern. With the failure of each specific transmitting and receiving element an infinitesimal amount of degradation occurs in the signal. Such degradation may not even be apparent to the operator until a significant number of the element failures has occurred.

Thus, although there may be some loss in the range and/or resolution capability of the radar system, the total system continues to remain operational. Obviously with this type of redundancy preventive maintenance must be scheduled to replace periodically the defective elements (redundancy-with-repair) and restore the system to peak performance. A significant advantage of this form of redundancy is the ability to defer maintenance or at least schedule it, for example, during a period of minimum traffic where shutdown of the radar will cause a minimum of disruption to the air-traffic-control system.

Redundancy Summary

Both the advantages and disadvantages of redundancy must be considered prior to its incorporation in a design. The previously mentioned major disadvantages of using redundancy to solve a reliability problem are weight, cost, and complexity. Addition of backup systems and/or lower-level items increases the weight and cost of the hardware. This weight and cost may be reduced by the application of redundancy to the lower levels of the hardware breakdown structure (for example, the parts) rather than assemblies. A more harmful effect may be increased complexity which would negate the search for reliability improvement. For example, sensing, activation, and switching hardware added for backup item energization may reduce the overall reliability below that of the primary item.

In many applications redundancy provides reliability improvement with cost reduction. However, simple backup redundancy is not necessarily the most cost-effective way to compensate for inadequate reliability. The designer has the responsibility to determine what balance of redundancy alternatives is the most effective. In the tradeoff process, it may be determined that redundancy, by the dublication of hardware, may impact the cost of preventive maintenance. This is a significant factor in total life-cycle cost considerations. Redundancy may be easy to incorporate if an already-designed item is available, and it may be cheaper if the item is economical in comparison to redesign. Redundancy may be too expensive if the item is costly, or too heavy if the weight limitations are exceeded, and so on. These are some of the factors which the designer must consider. In any event, the designer should consider redundancy for reliability improvement of critical items (of low reliability) for which a single failure could cause loss of a system or of one of its major functions, loss of control, unintentional actuation of a function, or could present a safety hazard.

The incorporation of redundancy into a design must take into account "verifiable redundancy." Due to redundancy inclusion, it may not be possible to check some circuits prior to the start of a mission. So prior to mission start it can only be assumed that an item with nonverifiable redundancy is functional. This does not assure that all of the redundant elements are operational. In this sense, pre-mission failures could be masked by a redundant item. This appears contradictory to the basic purpose of adding redundancy, that is, to improve reliability. If the status of redundant elements is not known prior to the mission, then the purpose of redundancy is defeated. The possibility exists of starting a mission without the redundancy designed for preventing a reliability loss. The designer must take this into account for built-in test planning, inclusion of test points, packaging, and so on, when redundancy is used in design.

Some General Guidelines for Redundancy

1. The ability to repair redundant systems after failure provides large reliability gains.
2. Standby redundancy has a slight mathematical advantage over parallel redundancy (excluding switching failure rates).
3. One or more redundancy levels can result in significant reliability increases over a single equipment.

7

Failure Mode, Effects and Criticality Analysis

7.0 INTRODUCTION

The failure mode, effects and criticality analysis (FMECA) is composed of two separate analyses, the failure mode and effects analysis (FMEA) and the criticality analysis (CA). The FMEA documents all possible part failures in a system design using specified ground rules. It determines, by analysis of those failure modes, the effect of each potential failure on system operation. It also identifies single-point failures, that is, those failures critical to mission success or operator safety. In contrast, the CA ranks each potential failure mode according to the criticality of the failure effect and the probability of its occurrence. In performing the FMEA or the CA each potential failure studied is considered to be the only failure in the system, that is, a single-failure analysis. An FMEA can be accomplished without a CA, but a CA requires that an FMEA first identify the critical failure modes of the various items in the design.

The FMECA should be iterative in nature. It should be scheduled and completed concurrently with the design effort so that the design will reflect the analysis conclusions and recommendations. The FMECA results and the current status of the FMECA should be coordinated with other program tasks. These results should be utilized as inputs to system interfaces, design tradeoffs, reliability engineering, safety engineering, maintenance engineering, maintainability, logistic support analysis, test equipment design, test planning activities, and so on. Each failure mode should be explicitly defined and should be addressed at each respective indenture level.

FMEA Versus Fault Tree

The FMEA utilizes inductive logic or a "bottom up" approach. It begins at the lowest level of the system hierarchy (for example, component part), and using a knowledge of the failure modes of each part, it traces up through the system hierarchy to determine the effect that each potential failure mode will have on system performance. With the FMEA the focus is on the parts of which the system is composed. Major benefits derived from an FMEA are as follows:

1. A method for selecting a design with a high probability of operational success and adequate safety.
2. A documented uniform method of assessing potential failure modes and their effects on operational success of the system.
3. Early visibility of system interface problems.
4. A list of potential failures which can be ranked according to the seriousness of their effects and the probability of their occurrence.
5. Identifies single-point failures critical to mission success or to personnel safety.
6. Criteria for early planning of necessary tests.
7. Quantitative, uniformly formatted input data for the reliability prediction, assessment, and safety models.
8. A basis for troubleshooting procedures and for the design and location of performance monitoring and fault-sensing devices and built-in automatic test equipment.
9. An effective tool for the evaluation of a proposed design, together with any subsequent operational or procedural changes and their impact on mission success and personnel safety.

Items 5 and 8 are two of the most important functions performed by an FMEA.

A principal advantage of the FMECA is its procedural simplicity. The analytical process is straightforward and permits complete and orderly evaluation of the design.

The FMEA approach contrasts with the fault tree analysis (discussed in Chapter 8) which utilizes deductive logic or a "top down" approach. In fault tree analysis, a system failure is assumed, and is traced down through the system hierarchy to determine the event or series of events that could cause such a failure. The fault tree is the "Sherlock Holmes" approach. With the fault tree the focus is on the total system.

An FMEA may be accomplished before the reliability prediction is completed to provide initial modeling and prediction information.

When performed as an integral part of the early design process, it should be periodically updated to reflect design changes as they are incorporated. During the early design stages, a detailed knowledge of the component parts may not be available; however, some knowledge of the "black boxes" which make up the system is usually available. Thus an FMEA might start at the "black box" level and subsequently be expanded as more detailed knowledge becomes available. The results of the most current FMEA should be a major contribution to each design review.

FMEA Computer Analysis

The computer can also be helpful in performing an FMEA, since a large number of computations and a large amount of record-keeping are required even for medium-sized systems.

In the failure-mode effects portion of the analysis the computer is used primarily for bookkeeping purposes: however, function evaluation, using performance models is also possible. If computer programs containing the design equations which relate system outputs to various design parameters are available, each item may be hypothesized to fail in each one of its modes, and the effect on the system computed. Several such computer programs are available for circuit evaluation.

Circuit analysis programs such as ECAP (Electronic Circuit Analysis Program; see Chapter 5.1) accept a topological input description of the circuit and then synthesize the circuit equations. These can then be used to evaluate potential failure effects. Computer running time, however, can become excessive, since the circuit equations must be generated over again for each run. For extreme failure modes such as an open or a short of a part, the circuit configuration is changed and a completely new solution is required.

7.1 FAILURE MODE AND EFFECTS ANALYSIS

The purpose of the FMEA is to identify potential hardware design deficiencies including undetectable failure modes and single-point failures. This is done by a thorough, systematic, and documented analysis of the ways in which a system can fail, the causes for each failure mode, and the effects of each failure. Its primary objective is the identification of catastrophic and critical failure possibilities so that they can be eliminated or minimized through design change. The FMEA results may be either qualitative or quantitative. The depth of the detail addressed by the FMEA should be tailored to meet overall program requirements.

The initial phase of the FMEA may be performed during the conceptual design stage. The next phase is performed in parallel with the detail design effort. The analysis is periodically updated throughout the development program as design changes occur. During each subsequent design phase, the results of the earlier FMEAs are revised and updated to reflect all applicable design changes. The final FMECA is performed for the Critical Design Review (see Section 15.3) just before, or concurrently with, the release of detail drawings for production.

Steps in Performing an FMEA

There are five essential steps in the performance of an FMEA. They are the same for both the initial FMEA and for all subsequent updates of the FMEA. These five steps are:

1. Reliability block diagram construction: A reliability block diagram is constructed (or updated), indicating the functional dependencies among the various elements of the system. It defines and identifies each required subsystem.
2. Failure definition: Rigorous failure definitions must be established for the system, subsystem, and all lower equipment levels. A properly executed FMEA provides documentation of all critical components in a system.
3. Failure effect analysis: A failure effect analysis is performed on each item in the reliability block diagram. This takes into account each different failure mode of the item and indicates the effect of that item's failure upon the performance of the next higher level in the block diagram. (Table 7.1 shows a sample group of failure mode ratios for various electronic parts. They may be used as a guide; however, they are only estimates and should be revised based on user experience.)
4. Bookkeeping task: The system and each sub-item must be properly identified and indexed.
5. Critical items list: The critical items list is generated or updated based on the findings in steps 1, 2, and 3.

Digital Microcircuit Failure Modes

There are any number of failure modes for each integrated circuit, but only the most common and logical ones are considered and these are grouped by logical function. Outputs are the only sections of an integrated circuit that are normally analyzed. If all outputs are considered then all inputs will also be covered.

Table 7.1 Failure-Mode Distribution of Parts

Part	Failure mode	Approximate %
Blowers	Winding failure	35
	Bearing failure	50
	Sliprings, brushes, and commutators	5
	Other	10
Capacitors—fixed ceramic dielectric	Short circuit	50
	Change of value	40
	Open circuit	5
	Other	5
Capacitors—fixed aluminum electrolytic	Open circuit	40
	Short circuit	30
	Excessive leakage current	20
	Decrease in capacitance	10
Capacitors—fixed mica or glass dielectric	Short circuit	70
	Open circuit	15
	Change of value	10
	Other	5
Capacitors—fixed metallized paper or film	Open circuit	65
	Short circuit	30
	Other	5
Capacitors—fixed tantalum electrolytic	Open circuit	35
	Short circuit	35
	Excessive leakage current	10
	Decrease in capacitance	5
	Other	15
Circuit breakers	Mechanical failure of tripping device	70

Table 7.1 (Continued)

Part	Failure mode	Approximate %
Circuit breakers	Other	30
Clutches—magnetic	Bearing wear	45
	Loss of torque due to internal mechanical degradation	30
	Loss of torque due to coil failure	15
	Other	10
Coils	Insulation deterioration	75
	Open winding	25
Connectors	Shorts (due to poor seal)	30
	Solder joint failure	25
	Degradation of insulation resistance	20
	High contact resistance	10
	Miscellaneous mechanical failures	15
Crystal units	Opens	80
	No oscillations	10
	Other	10
Diodes—silicon	Short circuits	75
	Intermittent circuits	20
	Open circuits	5
Insulators	Mechanical breakage	50
	Deterioration of plastic material	50
Microcircuits	Short High	33
	Short Low	33
	Open	33

Table 7.1 (Continued)

Part	Failure mode	Approximate %
Motors, drive, and generators	Winding failures	20
	Bearing failures	20
	Sliprings, brushes, and commutators	5
	Other	55
Motors, servo, and tachometers	Bearing failures	45
	Winding failures	40
	Other	15
Relays—electro-mechanical	Contact failures	75
	Open coils	5
	Other	20
Resistors—fixed carbon, and metal film	Open circuits	80
	Change of value	20
Resistors—fixed composition	Change of value	95
	Other	5
Resistors—variable composition	Erratic operation	95
	Insulation failure	5
Resistors—variable wirewound	Erratic operation	55
	Open circuits	40
	Change of value	5
Resistors—variable wirewound, precision	Open circuits	70
	Excessive noise	25
	Other	5

Table 7.1 (Continued)

Part	Failure mode	Approximate %
Switches, rotary	Intermittent contact	90
	Other	10
Switches, toggle	Spring breakage (fatigue)	40
	Intermittent contact	50
	Other	10
Synchros	Winding failures	40
	Bearing failure	30
	Slipring or brush failure	20
	Other	10
Thermistors	Open circuits	95
	Other	5
Transformers	Shorted turns	80
	Open circuits	20
Transistors—silicon	High collector-to-base leakage current	60
	Low collector-to-emitter breakdown voltage (BVceo)	35
	Open circuit	5
Varistors	Open circuit	95
	Other	5

Source: *Engineering Design Handbook: Design for Reliability*, AMCP 706-196.

Digital microcircuit output failures can be categorized into three modes:

Logic stuck high
Logic stuck low
Open or undefined (that is, between the levels)

These three failure modes simulate the three logic states used by the designer in designing the circuit. Simple integrated circuits are relatively easy to analyze and can be categorized in a straightforward manner as shown in Figure 7.1.

Microprocessors, LSI and VLSI circuits must usually be broken down first into functional output categories. These will differ greatly from one part to another. The analyst then must consider the basic function of the device. Figure 7.2 shows the failure mode breakdown for a microprocessor. A thorough knowledge of these devices is essential to the performance of the FMECA since they are complex self-contained systems. Interface circuits are analyzed in the same manner.

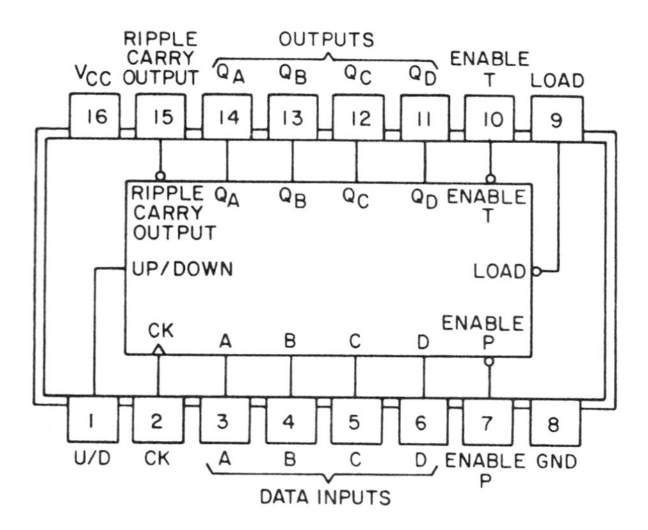

Figure 7.1 Digital microcircuit failure modes. Failure modes include: pins 11, 12, 13, 14, 15, outputs only; short high, short low, open.

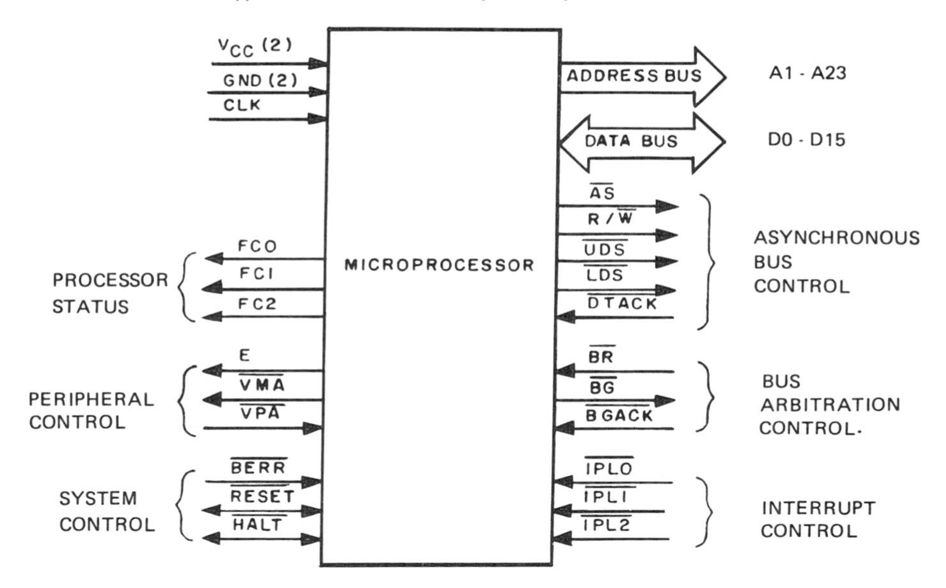

Figure 7.2 Microprocessor failure modes. Failure modes include: address bus (pins 29-48, 50-52); data bus (pins 1-5, 54-64); processor status (26-28); peripheral control (pins 19, 20); system control (pins 17, 18); async bus control (pins 6-9); bus arbitration control (pin 11); short high, short low, open, for all pins.

Analog Microcircuit Failure Modes

The nature of analog microcircuits allows them to be analyzed in a manner similar to that of digital microcircuits. Figure 7.3 illustrates the failure mode analysis of a common analog microcircuit.

The individual performing the failure mode analysis should use personal judgement regarding which failure modes to consider. In the future, as more complex analog devices become available, the analyst also may have to first break down the chip into functional blocks and start the analysis from there.

FMEA Format

Figure 7.4 illustrates one possible format for documenting a failure mode effect analysis. For each component in the system, appropriate information is entered in the applicable column.

A numerical reference for all items in the reliability block diagram should be provided using a standard coding system. One such

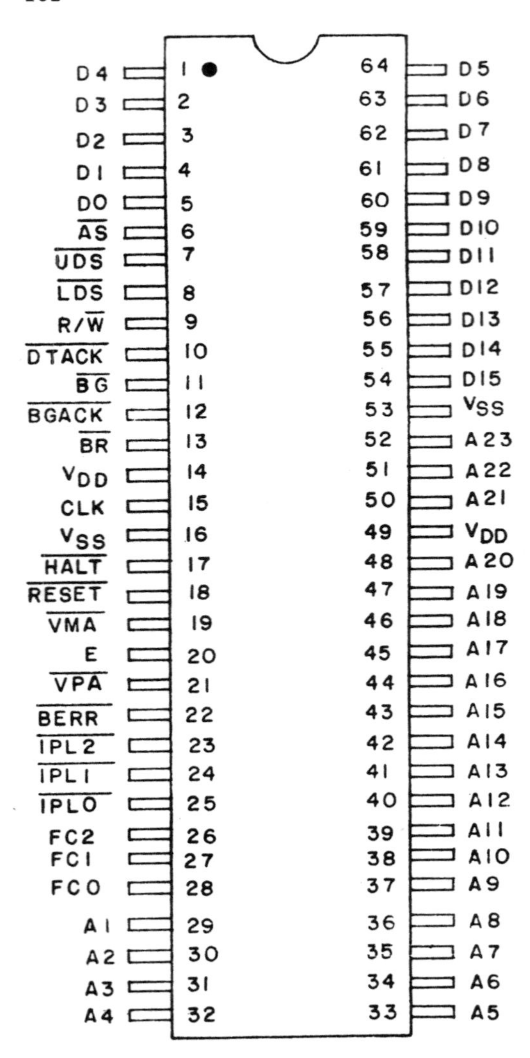

Figure 7.2 (Continued)

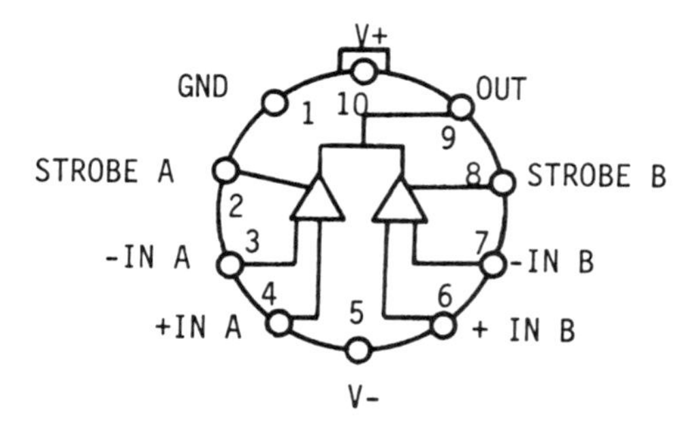

Figure 7.3 Analog microcircuit failure modes. Failure modes include: for pin 9, short high (V+), short low (V−), and open.

scheme is that documented in MIL-STD-1629, "Procedures for Performing a Failure Mode, Effects and Criticality Analysis." Items can be assigned a simple code such as: the system is assigned a letter, and then the subsystems, sets, and groups are assigned numbers in a specifically ordered sequence. For example, the code S-23-01 could designate the first group of the third set in the second subsystem in system "S". The exact coding system used is not as important as making sure that each block has a unique number. Identical items used in different systems, or in the same system but used in different applications, should be assigned different code numbers.

During the course of the FMEA, a number of changes to the block diagrams may be required. Therefore, to minimize the number of changes in the coding system, it is usually recommended that the FMEA be completed before assignment of code numbers is finalized.

The principles of FMEA are straight forward and easy to grasp. However, performing an FMEA can be tedious and time-consuming. Nevertheless, properly performed, it can be cost-effective. The bookkeeping aspects of the task, namely keeping track of each item and its place in the hierarchy, are very important. Mistakes are easy to make. Fortunately, this aspect of the task lends itself superbly to computer automation.

Identification of Critical Failures

Based on the failure effects analysis, a list of critical items is prepared. This list contains those items whose failure can result in a possible loss, probable loss, or certain loss of the next higher level in the reliability block diagram. All items that can cause system loss should be identified clearly by their inclusion in the critical items list.

Item	Code	Function	Failure Mode	Failure Effect	Loss Probability
Item name	Item or circuit designation code	Concise statement of the item's function	Concise statement of the mode(s) of item failure	Explanation of the effect of each failure mode on the performance of the next higher level in the symbolic logic block diagram	Numerical index indicating the probability of system loss if the item fails in the mode indicated

Figure 7.4 Sample failure effects analysis form.

7.2 SEVERITY/CRITICALITY ANALYSIS

After the critical items have been identified by the FMEA the criticality analysis (CA) is performed to rank the items on this list so that they can be addressed in a hierarchical order, eliminating by design change the most serious ones first and then the less serious ones.

The criticality analysis may be either qualitative or quantitative in nature depending upon the data available and the depth of analysis desired. A quantitative approach will require specific failure-rate data, whereas the qualitative approach will not.

Qualitative Approach

The qualitative approach determines criticality simply on the basis of the probability of occurrence of a specific failure mode. Simple qualitative probability of occurrence levels may be defined as follows:

1. Level A—frequent: A high probability of occurrence during the item operating time interval, defined as a single failure-mode probability greater than 0.20.
2. Level B—reasonably probable: A moderate probability of occurrence during the item operating time interval, defined as a single failure-mode probability of occurrence which is greater than 0.10 but less than 0.20.
3. Level C—occasional: An occasional probability of occurrence during item operating time interval defined as a single failure-mode probability of occurrence which is greater than 0.01 but less than 0.10.
4. Level D—remote: An unlikely or remote probability of occurrence during item operating time interval, defined as a single failure-mode probability of occurrence which is greater than 0.001 but less than 0.01.
5. Level E—extremely unlikely: A failure whose probability of occurrence is essentially zero during item operating time interval, defined as a single failure-mode probability of occurrence less than 0.001.

Quantitative Approach

The quantitative criticality analysis is implemented as follows:

1. The reliability block diagram, FMEA system coding, and critical items list are reviewed and updated as necessary.
2. Quantitative criticality is then determined based upon four factors:

(a) the probability of system loss
(b) the item failure mode
(c) the failure-mode distribution ratio
(d) the unreliability of the item

Item criticality, CR_i is defined by the equation:

$$(CR)_{ij} = \alpha_{ij}\beta_{ij}\lambda_i \tag{7.1}$$

where:

α_{ij} = failure-mode frequency ratio of item i for failure mode j (see Table 7.1), that is, the ratio of failures of the type being considered to all failures of the item.

β_{ij} = loss probability of item i for failure mode j (that is, the probability of system failure if the item fails). A suggested scale might be: certain loss = 1.00; probable loss = 0.1 to 1.0; possible loss = 0 to 0.10; no effect = 0.0.

λ_i = failure rate of item i.

$(CR)_{ij}$ = system failure rate due to item i's failing in its mode j.

System criticality is defined by the equation:

$$(CR)s = \sum_{i=1} \left[\sum_{j=1} (CR)_{ij} \right] \tag{7.2}$$

where:

$(CR)s$ = system criticality (failure rate)

Σ_j = sum over all failure modes of item i

Σ_i = sum over all items

Criticality Format

One possible format for recording the criticality analysis is shown in Figure 7.5. This form is a modification of Figure 7.4 to include also the failure-mode frequency ratio and the failure rate.

Criticality Matrix

The criticality matrix provides a means of identifying and comparing each failure mode to all other failure modes with respect to severity.

(1) Item	(2) Code	(3) Function	(4) Failure Mode	(5) Failure Effect	(6) Loss Probability (β)	(7) Failure-Mode Frequency Ratio (α)	(8) Failure Rate (per 10^6 hrs)	(9) Criticality	(10) Comments

Figure 7.5 Sample failure effects analysis form. Description of columns: (1) item name; (2) item identification or circuit designation code; (3) concise statement of the item's function; (4) concise statement of the mode(s) of item failure; (5) explanation of the effect of each failure mode on the performance of the next higher level in the symbolic logic block diagram; (6) numerical index indicating the probability of system loss if the item fails in the mode indicated; (7) that fraction of the part failure rate (λ_p) related to the particular failure mode under consideration; (8) λ_p from MIL-HDBK-217 or other failure rate source; (9) calculated item criticality per equation 7.1; (10) self explanatory. [From *Electronic Reliability Design Handbook* (MIL-HDBK-338), Naval Publications and Forms Center, Philadelphia.]

The matrix is constructed by inserting item or failure-mode identification numbers in the applicable matrix locations representing the severity classification category and either the probability of occurrence level or the criticality number for the failure modes of that item. The resulting matrix shows the distribution of criticality of item failure modes and provides a tool for assigning corrective action priorities. As shown in Figure 7.6, the further along the diagonal line from the origin the failure mode is recorded, the greater its criticality and the more urgent the need for implementing corrective action. The example criticality matrix in Figure 7.6 is constructed to show either criticality number or probability of occurrence level. Either parameter can be used for the vertical axis.

As shown in Figure 7.6 potential failure modes which fall into the areas marked with the double cross-hatch should be the first ones eliminated by redesign or other corrective action. Potential failure modes in the areas marked with the single cross-hatch would be attacked next and then, depending upon costs, delivery, and other constraints, the remaining potential failure modes would be addressed.

Criticality Analysis Outputs

The criticality analysis, regardless of the approach used, should provide at least the following outputs:

1. A prioritized list of reliability critical items.
2. Suggested indenture level redesigns.
3. A list of unacceptable failure modes, that is, those items with criticalities too high to meet the design requirements.

Criticality Reduction

If the calculated criticality for a given item is excessive, one or more of the following design actions would be required to reduce its criticality:

1. Reduce the item failure rate by increased part derating.
2. Decrease the item failure rate by incorporating additional cooling and reducing internal temperatures.
3. Change the failure-mode distribution ratio by the selection of a different type or style of part.
4. Reduce the probability of loss by incorporating redundancy.

MIL-STD-1629 contains detailed procedures and suggested formats for performing a criticality analysis.

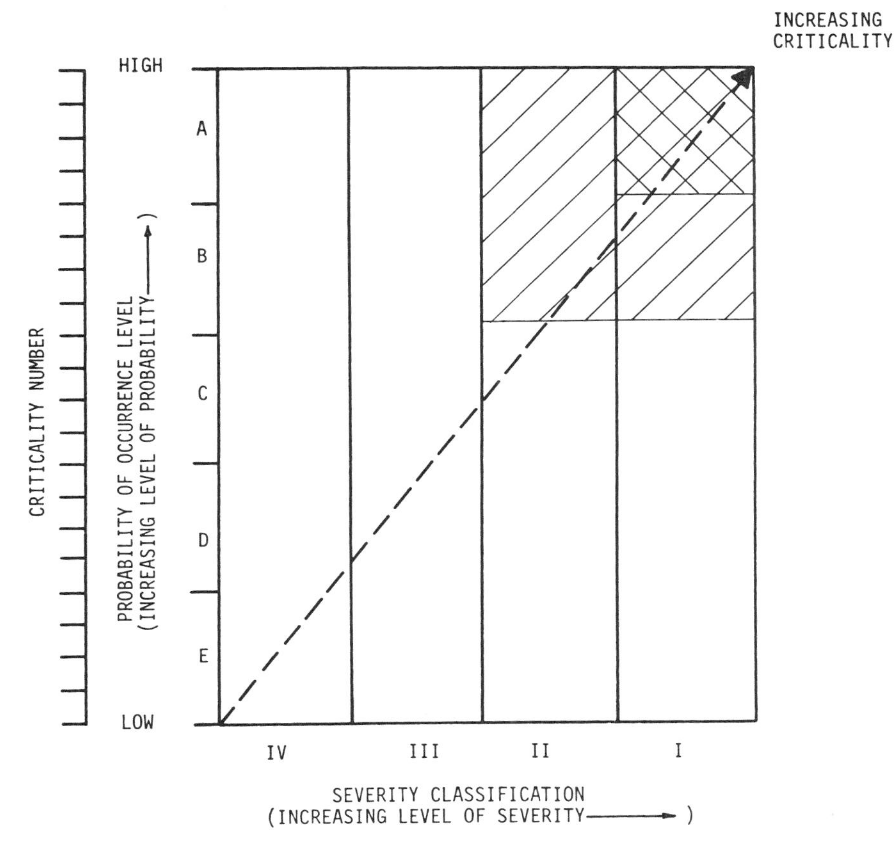

Figure 7.6 Criticality matrix. *Note*: Both criticality number and probability of occurrence level are shown for convenience.

7.3 FMECA EXAMPLE

For the sake of illustration, an abbreviated FMECA will be performed on a hypothetical radar system. This radar system is composed of five subsystems; transmitter, receiver, antenna, display, and power supply as shown in Figure 7.7. Of specific interest in this example is the receiver subsystem (block 20) which itself is composed of seven units (blocks 20A1 through 20A7).

Receiver failure is defined as cessation of operation. The FMEA is performed by filling in work sheets for all units in the receiver subsystem. (In an actual FMEA of course this procedure

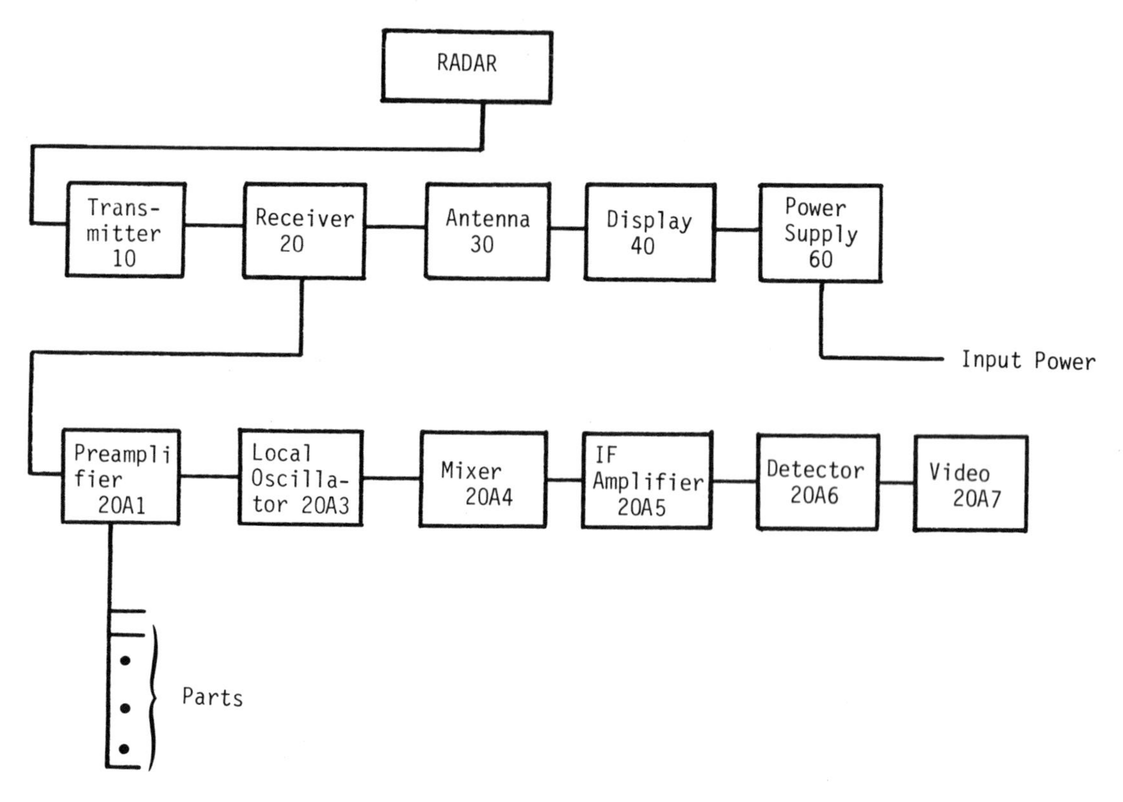

Figure 7.7 Radar block diagram. (From *Engineering Design Handbook: Design for Reliability*, AMCP 706-196.)

would be repeated for each subsystem.) Critical items are identified next, that is, those whose failure effect is identified as something other than "no effect." The criticality analysis is then performed, in this example only for the preamplifier, as shown in Table 7.2. Qualitative estimates of the loss probability (β) value (column 6 of Table 7.2) are then made for each part. For this system analysis, the following values of β are assumed: certain loss = 1.0; possible loss = 0.1; and no effect = 0. The failure-mode frequency ratio (α) for each failure mode is then determined for each part and entered in column 7 of Table 7.2. Resistor 20A1R1 is a fixed film with two failure modes as shown in Table 7.1: open = 0.8 and drift = 0.2. The failure rate (λ) for each component is then derived from MIL-HDBK-217 or some other suitable source and tabulated in column 8 of Table 7.2. Here, λ(20A1R1) = 1.5 failures per 10^6 hr. The criticality (CR) for each failure mode of each part is then computed using Equation 7.1 (ignoring values with greater than three decimal places).

$$\text{CR(20A1R1: open)} = 0.80 \times 1.00 \times 1.5 \times 10^6 \text{ hr} = 1.2 \text{ failures per } 10^6 \text{ hr}$$

$$\text{CR(20A1R1: drift)} = 0.20 \times 0.10 \times 1.5 \times 10^6 \text{ hr} = 0.030 \text{ failures per } 10^6 \text{ hr}$$

The total CR for the unit is then computed, using Equation 7.2. For the preamplifier unit, CR = 5.690 failures per 10^6 hr.

FMECA Summary

The FMECA does not replace the need for sound design engineering judgment. It will not in itself guarantee a good design. It is simply a logical way of establishing bookkeeping that can be systematically analyzed for design reliability.

Its purpose is to define and focus attention on problems and to indicate the need for a design solution. The analysis is practical in determining many significant details which may otherwise require separate, individual studies. Like other design tools, the FMECA has its limitations. However, if it is used for its intended purpose the FMECA can be a significant aid to better design.

Table 7.2 Preamplifier Criticality

(1) Item	(2) Code	(3) Function	(4) Failure Mode	(5) Failure Effect
Resistor	20A1R1	Voltage divider	Open	No output
Resistor	20A1R1	Voltage divider	Value change	Wrong output
Resistor	20A1R2	Voltage divider	Open	No output
Resistor	20A1R2	Voltage divider	Value change	Wrong output
Capacitor	20A1C3	Decoupling	Open	No effect
Capacitor	20A1C3	Decoupling	Short circuit	No output
Capacitor	20A1C3	Decoupling	High leakage current	No effect
Capacitor	20A1C3	Decoupling	Capacitance decrease	No effect
Diode	20A1CR3	Voltage divider	Short circuit	No output
Diode	20A1CR3	Voltage divider	Intermittent circuit	No output
Diode	20A1CR3	Voltage divider	Open circuit	No output
Transistor	20A1Q4	Amplifier	High leakage current	No output
Transistor	20A1Q4	Amplifier	Low break-down voltage	No output
Transistor	20A1Q4	Amplifier	Open terminals	No output
Transformer	20A1T5	Coupling	Shorted turns	Wrong output
Transformer	20A1T5	Coupling	Open circuit	No output
Resistor	20A1R6	Bias	Open circuit	No output

(6) Loss probability (β)	(7) Failure-mode frequency ratio (α)	(8) Failure rate (per 10^6 hrs)	(9) Criticality	(10) Comments
1.00	0.80	1.5	1.200	Film resistor
0.10	0.20	1.5	0.030	Film resistor
1.00	0.80	1.5	1.200	Film resistor
0.10	0.20	1.5	0.030	Film resistor
0.00	0.35	0.22	0.000	Tubular tantalum
1.00	0.35	0.22	0.077	Tubular tantalum
0.00	0.10	0.22	0.000	Tubular tantalum
0.00	0.05	0.22	0.000	Tubular tantalum
1.00	0.75	1.0	0.750	
1.00	0.20	1.0	0.200	
1.00	0.05	1.0	0.050	
1.00	0.60	3.0	1.800	
1.00	0.35	3.0	0.105	
1.00	0.05	3.0	0.150	
0.10	0.80	0.3	0.024	
1.00	0.20	0.3	0.06	
1.00	0.05	0.006	0.000	Composition resistor

Table 7.2 (Continued)

(1) Item	(2) Code	(3) Function	(4) Failure mode	(5) Failure effect
Resistor	20A1R6	Bias	Value change	No effect
Capacitor	20A1C7	Bypass	Open circuit	No effect
Capacitor	20A1C7	Bypass	Short circuit	Wrong output
Capacitor	20A1C7	Bypass	High leakage current	No effect
Capacitor	20A1C7	Bypass	Capacitance decrease	No effect

(6) Loss probability (β)	(7) Failure-mode frequency ratio (α)	(8) Failure rate (per 10^6 hrs)	(9) Criticality	(10) Comments
0.00	0.95	0.005	0.000	Composition resistor
0.00	0.40	0.48	0.000	Aluminum electrolytic
0.10	0.30	0.48	0.014	Aluminum electrolytic
0.00	0.20	0.48	0.000	Aluminum electrolytic
0.00	0.10	0.48	0.00	Aluminum electrolytic
	Criticality total for unit =		5.690	

8

Fault Tree Analysis

8.0 INTRODUCTION

Fault tree analysis (FTA) is a graphical method of risk analysis. Its purpose is to identify critical failure modes within a system or equipment. Utilizing a pictorial approach, it identifies critical faults in constituent lower level elements and determines, in a logical way, which failure modes at one level produce critical failures at a higher level in the system. The technique is particularly useful in safety analysis where the block diagramming discipline helps prevent oversights.

As mentioned in Chapter 7, the FMEA is considered to be a "bottom up" analysis, whereas FTA is considered to be a "top down" analysis. FMEAs and FTAs are complimentary and basically equivalent methods of risk analysis. The choice between these two methods depends on the nature of the risk to be evaluated. There are important differences, however, between the two techniques. A major advantage of the FTA is its ability to address human errors, which the FMECA can not address.

The Fault Tree

The fault tree is based upon deductive reasoning, that is, reasoning from the general to the specific. A specific fault is postulated, and then an attempt is made to find out the modes of system or component behavior that contributed to this failure. This is frequently referred to as the "Sherlock Holmes" approach. Holmes, faced with given evidence, had the task of reconstructing the events leading up to the crime. All successful detectives and fault tree analysts must be experts in deductive reasoning.

Fault tree analysis focuses on one particular undesired event at a time and determines all possible causes of that event. The undesired event is the top event in that fault tree diagram. It is generally a complete, or catastrophic failure rather than a drift type of failure. Careful definition of the top event is extremely important to the success of the analysis. If the top event is too general, the analysis becomes unmanageable; if it is too specific, the analysis does not provide a sufficiently broad view of the system. Fault tree analysis can be an expensive and time-consuming exercise, and its cost must be measured against the cost associated with the occurrence of that specific undesired event.

Because of the deductive nature of the FTA, it requires greater skill on the part of the analyst than does the FMECA. Fault tree analysis is particularly useful in studying highly complex functional paths in which the outcome of one or more combinations of noncritical events may produce an undesirable critical event. Typical FTA candidates are functional paths or interfaces that could have a critical impact upon safety of the general public, or safety to operating and maintenance personnel or to the probability of producing an error-free command in an automated system with a multiplicity of redundant, overlapping outputs.

The fault tree provides a concise and orderly description of the various combinations of possible occurrences within the system which can result in a predetermined critical output event. Performance of a FTA does require considerable engineering time, and the quality of the results is only as good as the validity of input data and the accuracy of the fault tree logic.

Fault tree methods can be applied in the early design phase, and then progressively refined and updated as the design evolves to track the probability of an undesired event. Initial fault tree

Table 8.1 Potential Applications for FTA Results

1. Allocation of critical failure mode probabilities among lower levels of the system
2. Comparison of alternative design configurations from a safety point of view
3. Identification of critical fault paths and design weaknesses for subsequent correction action
4. Evaluation of alternative corrective action approaches
5. Development of operational, test, and maintenance procedures to recognize and accommodate unavoidable critical failure modes

Table 8.2 Input Requirements for an FTA

1. Definition of events and interconnections
2. Definition of the principle fault and modes of failure
3. Definition of applicable human errors
4. Equipment design information
5. Definition of the maintenance concept for the equipment
6. Definition of the equipment operating conditions
7. Definition of the equipment use

diagrams might represent functional blocks (for example, units, or equipments), becoming more definitive at lower levels as the design materializes in the form of specific parts and materials. Potential applications for the results of a fault tree analysis are shown in Table 8.1. The input data requirements for performing an FTA are summarized in Table 8.2.

8.1 FAULT TREE CONSTRUCTION

The fault tree itself is a graphic model of the various parallel and sequential combinations of faults that will result in the occurrence of the predefined undesired event. The faults can be events associated with component hardware failures or human errors, or any other pertinent events which can lead to the undesired event. A fault tree thus depicts the logical interrelationships of basic events that lead to the undesired event, that is, the top event of the fault tree.

A fault tree does not model all possible system failures or all possible causes for system failure. It focuses on one top event that corresponds to a particular system failure mode. Thus it contains only those faults that contribute to that top event. These faults are not exhaustive; they cover only the most credible faults as assessed by the analyst.

A fault tree may be viewed as a complex of logic gates which serve to permit or inhibit the passage of a fault up the tree. The gates show the relationships of events needed for the occurrence of the "higher" event. The "higher" event is the output of that gate. The "lower" events are the inputs to that gate. The gate symbol denotes the type of relationship between the input events and the output event.

The symbols commonly used in fault tree diagramming are illustrated in Figure 8.1. These symbols are used to show basic functional relationships in the block diagrams and to build the equivalent fault tree diagrams depicting successful operation. The fault tree is constructed using the symbols shown in Figure 8.1 in accordance with the construction rules shown in Table 8.3.

Constructing the Fault Tree

Before the fault tree can be constructed we must have a thorough functional block diagram of the system. The functional reliability block diagram for the system and equipment must clearly indicate the functional paths in which the critical failure mode to be circumvented or eliminated is located, and it must clearly define the critical failure mode in terms of the system-level malfunction or symptom to be avoided.

The fault tree logic diagram is constructed relating all possible sequences of events whose occurrence would produce the undesired

Table 8.3 Fault Tree Construction Rules

1. Fault events (rectangle symbols) indicate the fault state and when this state occurs.
2. The fault event describes the state of either the system or a component.
3. The state of the system event may use the OR, AND, INHIBIT, or NO gate.
4. The state of a component uses only an OR gate. Component failures include:

   Primary—failure within normal parameters

   Secondary—failure due to excessive stress

   Command—failure due to a command element
5. Gate to gate connections are not allowed. A fault event must be used between gates.
6. Normal, not bizarre, events are assumed when a fault occurs.
7. Fault events under a gate may restate the output event.
8. The fault event is repeated on both sides of a transfer.
9. Basic faults indicate the fault state but not when this state occurs.

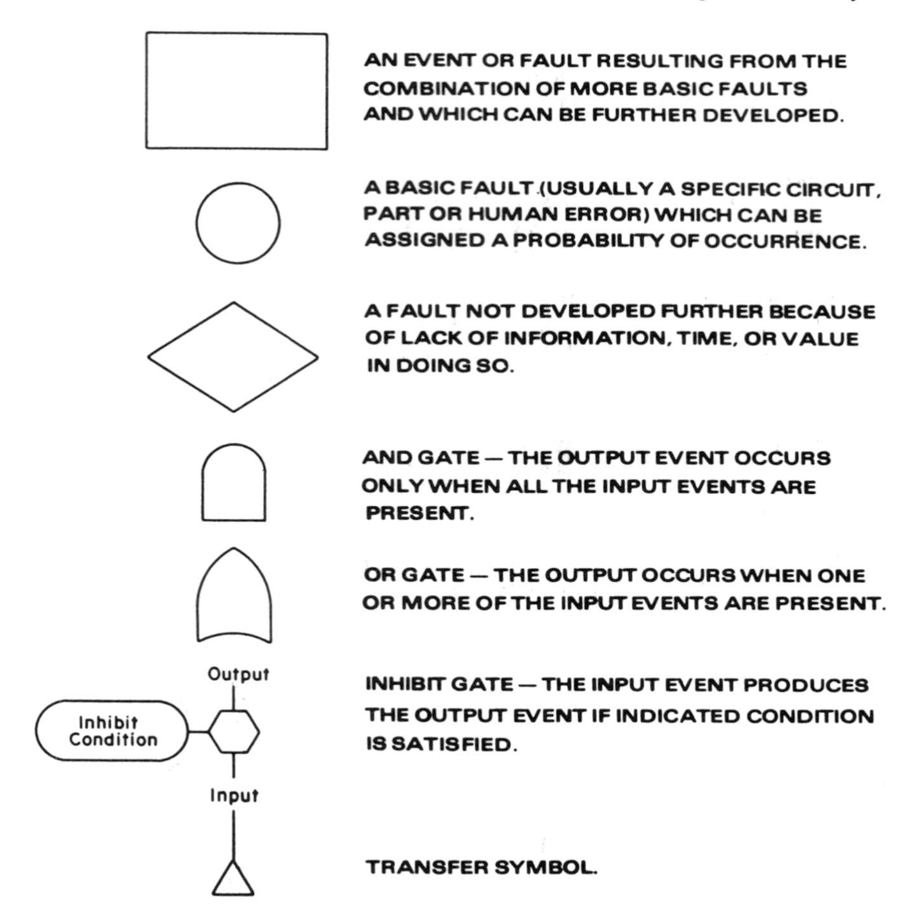

Figure 8.1 Fault tree symbols. (From *Fault Tree Handbook*, NUREG-0492, U.S. Nuclear Regulatory Comm.)

events identified in the functional block diagram. The fault tree should depict the paths that lead to each succeeding higher level in the functional configuration.

In constructing the fault tree, consideration must be given to the time-sequencing of events and functions during the specified mission profile. This must be done for each functional path or interface within the reliability model. Very often the operational sequence involves one or more changes in hardware configuration, functional paths, critical interfaces, or application stresses. When such conditions apply, it may be necessary to construct a separate fault tree for each operating mode, function, or mission event in the mission sequence.

The construction of the fault tree itself is by far the most difficult portion of this task. The accuracy of the fault tree is entirely dependent upon the skill of the analyst. Unlike the parts list available in the FMEA, there is no means of checking back to make sure that a significant potential failure contribution has not been overlooked. Because of the intuitive nature of fault tree construction, computer automation is of little help during this important portion of the task. Thus, fault tree construction, more than most other reliability tasks, requires the services of senior engineering personnel highly skilled in the art.

8.2 FAULT TREE ANALYSIS: QUALITATIVE AND QUANTITATIVE EVALUATION

Fault Tree Evaluation Techniques

Once the fault tree has been constructed it is evaluated to obtain qualitative and/or quantitative results. Qualitative analysis results include definition of: (1) the minimum cut sets of the fault tree, (2) quantitative component importances, and (3) minimum cut sets potentially susceptible to common-cause (common-mode) failures. The minimum cut set gives all the unique combinations of component failures that cause system failure. The qualitative importances give a qualitative ranking for each component with regard to its contribution to system failure. The common-cause or common-mode evaluations identify those minimum cut sets consisting of multiple components which, because of a common susceptibility, can all potentially fail due to a single failure cause.

Quantitative results which can be obtained from the evaluation include: (1) numerical probability of occurrence, (2) quantitative importances of components and minimal cut sets, and (3) sensitivity and relative probability evaluation. The quantitative importances give the percentage of the times that system failure is caused by a particular minimal cut set or a particular component failure. The sensitivity and relative probability evaluations determine the effects of changing maintenance and checkout times, implementing design modifications, and changing component reliabilities. Also included in the sensitivity evaluations are error analyses that determine the effects of uncertainties in failure-rate data. Table 8.4 is a summary of the type of results obtained from a fault tree evaluation.

Qualitative Evaluations

The fault tree is not in itself a quantitative model. It is a qualitative model that can be evaluated quantitatively, and often is. This qualitative aspect, of course, is true of virtually all varieties of system models. The fact that a fault tree is a particularly convenient model to quantify does not change the qualitative nature of the model itself.

For qualitative evaluations, the minimum cut sets are obtained using Boolean reduction of the fault tree. The minimal cut sets obtained are used not only in the subsequent qualitative evaluations but in all of the quantitative evaluations as well.

Cut Sets

Cut sets are commonly used in the analysis of fault trees. A cut set is defined as any basic event or combination of basic events whose occurrence will cause the top event to occur. Finding the cut sets for a given fault tree is a simple but repetitious task. There are just two simple rules to follow:

1. An "AND" gate increases the size of a cut set.
2. An "OR" gate increases the number of cut sets.

The cut sets so determined may not represent the simplest configuration, so the next step is to eliminate all redundant items and reduce these cut sets to the "minimum" cut set.

A minimal cut set is defined as the smallest combination of events which, if they all occur, will cause the top event to occur. Thus, by definition, a minimal cut set is that combination of primary events sufficient to cause the top event. This combination is the "smallest" combination in that all the failures are needed for the top event to occur; if any one of the failures in the cut set does not occur, then the top event will not occur (at least not by this combination).

Any fault tree will consist of a finite number of minimal cut sets which are unique for that top event. The one-component minimal cut sets, if there are any, represent those single failures which will cause the top event to occur. The two-component

Table 8.4 Results Obtainable from Fault Tree Evaluation

Qualitative results	
Minimum cut sets	Combination of component failures causing system failure
Qualitative importances	Qualitative rankings of contribution to system failure
Common-cause potentials	Minimal cut sets potentially susceptible to a single failure cause
Quantitative results	
Numeric probabilities	Probabilities of system and cut set failures
Quantitative importances	Quantitative rankings of contributions to system failure
Sensitivity evaluations	Effects of changes in models and data, error determinations

minimal cut sets represent the double failures which together will cause the top event to occur. For any minimal cut set, all "n" components in the cut set must fail in order for the top event to occur.

Qualitative Importances

After obtaining the minimal cut sets, some idea of failure importance can be obtained by ordering the minimal cut sets according to their size. Single-component minimal cut sets (if any) are listed first, then double-component minimal cut sets, then triple-component, and so on.

Since the failure probabilities associated with the minimal cut set decrease by orders of magnitude as the size of the cut set increases, the ranking according to size gives a gross indication of the importance of that specific minimal cut set. For example, if individual component failure probabilities are of the order of 10^{-3}, a single-component cut set will be of the order of 10^{-3}, and a double-component cut set 10^{-6}, a triple-component 10^{-9}, and so on. Component failure probabilities are generally different and depend on maintenance or testing intervals, downtimes, and so on. Therefore the ranking of minimal cut sets according to size gives only a general indication of their importance.

Common-Cause Susceptibilities

Primary failures (that is, component failures) on a fault tree do not necessarily have to be independent. A single, more basic cause may result in multiple failures that cause the system to fail. Multiple failures that can cause the system to fail and that originate from a common cause are termed common-cause (or common-mode) failures.

In evaluating a fault tree, we do not know which failures will be common-cause failures. However, we can indicate the susceptibility that component failures may have to a common cause. By definition, the top event occurs, that is, system failure occurs, if all the primary failures in a minimal cut set occur. Therefore we are interested only in those common causes that can trigger all of the primary failures in a minimal cut set. A cause that does not trigger all the primary failures in a minimal cut set will not by itself cause system failure.

To identify minimal cut sets susceptible to common-cause failures we can first define common-cause categories, for example, general areas that can cause component dependence. Examples of common-cause categories include a common manufacturer, environment, energy sources (not usually explicitly shown in the fault tree), and the human operator.

Quantitative Evaluations

Once the minimal cut sets are defined, if quantitative results are desired, probability evaluations can be performed. The quantitative evaluations are most easily performed in a sequential manner, first by determining the component failure probabilities, then the minimal cut set probabilities, and finally the system (that is, top event) probability. Quantitative measures of the importance of each cut set and of each component can also be obtained in this process.

Quantitative results require additional models and data beyond that required for the qualitative evaluations. First we must develop a failure probability model; this mathematical model of the fault tree is used for computation of the probability of critical event occurrence based on the failure modes identified in the fault tree diagram.

Then we must collect fault and/or failure data. Failure rates for most standard electronic and electromechanical parts are available from MIL-HDBK-217. When necessary, failure-rate values for mechanical parts may also be obtained from sources such as the *Nonelectronic Parts Reliability Data Book* (NPRD-2) (published by the Reliability Analysis Center, see Chapter 15.4), or IEEE-STD-500 *Electrical, Electronic, Sensing Component and Mechanical Equipment Reliability Data for Nuclear Power Generating Stations*. Failure rate data for new parts and recently developed parts may not always be available from these sources. In such cases, it may be necessary to draw on vendor data or to perform special studies to obtain such data.

Other data needed at this point may include, failure-mode distributions (see Table 7.1) for critical parts, operating time, human error rates, and so on.

Next we may determine the failure probabilities for the identified failure modes, that is, determine the probability of occurrence (probability of failure) for each event or failure mode identified in the model. We may also compute safety parameters at the system level using the previously derived models and failure data.

In the absence of complete and validated failure-rate and failure-mode data for all inputs, a preliminary fault tree analysis can be performed using conservative estimates of failure rates for the critical failure modes. This preliminary analysis will identify those input values which have little effect, as well as those having a critical effect on system performance. The critical inputs can then later be investigated, in greater depth if necessary. Evaluation of the fault tree model may reveal that the conservatively estimated values are sufficient to satisfy the performance goal.

Delphi Methods

In quantitative risk assessment, the unknown frequencies (probabilities) of low-probability events must often be estimated, usually with very little objective data. In these cases subjective data such as expert opinions must often be used. Estimates may be obtained using *Delphi* methods from a large group of experts in the field. Each expert provides an anonymous risk estimate plus a measure of its uncertainty. These estimates are then combined to give a usable risk estimate.

The Delphi method establishes a feedback system for the written communication of data and information among a group of professionals. It involves the following requirements:

1. A group of experts on the topic of concern and an administrative team to carry out the exercise
2. A structured methodology that focuses on the nature of the problem being considered
3. A series of questionnaires that utilize this structure in terms of specific issues or data
4. An opportunity for the experts to revise their judgements based upon summaries of the group's previous response as prepared by the administrative team
5. Successive iterations through items 3 and 4 until the revisions of the questions do not change

Monte Carlo Evaluation

The most common method for finding a quantitative solution to the fault tree is by use of the Monte Carlo method. Monte Carlo methods are simple in principle but in practice can be expensive. The Monte Carlo method uses a digital computer to provide repetitive mathematical simulations of the system. Probability data are provided as inputs, and the simulation program represents the fault tree providing quantitative results. In this manner, thousands or millions of trials can be simulated. A typical simulation program involves the following steps:

1. Assign failure data to input fault events within the tree and, if desired, repair data.
2. Represent the fault tree in a computer program to provide quantitative results for the overall system performance, subsystem performance, and the basic input-event performance.
3. List the failures that lead to the undesired event and identify minimal cut sets contributing to the failure.
4. Compute and rank basic input-failure and availability performance results.

In performing these steps, the computer program simulates the fault tree and, using the input data, randomly selects the various parameter data from assigned statistical distributions. It then tests whether or not the top event occurred within the specified time period. Each test is a trial, and a sufficient number of trials is run to obtain the desired quantitative resolution. Each time the top event occurs, the contributing effects of input events and the logical gates causing the specified top event are stored and listed as a failure output. The output provides a detailed perspective of the system under simulated operating conditions and provides a quantitative basis to support objective decisions.

The last step then is to identify critical fault paths, that is, to identify those paths where the probability of an unsafe failure mode at the system level exceeds the specification requirements. These are the critical paths which contribute most significantly to the problem.

A number of computer programs have been developed for fault tree analysis. Some of these are listed in Figure 8.2. One of the more comprehensive documents available on fault tree analysis is the *Fault Tree Handbook* (NUREG-0492).* It contains a more detailed description of most of the computer programs shown in Figure 8.2.

---

*Available from the *Division of Technical Information and Document Control, U.S. Nuclear Regulatory Commission, Washington, DC 20555.*

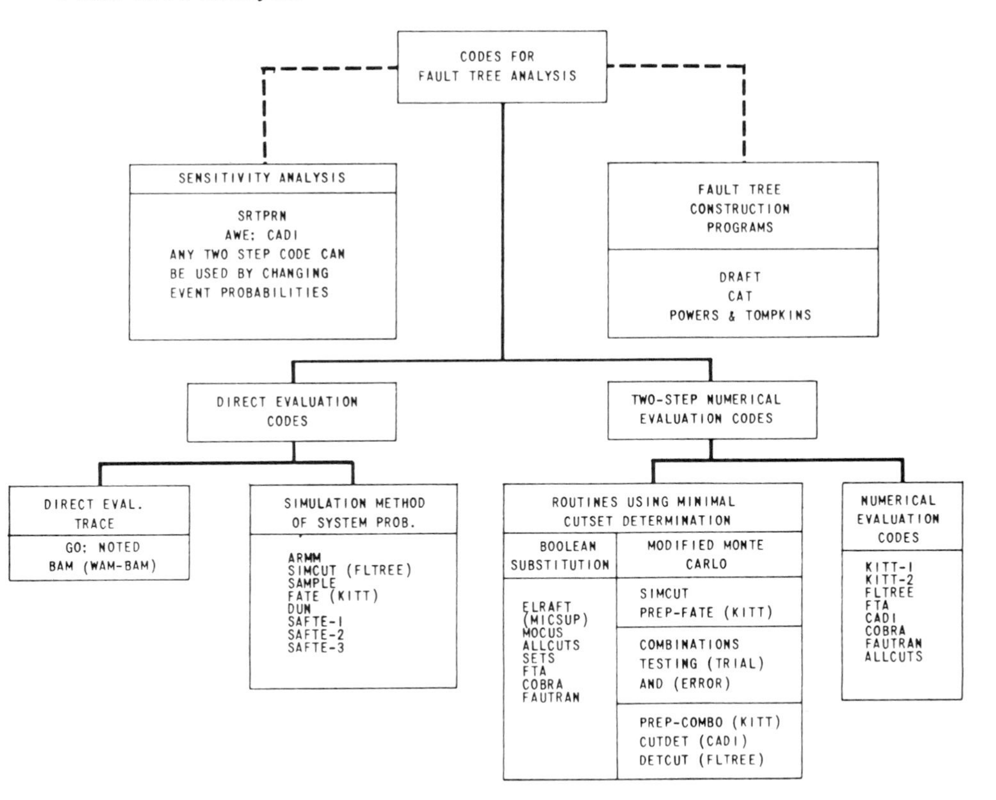

Figure 8.2 Fault tree analysis programs. Most codes are available from: *Argonne Code Center, Building 221, Argonne National Laboratory, 9700 S. Case Avenue, Argonne, IL 60439.*

A specific example of one of these computerized techniques is the *graphic reliability analysis package* (GRAP).* This is an interactive program to construct, modify, and quantify fault trees. The program automates many of the clerical tasks involved in fault tree analysis and facilitates more effective use of the fault tree analyst's time. It also contains code sets for deriving the minimum cut set associated with the fault tree.

---

*This package is available from *Babcock and Wilcox, Computer Services Center, P.O. Box 1260, Lynchburg, VA 24505.*

8.3 FAULT TREE EXAMPLE

First of all review again the symbols used in constructing a fault tree as shown in Figure 8.1. The basic relationships between functional reliability (success) block diagrams and the equivalent fault tree diagrams, using some of these symbols, are illustrated in Figures 8.3 and 8.4.

Success of a simple two-element series system comprised of blocks A and B is given by R = AB; the probability of system failure (that is, of unsuccessful or unsafe performance) is given by $\bar{R} = 1 - R = 1 - AB$. When individual element unreliability ($\bar{R}i$) is less than 0.1, the following approximations may be used to simplify computations for the fault tree logic diagram with little error (10%):

$$\bar{R} = 1 - AB = 1 - (1 - \bar{A})(1 - \bar{B})$$
$$= \bar{A} + \bar{B} - \bar{A}\bar{B} \approx \bar{A} + \bar{B}$$

The two-element block diagrams of Figure 8.3(a) and 8.3(b) are reconfigured as a simple parallel redundant system in Figure 8.3(c) to illustrate the treatment of parallel redundant elements in the fault tree logic diagram. Note that AND gates for the combination of successes (R's) become OR gates for the combination of failures ($\bar{R}$'s); and OR gates for R's become AND gates for $\bar{R}$'s. This is further illustrated in the series parallel network of Figure 8.4.

Accidental Ignition Example

To study the fault tree analysis in more depth, consider a hypothetical example: The undesired event to be assessed is the accidental ignition of a rocket engine. The fault tree analysis of this critical failure mode will be performed as illustrated in the following steps.

Step 1: Develop Functional Reliability Block Diagram

Develop a reliability block diagram for the functional paths in which the critical failure mode is to be circumvented or eliminated. Define the critical failure mode in terms of the symptom to be avoided. In this case the firing circuit of Figure 8.5 is designed to ignite a rocket motor in the following sequence:

1. Shorting switch S1 is opened to release the launcher.
2. Firing switch S2 is closed, energizing relay R1.
3. Relay R1 activates guidance and control (G&C) function.
4. Relay R2 is activated by a signal from G&C closing the firing circuit, thus igniting the rocket motor.

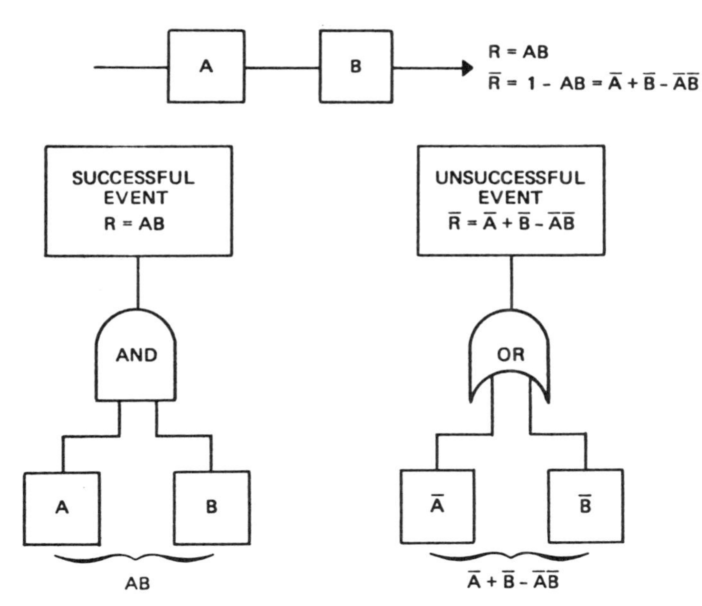

Figure 8.3 Block diagram to fault tree transformation: two series elements. (From "Reliability Data Analysis and Interpretation," *Reliability Guides*, vol. 4, NAVORD OD 44622, copyright 1974 by Bird Engineering-Research Association, Inc., Vienna, VA.)

The rocket motor can be inadvertently ignited by (1) electronic failure, (2) electromagnetic interference (EMI), or (3) by external factors such as shock, elevated temperature, and so on. These are the events to be studied in the fault tree analysis.

Step 2: Construct the Fault Tree

Develop the fault tree logic diagram relating all possible sequences of events whose occurrence would produce the undesired event. The fault tree depicts the paths leading to each higher level in the functional configuration. Figure 8.6 illustrates the construction of just one branch of the fault tree for the ignition circuit.

In constructing the fault tree for each functional path or interface within the reliability model, consideration must be given to the time-sequencing of events and functions during the operational profile. Very often the operational sequence involves one or more changes in hardware configuration, functional paths, critical interfaces, or application stresses. When such conditions are found it

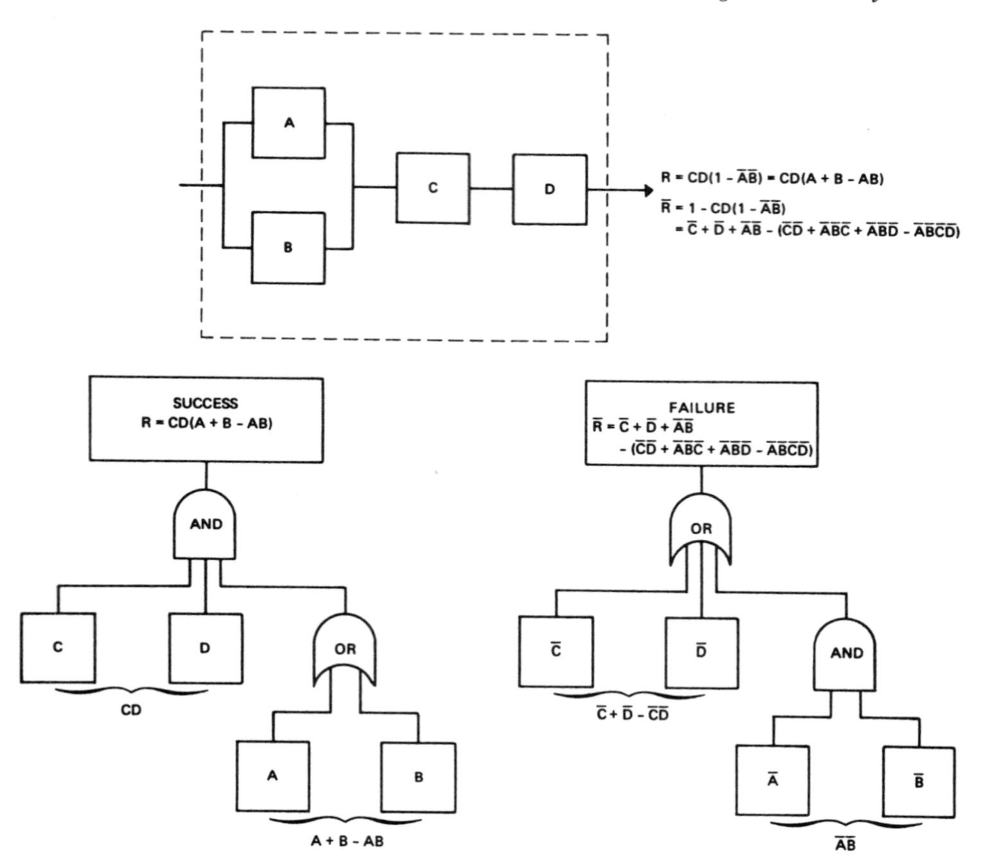

Figure 8.4 Block diagram to fault tree transformation: series/parallel network. (From "Reliability Data Analysis and Interpretation," *Reliability Guides*, vol. 4, NAVORD OD 44622, copyright 1974 by Bird Engineering-Research Association, Inc., Vienna, VA.)

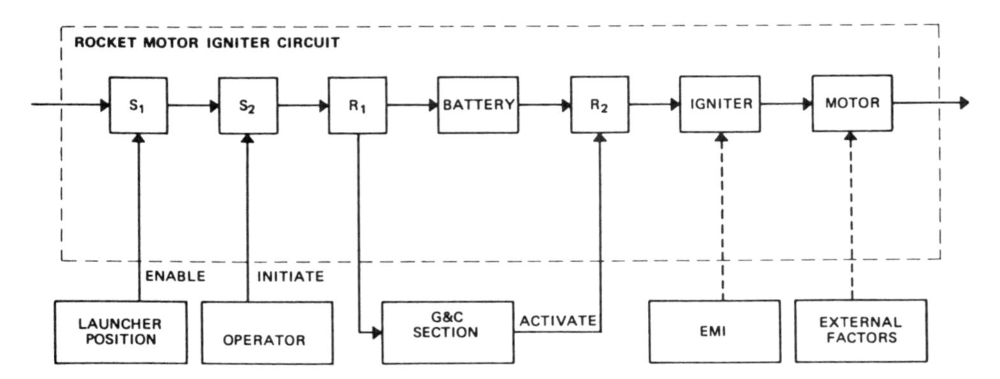

Figure 8.5 Block diagram rocket motor firing circuit. (From "Reliability Data Analysis and Interpretation," *Reliability Guides*, vol. 4, NAVORD OD 44622, copyright 1974 by Bird Engineering-Research Association, Inc., Vienna, VA.)

may be necessary to develop a separate fault tree for each operating mode, function, or mission event in the sequence.

Step 3: Develop Failure Probability Model

Develop the mathematical model of the fault tree for computation of the probability of critical event occurrence based on the failure modes identified. For example, the undesired critical failure mode identified in Figure 8.6 and given by the top-level model is:

$$\bar{A} = \bar{B} + \bar{C} - \bar{B}\bar{C}$$

As indicated in the figure, $\bar{C}$ represents the probability of accidental rocket motor firing due to premature ignition via the firing circuit either because of hardware failure ($\bar{F}$) or electromagnetic interference ($\bar{G}$), that is:

$$\bar{C} = \bar{F} + \bar{G} - \bar{F}\bar{G}$$

Considering hardware failures only, the probability of premature ignition due to hardware failure is:

$$\bar{F} = \bar{H}\bar{J}$$

where:

$$\bar{H} = \bar{K}\bar{L}\bar{M}$$

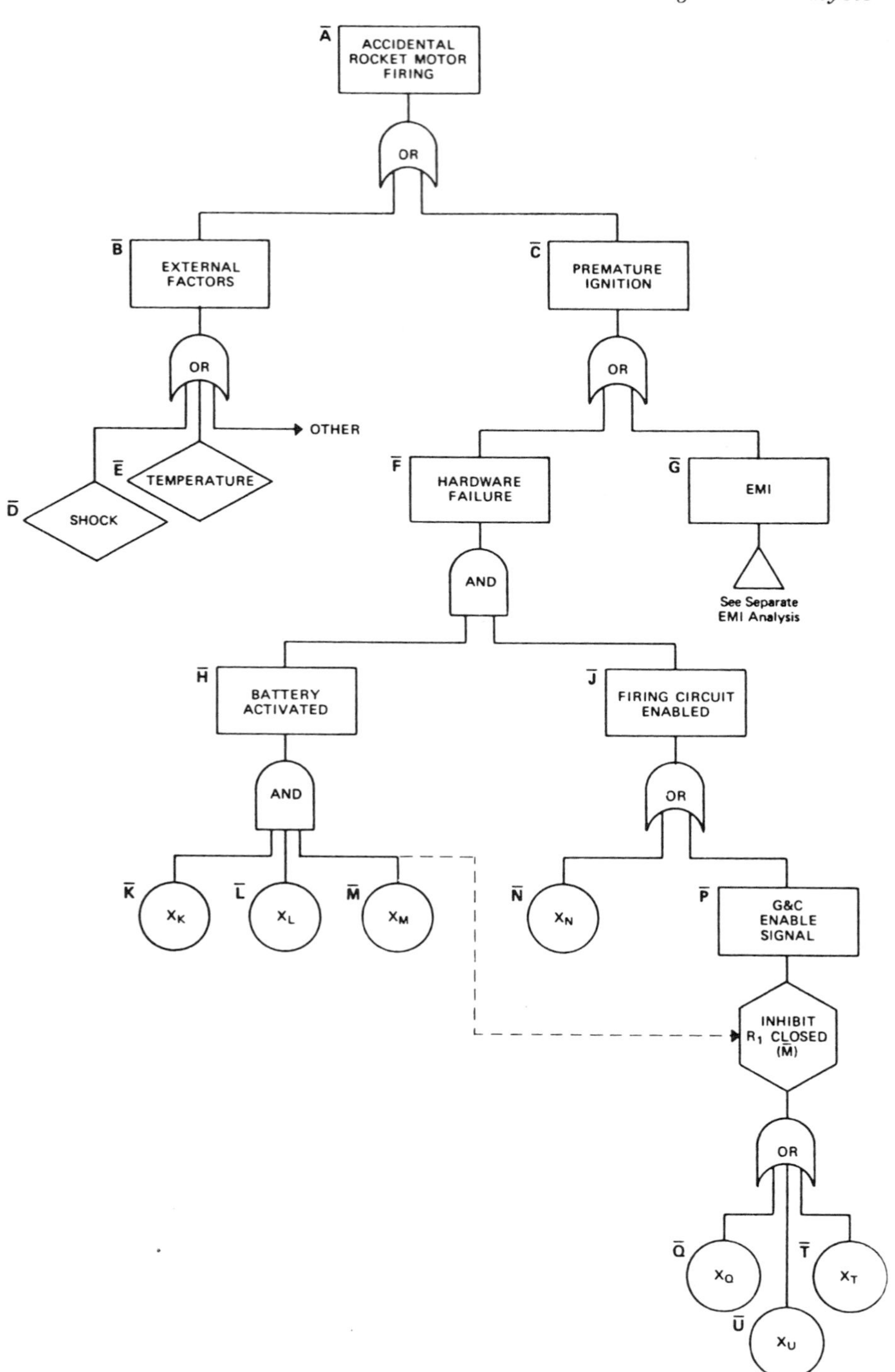
A
ACCIDENTAL ROCKET MOTOR FIRING
OR
B
EXTERNAL FACTORS
C
PREMATURE IGNITION
OR
OR
OTHER
E
TEMPERATURE
D
SHOCK
F
HARDWARE FAILURE
G
EMI
AND
See Separate EMI Analysis
H
BATTERY ACTIVATED
J
FIRING CIRCUIT ENABLED
AND
OR
K
X_K
L
X_L
M
X_M
N
X_N
P
G&C ENABLE SIGNAL
INHIBIT R_1 CLOSED ($\bar{M}$)
OR
Q
X_Q
T
X_T
U
X_U

$$\bar{J} = \bar{N} + \bar{P} - \bar{N}\bar{P}$$

$$\bar{P} = \bar{Q} + \bar{T} + \bar{U} = (\bar{Q}\bar{T} + \bar{Q}\bar{U} + \bar{T}\bar{U} - \bar{Q}\bar{T}\bar{U})$$

Step 4: Determine Failure Probabilities of the Failure Modes

Determine the probability of occurrence (that is, probability of failure) for each event or failure mode identified in the model. Compute safety parameters at the system level by applying the failure data in the models derived in step 3.

Assume, for example, the following failure probabilities in the premature ignition branch of the fault tree:

$$\bar{K} = 50 \times 10^{-3}$$
$$\bar{L} = 100 \times 10^{-3}$$
$$\bar{M} = 40 \times 10^{-3}$$
$$\bar{N} = 5 \times 10^{-3}$$
$$\bar{Q} = 2 \times 10^{-3}$$
$$\bar{T} = 1 \times 10^{-3}$$
$$\bar{U} = 0.5 \times 10^{-3}$$

Starting at the bottom, combine these data in the failure probability models previously developed in step 3, and estimate the system-level probability as follows:

$$\bar{P} = \bar{Q} + \bar{T} + \bar{U} - (\bar{Q}\bar{T} + \bar{Q}\bar{U} + \bar{T}\bar{U} - \bar{Q}\bar{T}\bar{U})$$
$$= (2 + 1 + 0.5)10^{-3} - [(2 + 1 + 0.5)10^{-6} - 1(10^{-9})]$$
$$\approx 3.5 \times 10^{-3}$$

Higher order (product) terms in the model were dropped in the probability model since the values of individual terms are much less than 0.10. Combining $\bar{P}$ with $\bar{N}$ to find J yields:

Figure 8.6 Fault tree for rocket motor firing circuit. Failure modes (X): X_k = switch S_1 shorted; X_L = switch S_2 shorted; X_M = relay R_1 failed closed; X_N = relay R_2 failed closed; X_Q = high leakage current; X_T = transistor Q-2 open; X_U = connector short to B^+. (From "Reliability Data Analysis and Interpretation," *Reliability Guides*, vol. 4, NAVORD OD 44622, copyright 1974 by Bird Engineering-Research Association, Inc., Vienna, VA.)

$$\bar{J} = \bar{N} + \bar{P} - \bar{N}\bar{P}$$

$$= 5 \times 10^{-3} + 3.5 \times 10^{-3} - 17.5 \times 10^{-6}$$

$$\approx 8.5 \times 10^{-3}$$

This is the probability of accidental firing circuit operation conditional on relay R1 having failed in the closed position (that is, in $\bar{M}$) in the battery branch of the fault tree. Also in the battery branch, the battery can be accidentally activated only if switches S1 and S2 both fail in the shorted mode, and if relay R1 fails in the closed position. Thus:

$$\bar{H} = \bar{K}\bar{L}\bar{M}$$

$$= (50 \times 10^{-3})(100 \times 10^{-3})(40 \times 10^{-3})$$

$$= 200 \times 10^{-6}$$

The probability of premature ignition due to hardware failure is then estimated from:

$$\bar{F} = \bar{H}\bar{J} = (200 \times 10^{-6})(8.5 \times 10^{-3})$$

$$= 1.70 \times 10^{-6}$$

Assume that the EMI analysis disclosed a probability of accidental ignition of $\bar{G} = 5 \times 10^{-6}$. Thus the probability of premature ignition due to either hardware failure or EMI exposure is:

$$\bar{C} = \bar{F} + \bar{G} - \bar{F}\bar{G}$$

$$\approx (1.70 \times 10^{-6}) + (5 \times 10^{-6}) - [(1.70 \times 10^{-6})(5 \times 10^{-6})]$$

$$\approx 6.70 \times 10^{-6}$$

Failure data accrued during rocket motor qualification tests indicates $\bar{D} = 2.5 \times 10^{-6}$ and $\bar{E} = 12.5 \times 10^{-6}$. Under these circumstances:

$$\bar{B} = \bar{D} + \bar{E} - \bar{D}\bar{E}$$

$$= (2.5 \times 10^{-6}) + (12.5 \times 10^{-6}) - [(2.5 \times 10^{-6})(12.5 \times 10^{-6})]$$

$$= 15 \times 10^{-6}$$

Thus the probability of accidental rocket motor firing during the handling and loading sequence is:

$$\bar{A} = \bar{B} + \bar{C} - \bar{B}\bar{C}$$

$$\approx (15 \times 10^{-6}) + (6.70 \times 10^{-6}) - [(15 \times 10^{-6})(6.75 \times 10^{-6})]$$

$$\approx 21.7 \times 10^{-6}$$

That is, there is a risk of approximately 22 premature rocket motor firings per million missile load/launch attempts.

Step 5: Identify Critical Fault Paths

When the probability of an unsafe failure mode at the system level exceeds the specification requirements, then it is necessary to identify the critical paths which contribute most significantly to the problem. In this case, both paths $\bar{B}$ and $\bar{C}$ contributed about equally to the total risk.

Part IV
System Design and Analysis

9

System Reliability Analysis

9.0 INTRODUCTION

Chapter 1 of this text laid the theoretical and mathematical foundation for the reliability engineering discipline. In contrast, this chapter emphasizes the practical aspects of specifying, allocating, and predicting equipment and system reliability. This chapter deals with the numerical quantification of reliability. Numerical quantification of reliability is necessary to establish demonstrable reliability requirements, to assess the reliability of the current design, and to allow engineering tradeoffs to be made between different design approaches thereby optimizing the design.

The specification, prediction, and verification of equipment and system reliability are based upon the system reliability analysis techniques developed in this chapter. Thus the importance of and the need for a thorough understanding of the techniques themselves cannot be overemphasized.

System Definition and Specification

The first step in the reliability engineering process is to specify the numerical requirement(s). This is the reliability that the equipment or system must be designed to achieve. The essential elements of an equipment or system reliability specification are shown in Table 9.1.

Quantitative Reliability Requirements

A reliability requirement must be specified quantitatively to be meaningful. Four commonly used ways of quantitatively defining a

Table 9.1 Essential Elements of a Reliability Specification

1. A quantitative statement of the reliability requirement
2. A full description of the environment in which the equipment/system will be stored, transported, operated, and maintained
3. The time measure or mission profile
4. A clear definition of what constitutes an equipment/-system failure
5. A detailed description of the test procedure and the accept/reject criteria that will be used to demonstrate the specified reliability

reliability requirement are *mean-time-between-failures*, *probability of survival*, *probability of success*, and *failures per million hours*.

Mean-Time-Between-Failures (MTBF)

This definition is useful for long-life equipments and systems where the planned mission lengths are typically short relative to the specified MTBF. Although this definition is adequate, it gives no positive assurance of a specified level of reliability in early life, except as the assumption of an exponential distribution can be proven to be valid.

Probability of Survival, R(t) (for a Specified Period of Time)

This definition is useful for defining reliability when a high reliability is required during the mission period but an MTBF greater than the mission period is of little consequence except as it influences availability.

Probability of Success, P(s)

This definition, independent of time, is used for specifying the reliability of one-shot devices, such as the flight reliability of missiles, and so on. It is also specified for items which are cyclic in nature, such as the launch reliability of a missile from its launcher.

Failures per Million Hours (fpmh)(λ)

The definition for failure rate (λ) over a specified period of time is used to specify the reliability of parts, units, and assemblies whose reliability for the time period of interest approaches unity. For equipment exhibiting an exponential failure distribution it is the reciprocal of the MTBF.

The specified reliability requirement may be defined in one of two ways:

1. As a *nominal* or design value. This is the average value with which the customer would be satisifed.
2. As a *minimum* acceptable value below which the customer would find the system totally unacceptable and which he could not tolerate in an operational environment. This value is based upon the operational requirements.

When a formal reliability demonstration test is required in accordance with MIL-STD-781, the minimum acceptable MTBF is the specified MTBF. It is stated as the *lower test MTBF*; however, the equipment would be designed to meet the *upper test MTBF* to assure that it will pass the formal demonstration test. The difference between the lower test MTBF and the upper test MTBF is known as the discrimination ratio. This concept will be discussed in greater detail in Chapter 12.

The second of the two methods for specifying a reliability requirement is preferred since it automatically establishes the design goal at a clearly defined value above a known minimum.

Table 9.2 summarizes the appropriate methods of stating the reliability requirements for various functions, usage, and maintenance conditions.

The reliability requirement must always be realistic in terms of actual need. It must also be consistent with the current state of the art. Otherwise, the requirement may be unattainable or attainable only with excessive expenditures of money and time.

Although probability of success, probability of survival, mean-time-between-failures, and its reciprocal, failures per million hours are the most common methods of specifying reliability, they are by no means the only ways. Other reliability parameters frequently used in the commercial and military specifications include:

Mean-miles-between-failures
Mean-time-between-maintenance-actions
Mean-cycles-to-failure
Mean-number-of-copies-between-failures
Failures per 10^9 hours (FITs)

Thus the reliability requirements must be quantitative in nature and must be specifically tailored to the needs and peculiarities of each specific program.

Table 9.2 Methods of Specifying Reliability Considering Complexity and Use

	Conditions of use			
	Repairable		Nonrepairable	Time dependent
Level of complexity	Continuous duty long life	Intermittent duty short missions	Continuous or intermittent	One-shot
Complex systems	R(t) or MTBF	R(t) or MTBF	R(t) or MTBF	P(S) or P(F)
Systems or subsystems	R(t) or MTBF	R(t) or MTBF	R(t) or λ	P(S) or P(F)
Units, assemblies, subassemblies, and parts	λ	λ	λ	P(F)

Note: R(t) = reliability for specified mission, or period of time (t), MTBF = mean-time-between-failures, P(S) = probability of success, P(F) = probability of failure, λ = failure rate.

Source: *Reliability Engineering Handbook*, NAVWEPS 00-65-502, copyright 1970 by Bird Engineering-Research Association, Inc., Vienna, Va.

Table 9.3 Performance Parameters, Limits, and Failure Criteria

Performance parameter	Units of measure	Specified requirement	Failure classification in terms of performance limits
Power output (P)	Kilowatts	500 ± 20%	Major: $200 < P < 400$ Critical: $P < 200$
Channel capacity (n)	Number of channels	$n = 48 + 0$	Major: $24 < n < 40$ Critical: $n < 24$
Voltage gain (A)	Decibels	$A = 40 \pm 3$ dB	Major: $30 < A < 37$ Critical: $A < 30$
Detection range (R)	Nautical miles	$R = 300 + 0 - 50$	Major: $150 < R < 250$ Critical: $R < 150$

Source: *Reliability Engineering Handbook*, NAVWEPS 00-65-502, copyright 1970 by Bird Engineering-Research Association, Inc., Vienna, Va.

Definition of Failure

The definition of failure may well include specific performance parameters together with their allowable limits as shown in Table 9.3 for a hypothetical radar system.

9.1 RELIABILITY ALLOCATION OR APPORTIONMENT

The next step in the design process is to translate the overall system reliability requirement into reliability requirements for each of the subsystems and lower-tier items. This process is known as reliability apportionment or allocation. Given a set of system reliability requirements it is necessary to determine requirements for each of the various individual subelements of the system. The problem is to establish a procedure by which consistent and reasonable reliability allocations may be made.

The reliability parameters apportioned to the subsystems are used first as guidelines to determine design feasibility. If the allocated reliability for a specific subsystem cannot be achieved with the current technology, then the system design must be modified and new allocations defined. This is repeated until an allocation is achieved that satisfies the system-level requirement with all of its constraints, and defines subsystems that can be designed with the current state of the art. The allocation process is performed at each of the lower-levels of the system hierarchy, for example at the subsystem, equipment, module, component, and so on, and in turn provides requirements for subcontracted items.

Simple Apportionment Example

Apportionment may be done by converting the MTBF of the system to its equivalent failure rate, λ, and then allocating this λ out to the various elements such that:

$$\lambda_{system} = \lambda_a + \lambda_b + \lambda_c + \cdots + \lambda_x$$

A convenient approach in the refinement process is to allocate some failure probability to a reserve. This technique recognizes that new functions may be added, and thereby precludes the need for continual reallocation to accommodate additional design definitions later. It is common practice to apportion out only 90% of the total failure rate initially.

The apportionment is made by the reliability engineer in conjunction with the designer of the specific item, taking into account the complexity of the item, the type of parts used, their quality grade, and so on. The remaining 10% of the system failure rate may be held in reserve to support any future reapportionments which become necessary as the details of the design are better known. The apportionment is normally performed initially during the conceptual design of the equipment, and is updated (reapportioned) as the design becomes better known. A simple reliability apportionment example is shown in Table 9.4.

Apportionment of the reliability requirement, however, is not a simple task. It draws heavily upon the previous experience of the reliability engineer, his knowledge of similar systems and their reliability experience to assure an equitable distribution of the available λ to the various subelements. Included in this process should be a comparison of the emerging requirements with empirical data for similar hardware to determine the realism of the allocation in terms of the expectations. If some of the requirements appear to be unreasonably difficult to achieve, than this analysis becomes the basis for performing design tradeoffs among the subsystems to reallocate the system requirement. This total process—gross allocations, comparisons with empirical data, tradeoffs, reiterating as required—eventually results in the apportioned subsystem requirements.

Reapportionment

When reallocation is necessary to increase the λ available for a given element it will be necessary to tighten up the λ requirement for other elements because the overall system requirement must still be met.

Table 9.4 Simple Reliability Apportionment Example

System MTBF requirement = 1000 hr

Equivalent failure rate λ = 1/MTBF = 1000 fpmh

90% of the equivalent failure rate = 900 fpmh

Item A:	λ_A = 400 fpmh	$MTBF_A$ = 2500 hr
Item B:	λ_B = 200 fpmh	$MTBF_B$ = 5000 hr
Item C:	λ_C = 200 fpmh	$MTBF_C$ = 5000 hr
Item D:	λ_D = 100 fpmh	$MTBF_D$ = 10000 hr
	$\Sigma\lambda_{A-D}$ = 900 fpmh	

$$\text{Apportioned system MTBF} = \frac{1}{\Sigma\lambda_{A-D}} = \frac{1}{900} = 1111 \text{ hr}$$

As a last resort the 10% reserve may be distributed. Reallocation should be anticipated to occur at least once and probably two or three times as the design progresses through its various stages.

Allocation should be initiated early in the development phase, this offers the greatest flexibility in tradeoffs and redefinitions. Allocation should begin early to allow time to establish each of the lower level reliability requirements (system requirement allocated to subsystems, subsystem requirements allocated to assemblies, and so on). Eventually the reliability requirements must be frozen at some point to establish baseline requirements for the designers.

Where major system elements are to be subcontracted, the allocated values may well become the contractual requirements for the subcontracted item. At this time the allocated value becomes a firm requirement, hence it is not available for further reapportionment.

After the lower-level reliability requirements are defined, they are levied on the responsible equipment design engineers at each hardware level. Without the specific reliability requirements that must be achieved, reliability becomes a vague, undefined, general objective for which no one is responsible. The reliability requirements produced by allocation are essential inputs to other related activities. Maintainability, safety, quality engineering, logistics, and test planning are activities whose work will be facilitated by established reliability requirements.

Detailed Apportionment Techniques

A variety of techniques is available for reliability allocation. These techniques vary in complexity, depending upon the amount of subsystem definition available and the degree of rigor desired. Some of the available detailed apportionment techniques are:

1. Equal apportionment
2. AGREE apportionment
3. ARINC apportionment
4. Feasibility-of-objectives apportionment
5. Dynamic programming apportionment

Equal Apportionment Technique

In the absence of definitive system information, other than the fact that it contains n subsystems in series, equal apportionment to each subsystem would seem reasonable. In this case, the nth root of the system reliability requirement would be apportioned to each subsystem.

Equal apportionment assumes a series of n subsystems, each of which is to be assigned the same reliability goal. The primary weakness of this method is that subsystem goals are not assigned in accordance with the degree of difficulty associated with achievement of these goals. Since it does not take into consideration complexity and other design factors, extensive reallocation would probably be necessary.

The equal apportionment technique is based upon the formula:

$$R_i = (R_{system}) \exp \frac{1}{n} \tag{9.1}$$

$i = 1, 2, \ldots, n$

where:

R_i = reliability allocated to the ith item

R_{system} = required system reliability

n = total number of items

AGREE Apportionment Technique

This technique takes into consideration both the complexity and the importance of each subsystem. It assumes a series of K subsystems, each with exponential failure distributions. The apportioned reliability goal is usually expressed in terms of MTBF.

The AGREE apportionment is based upon the following formulas:

$$\theta_i = \frac{N w_i t_i}{n_i[-\ln R(t)]} \tag{9.2}$$

and

$$R_i(t_i) = \exp\left(\frac{-t_i}{\theta_i}\right) \tag{9.3}$$

where:

$i = 1, 2, \ldots, k$ = subsystem index

t = required system mission time

t_i = mission time of the ith item

w_i = importance factor of the ith item

n_i = number of modules of the ith item

$R(t)$ = system reliability requirement

$R_i(t_i)$ = the reliability apportioned to the ith subsystem for its mission time

θ_i = allocated MTBF for the ith item

$$N = \sum_{i=1}^{k} n_i = \text{total number of modules in the system}$$

The importance factors (w_i) for each subsystem are determined based upon the probability of system failure as a result of subsystem failure. The procedure used in this technique is as follows:

1. Determine the importance factor of each subsystem, that is, the probability of system failure as a result of subsystem failure.
2. Calculate the apportioned subsystem MTBFs using Equation 9.2 based on the system reliability requirements, importance factors, performance times, and the number of modules being used.
3. Calculate the apportioned reliability to each subsystem using Equation 9.3.

A group of electronic parts, a fictitious module is designated as the basic electronic building block for the purpose of partitioning an electronic system. This concept of a module accomplishes three goals: (1) it takes into account the relative inherent complexity; (2) it provides consistent minimum acceptable reliability values; and (3) it allows reliability requirements to be dynamic so that state-of-the-art changes can be incorporated as they occur.

ARINC Apportionment Technique

This method assumes: (1) series subsystems with constant failure rates, (2) that any subsystem failure causes system failure, and (3) that subsystem mission time is equal to system mission time. Reliability requirements are expressed in terms of failure rate, λ.

The technique utilizes the following steps:

1. The objective is to choose λ^*_i such that:

$$\sum_{i=1}^{n} \lambda^*_i \leqslant \lambda^*$$

where:

λ^*_i = the failure rate allocated to subsystem i

λ^* = the required system failure rate

n = the number of subsystems

2. Determine the subsystem failure rates (λ_i) from past observation or estimation.
3. Assign a weighting factor (w_i) to each subsystem according to the failure rates determined in step 2 such that:

$$w_i = \frac{\lambda_i}{\sum_{i=1}^{n} \lambda_i}$$

4. Allocate subsystem failure rate requirements such that:

$$\lambda^*_i = w_i \lambda^*$$

Feasibility-of-Objectives Apportionment Technique

This technique was developed primarily for allocating reliability in repairable electromechanical systems. Subsystem allocation factors are computed as a function of numerical ratings of system intricacy, state of the art, mission performance time, and environmental conditions. Each rating, on a scale from 1 to 10, is estimated using engineering judgment based upon experience. Rating values are assigned as follows:

1. System intricacy—this is based on the probable number of parts or components making up the system and the intricacy of the assembled parts or components. The least intricate system is rated 1, and the most intricate system is rated 10.
2. State-of-the-art—the present state of engineering progress in all fields is considered. The least developed design or method is assigned a value of 10, and the most highly developed a value of 1.
3. Mission performance time—a continuously operating element is rated 10, an element with minimum operating time during the mission is rated 1.
4. Environment—elements expected to operate in very severe and harsh environments are rated 10, those expected to encounter the least severe environments are rated 1.

Ratings are usually assigned by the cognizant design engineer based upon engineering knowledge and experience. However, they may also be determined by a group of engineers using a voting method such as the Delphi technique (see Chapter 8.2).

An estimate is made of the types of parts and components likely to be used in the new system and the effect of their expected use on their reliability. If particular components have proven unreliable in a particular environment, then the environmental rating is increased accordingly.

The four separate ratings for each subsystem are then multiplied together to give a composite rating for the subsystem. Each subsystem rating will be between 1 and 10^4. The subsystem ratings are usually normalized so that their sum is 1.

The basic equations are:

$$\lambda_s T = \Sigma \lambda_k T \tag{9.4}$$

$$\lambda_k = C_k \lambda_s \tag{9.5}$$

where

C_k = complexity of subsystem k

$$C_k = \frac{w_k}{W} \tag{9.6}$$

w_k = composite rating for subsystem k

$$w_k = r_{1k} r_{2k} r_{3k} r_{4k} \tag{9.7}$$

$$W = \sum_{k=1}^{N} w_k \tag{9.8}$$

and where:

λ_s = system failure rate

T = mission duration

λ_k = failure rate allocated to each subsystem

N = number of subsystems

r_{1k} = rating for each of the four factors for each subsystem

Feasibility-of-objective example A system consisting of six subsystems, A, B, C, D, E, and F, has a system reliability requirement of 0.90 for a 120 hr mission. Engineering estimates are then made for intricacy (r_1), state of the art (r_2), performance time (r_3), and environment (r_4).

The subsystems and their ratings are illustrated in Table 9.5, columns 1 thru 5. The allocated failure rate is then computed for each subsystem. The procedure is as follows:

Table 9.5 Feasibility-of-Objectives Example

(1)	(2)	(3)	(4)	(5)	(6)	(7)	(8)
Subsystem	Intricacy (r_1)	State of the art (r_2)	Performance time (r_3)	Environment (r_4)	Overall rating (w_k)	Complexity (C_k)	Allocated failure rate (per 10^6 hr)
1. A	5	6	5	5	750	0.097	85
2. B	7	6	10	2	840	0.109	96
3. C	10	10	5	5	2500	0.324	284
4. D	8	8	5	7	2240	0.298	255
5. E	4	2	10	8	640	0.083	73
6. F	6	5	5	5	750	0.097	85
Total					7720	1.000	878

Note: System reliability requirement = 0.98, mission time = 120 hr, λ_s = 878 per 10^6 hr.

1. Compute the product of the rating r_i for each subsystem using Equation (9.7) and their sums using Equation (9.8) and enter in Table 9.5, column 6:

$$w_1 = 5 \times 6 \times 5 \times 5$$
$$= 750$$
$$\vdots \qquad \vdots$$
$$w_6 = 6 \times 5 \times 5 \times 5$$
$$= 750$$
$$w = 750 + 840 + 2500 + 2240 + 640 + 750$$
$$= 7720$$

2. Compute the complexity factors C_k for each subsystem using Equation (9.6) and enter in Table 9.5, column 7:

$$C_1 = 750/7720$$
$$= 0.097$$
$$\vdots \qquad \vdots$$
$$C_6 = 750/7720$$
$$= 0.097$$

3. Compute system failure rate λ_s from system specifications: R_s = 0.90 and T = 120 hr:

$$\lambda_s = -\ln \frac{0.90}{120 \text{ hr}}$$
$$= 878.0 \text{ per } 10^6 \text{ hr}$$

4. Compute the allocated subsystem failure rate λ_k, using Equation (9.5) and enter in Table 9.5, column 8:

$$\lambda_1 = 0.097 \times (878.0 \text{ per } 10^6 \text{ hr})$$
$$= 85.17 \text{ per } 10^6 \text{ hr}$$
$$\lambda_2 = 0.109 \times (878.0 \text{ per } 10^6 \text{ hr})$$
$$\vdots \qquad \vdots$$
$$\lambda_6 = 0.097 \times (878.0 \text{ per } 10^6 \text{ hr})$$
$$= 85.17 \text{ per } 10^6 \text{ hr}$$

5. Round off failure rates to two significant figures, so that excessive accuracy is not implied; sum and compare with λ_s, step 3:

$$\lambda_k = 85 + 96 + 284 + 255 + 73 + 85$$
$$= 878 \leqslant 878$$

Dynamic Programming Apportionment Technique

If all subsystems are not equally difficult to develop dynamic programming provides an approach to reliability apportionment with minimum effort expenditure when the subsystems are subject to different, but identifiable effort functions.

An advantage of the dynamic programming approach is the fact that it can be implemented with a simple algorithm consisting of only arithmetic operations. Advantages of the dynamic programming approach include that:

1. Large problems can be solved with a minimum number of calculations (this minimum could be large for a complex system).
2. There is always a finite number of steps required in computing an optimum solution.
3. There are no restrictions on the form of the functional expression for computing reliability or the form of the cost-estimating equations. Nonlinear functions can be used if required.

Dynamic programming algorithms provide a guide through the maze of possible alternate calculations that may arise when large systems are being analyzed. The dynamic programming approach can also be applied to the problem of reliability optimization of redundant systems with repair. Use of the dynamic programming algorithm does not in any way remove the requirement for computing the reliability and cost for each system configuration. However, it minimizes the total number of calculations by rejecting those configurations that would result in a decreasing reliability or those with excessive costs.

Dynamic programming optimization also has application potential in other areas of reliability analysis. For example, models have been developed to determine an optimal number of redundant units subject to restraints such as weight, cost, volume, opposing failure modes, and so on.

The dynamic programming approach can be readily computerized, and a number of computer models are available.

9.2 RELIABILITY BLOCK DIAGRAMS AND MODELS

The reliability model of the system is the foundation upon which all of the other reliability tasks are based. A reliability model, or models, of the system, subsystem, or equipment is required for making numerical apportionments and is essential for evaluating complex series-parallel equipment arrangements. The rationale behind the reliability model must always be clearly documented.

Basic information for the reliability model is derived from schematic block diagrams depicting functional relationships between the subsystems and components available to provide the required performance. The reliability model reorients the diagrams into a series-parallel network showing reliability relationships among the various components and subsystems. Authenticity of the functional relationships should be verified by FMECA (see Chapter 7).

The reliability model should be developed as soon as program definition permits, even though numerical input data are not yet available. Careful review of the early models may reveal conditions where design or management action may be required. For example, single-point failures, which can cause premature mission termination or unacceptable hazards, can be identified for future elimination by alternate design approaches.

The reliability model or models should be iterated and continually expanded to a more detailed level as data becomes available. Details are added to the model as they become available so that technical evaluations may proceed along with the program decisions.

There are two fundamental types of reliability models, the *basic (series) reliability model* and the *mission reliability model*.

Basic (Series) Reliability Model

By definition, the basic reliability model is a series model consisting of all the elements of the item. It includes those elements related solely to redundancy and alternate modes of operation. The model includes both the reliability block diagram and its associated mathematical equation. The basic reliability model is used primarily to evaluate the maintenance and logistic support characteristics of the proposed design.

Mission Reliability Model

The purpose of the mission reliability model is to show the actual utilization of the various elements of the item in achieving mission success. The mission reliability model also consists of a reliability block diagram and its associated mathematical equation. Elements of the item intended for redundancy or alternate modes of operation are modeled in a series-parallel combination or similar configuration appropriate to the mission phase and mission application.

The mission reliability block diagram shows the interdependencies among the elements (subsystems, equipments, and so on) or functional groups of elements required for item success for each different service-use event. It uses concise visual shorthand, to show the various series-parallel block combinations (paths) that result in item success. A thorough understanding of the mission definition and the service-use profile are required to construct the mission reliability block diagram.

The reliability models together with duty cycle and mission duration information are used to develop mathematical equations which, with appropriate failure-rate and probability-of-success data, provide apportionment estimates and assessments of both basic reliability and mission reliability.

There are four principle methods by which the mission reliability model may be derived. These methods are: conventional probability, Boolean truth table, logic diagrams, and Monte Carlo simulation.

Conventional Probability

A reliability mathematical equation may be prepared from a reliability block diagram based upon conventional probability relationships. This method can be used for both single-functioned and multifunctioned systems.

The conventional probability method makes use of the equations developed for redundancy to handle series, parallel, and series-parallel combinations of equipments. For anything other than series-parallel configurations or complex configurations the following equation is used repeatedly:

$$P_S = P_S \text{ (if X is good) } R_X + P_S \text{ (if X is bad) } Q_X$$

where:

$$P_S = \text{reliability of mission}$$
$$P_S \text{ (if X is good)} = \text{reliability of mission if X is good}$$
$$P_S \text{ (if X is bad)} = \text{reliability of mission if X is bad}$$
$$R_X = \text{reliability of X}$$
$$Q_X = \text{unreliability of X} = 1 - R_X$$

Thus the reliability of the mission is equal to the reliability of the mission given that a specific portion of the system works times the probability that a portion of the system will work plus the reliability of the mission given that a specific portion of the system fails times the probability that the portion fails. The above formula can also be used to generate probability of success equations for series-parallel configurations.

Boolean Truth Table

A Boolean truth table may also be used to construct the reliability mathematical equation from a reliability block diagram based upon Boolean algebra. This method, applicable to both single-functioned and multifunctioned systems, is more tedious than the conventional probability method. However, it may be useful where the user is familiar with Boolean algebra.

The Boolean algebra approach lists all equipments in truth table form. The truth table has 2^n entries where n is the number of equipments in the system. The table has a one or zero entry in each column indicating success or failure respectively for each equipment. All possible combinations of all equipments working and failing are then listed. Each row of the truth table is examined and it is decided whether the combination of equipments working and equipments failed yields system success (S) or failure (F). A one is then inserted in the next column for each S. For each S entry, the respective probabilities for the indicated state of each equipment are multiplied to yield a P_S for that entry. A sample truth table is shown in Table 9.6.

Table 9.6 Boolean Truth Table

Trial number	Equipment			Success
	A	B	C	
1	1	0	0	0
2	0	1	0	0
3	0	0	1	0
4	1	0	1	1
5	0	1	1	1
6	1	1	0	0
7	1	1	1	1
8	0	0	0	0

Source: *Reliability Modeling and Prediction* (MIL-STD-765), Naval Publications and Forms Center, Philadelphia.

Logic Diagrams

The third method utilizes logic diagrams to prepare a reliability mathematical equation from a reliability block diagram. The logic diagram method can be used for both single-functioned and multi-functioned systems. This method is also more tedious than the conventional probability method, but it is quicker than the Boolean truth table approach for combining terms to simplify the mission reliability equation.

The logic diagram translates the reliability block diagram into a switching network. A closed contact represents equipment success, an open contact equipment failure. Each equipment required for each alternative mode of operation is identified by a contact along a path. All paths terminate at the same point (success). In the logic diagram all paths must be mutually exclusive. Thus, by using a few simple manipulations the amount of effort involved as compared with the Boolean truth table method can be significantly reduced.

For complex configurations the procedure is to reduce the reliability diagram to a series of series-parallel configurations by successively splitting the diagram into subdiagrams by removing one equipment and replacing it with a short circuit and an open circuit. Figure 9.1 shows an example of such a switching logic diagram.

Monte Carlo Simulation

The Monte Carlo simulation method synthesizes a system reliability prediction from a reliability block diagram by means of successive random sampling. The Monte Carlo simulation method is employed in instances where individual equipment probabilities (of equivalent reliability parameter) are known but the mission reliability model complexity is too great to derive a general equation for solution. The Monte Carlo simulation method does not result in a general probability-of-success equation but computes the system probability of success from the individual equipment probabilities and the reliability block diagram. A Monte Carlo simulation is performed by computer due to the large number of repetitive trials and calculations required to obtain a significant result. The Monte Carlo simulation method may be used for either single-functioned or multi-functioned systems.

The Monte Carlo simulation method is based on several principles of probability and on the techniques of probability transformation. A major underlying principle is the law of large numbers. It states that the larger the sample the more certainly the sample mean will be a good estimate of the population mean.

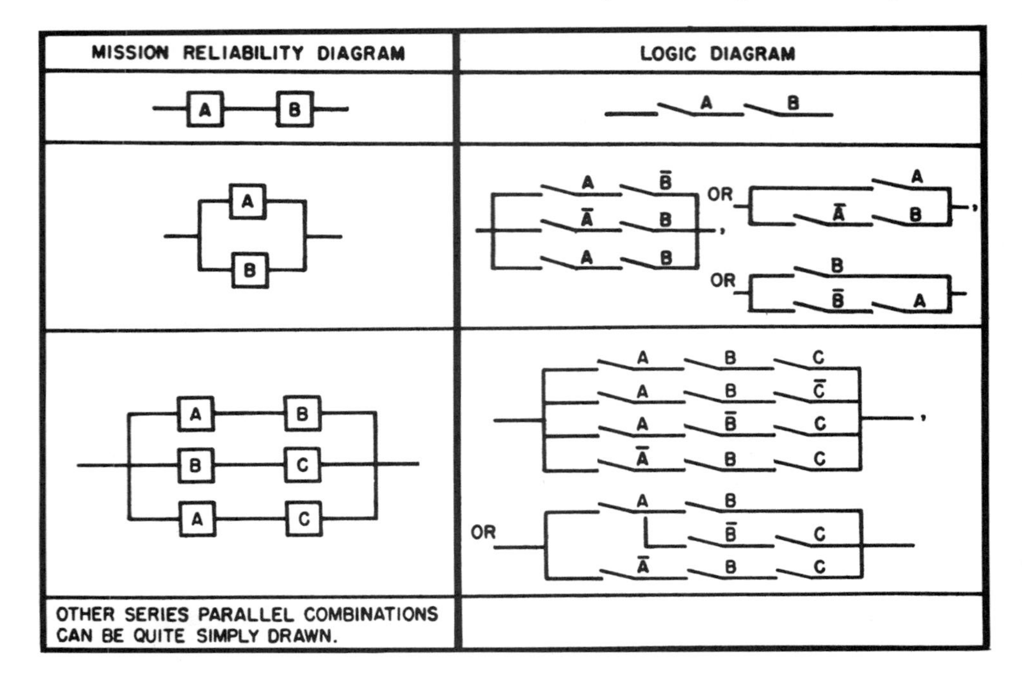

Figure 9.1 Switching logic diagrams. Note: When one logic switch A is open, all must be open and all $\bar{A}$ must be closed and similarly for B and C logic switches. [From *Reliability Modeling and Prediction* (MIL-STD-756), Naval Publications and Forms Center, Philadelphia.]

9.3 RELIABILITY PREDICTION

Reliability prediction is the process of quantitatively assessing an equipment design relative to its specified reliability requirement. The real value of the quantitative expression lies in the information conveyed with this value and the use which is made of that information. Predictions constitute decision criteria for selecting courses of action which affect reliability.

The primary objective of reliability prediction is to provide guidance relative to expected inherent reliability for a given design. Reliability predictions are most useful during the early system design, before hardware is constructed and tested.

During design and development, predictions serve as quantitative guides by which design alternatives can be judged relative to reliability. Reliability predictions assist in feasibility evaluation, comparison of alternate configurations, identification of potential problems, logistics support planning, determination of data deficiencies, tradeoff decisions, and allocation of reliability requirements. Predictions also provide criteria for planning and evaluating reliability growth and reliability demonstration testing. Some specific uses of reliability prediction are shown in Table 9.7.

Reliability prediction is an important key to system development. It allows reliability to become an integral part of the design process. To be effective, the prediction technique must relate engineering variables to reliability variables.

Two different types of reliability predictions must be addressed: the basic reliability prediction and the mission reliability prediction.

The basic reliability prediction utilizes the basic reliability model, that is, a series model. It assumes that every single part in the equipment is essential at all times and is used for estimating the demand for maintenance and logistic support.

The mission reliability prediction utilizes the mission reliability model and is intended to depict the intended utilization of the various

Table 9.7 Uses for Reliability Prediction

1. Establishment of reliability requirements, preliminary design specifications, and requests for proposals. They help to determine the feasibility of a proposed reliability requirement.
2. Comparison of established reliability requirements with state-of-the-art feasibility. They provide guidance in budget and schedule decisions.
3. Provide a uniform basis for proposal preparation, evaluation, and selection.
4. Evaluation of potential reliability through predictions submitted in technical proposals and reports.
5. Identification and ranking of potential problem areas and the suggestion of possible solutions.
6. Allocation of reliability requirements among the subsystems and lower-level items.
7. Evaluation of the choices between alternate parts, materials, units, and processes.
8. Conditional evaluation of the design for prototype fabrication during the development phase.
9. Provide a quantitative basis for design tradeoffs.

elements within the system to achieve mission success. It takes into account redundancy, duty cycles, and so on.

Both a basic reliability prediction and a mission reliability prediction should be prepared early in the program and updated whenever changes in design or data occur. Early predictions, while unrefined because of insufficient design detail, do provide useful feedback for establishing reliability requirements in the form of apportionments and for determining the feasibility of meeting reliability requirements.

As the item progresses from paper to hardware, predictions evolve into assessments as actual program test data become available and are integrated into the calculations. The validity of both predictions and assessments is a function of data quality and assumptions. Valid, timely analyses projecting or indicating deficient reliability provide the basis for corrective action. The sooner action is identified, the less its impact on the program, and the higher are the payoffs over the life of the item.

Prediction Methods

In general, there is a hierarchy of reliability prediction techniques available depending upon the depth of knowledge of the design and the availability of historical data on equipment and component part reliabilities. As the design proceeds from the conceptual through full-scale development to the production phase, data describing the system design evolves from a qualitative description of system functions to detailed specification and drawings suitable for hardware production. Therefore, a hierarchy of reliability prediction techniques have been developed to accommodate the different reliability study and analysis requirements and the availability of detailed data as the system design progresses. These techniques can be classified into three groups as follows:

1. Similar item or similar circuit prediction method
2. Parts count prediction method
3. Parts stress analysis prediction method

Similar Item or Similar Circuit Prediction Method

This prediction method utilizes specific experience on similar items or similar circuits. The quickest method of estimating reliability is to compare the item or the circuit under consideration with a similar item or circuit whose reliability is known. This method is most useful for items undergoing an orderly design evolution. Since the contemplated new design is similar to the old design, small differences can easily be isolated and evaluated. In addition, difficulties encountered in the old design serve as an alert to possible improvements in the new design.

Major factors of concern for a direct comparison of similar items or similar circuits should include:

1. Item or circuit physical and performance comparison
2. Design similarity
3. Manufacturing similarity
4. Service-use-profile similarity (logistic, operational and environmental)
5. Program and project similarity
6. Proof of previous reliability achievement

The validity of the similar item or similar circuit method is dependent upon the degree of equivalence between the items or circuits and not simply on the generic term used to describe the items or circuits. For example, comparing two power supplies, an existing ten-watt unit and a proposed one-kilowatt unit, and expecting comparable reliability values would not be reasonable. The much higher power of the proposed power supply will probably result in much lower reliability values due to design and stress differences. A comparison may be possible if there are scale factors to relate realistically reliability with specific item parameters such as power levels.

This reliability prediction technique permits very early estimation of the failure rate of a new equipment based on experience gained from operational equipments of similar design and function. The accuracy of the estimates, however, depends on the quality of historical data and the degree of similarity between the existing and new equipments.

Parts Count Prediction Method

The parts count method is used in the preliminary design stage when the number of parts in each generic part type class such as capacitors, resistors, and so on, are reasonably fixed, and the overall design complexity is not expected to change appreciably during later stages of development and production. The parts count method is based upon average failure rates for each component part type and it assumes the time-to-failure of the parts is exponentially distributed (that is, a constant failure rate).

The data needed to support the parts count method includes:

1. Generic part types (including device complexity for microelectronics)
2. Quantity of parts
3. Part quality levels (either known or assumed)
4. Item-use environment

The general expression for item failure rate with this method is:

$$\lambda_{ITEM} = \sum_{i=1}^{i=n} N_i \, (\lambda_G \pi_Q)_i$$

for a given equipment environment where:

λ_{ITEM} = total equipment failure rate (failure/10^6 hr)

λ_{Gi} = generic failure rate for the ith generic part (failures/10^6 hr)

π_{Qi} = quality factor for the ith generic part

N_i = quantity of ith generic part

n = number of different generic part categories

This equation applies if the entire equipment is being used in one environment. If the equipment comprises several units operating in different environments (such as avionics with units in both inhabited and uninhabited environments), then the equation should be applied to the portions of the equipment in each environment. The environment-equipment failure rates are then added to determine total equipment failure rate.

Generic (average) failure rates, λ_{Gi}, and quality factors, π_Q, are obtained from the latest version of MIL-HDBK-217, the basic document used for reliability predictions. MIL-HDBK-217 contains a number of tables of generic failure rates for various classes and types of parts and their associated quality factors.

Parts Stress Analysis Prediction Method

The parts stress analysis method is used to determine part failure rates in the detailed design stage when few assumptions are necessary about the parts used, their stress derating, their quality factors, their operating stresses, or their environment. The technique is based upon a detailed knowledge of the stresses to which the part will be subjected, for example, temperature, humidity, vibration, and so on, and the effect of those stresses on the part's failure rate. These factors should all be known or capable of being determined based upon the state-of-hardware definition for which the parts stress analysis method is applicable. Where unique parts are used, any assumptions regarding their failure-rate factors should be identified and justified. The parts stress analysis method is the most accurate method of reliability prediction other than measurement of reliability under actual- or simulated-use conditions. The parts stress analysis assumes that the time-to-failure of the parts is exponentially distributed (that is, a constant failure rate). The parts stress analysis prediction method was addressed in depth in Chapter 3.1.

The data needed to perform the parts stress analysis includes:

1. Specific part types (including device complexity for microelectronics)
2. Quantity of parts
3. Part quality levels
4. Item-use environment
5. Part operating stresses

The procedure for extracting failure-rate data from MIL-HDBK-217 differs according to part class and type. In general, however, the following steps are required:

1. A base failure rate is determined for each part. This value is established from the appropriate table in MIL-HDBK-217 and is a function of part type, environmental temperature, and the relative level of the more significant operational stresses.
2. The values of one or more multiplicative or additive factors are defined from applicable tables in MIL-HDBK-217. These factors define the relationship between the base failure rate and the predicted failure rate for the specific application of interest.
3. The part failure rate is calculated using the established base failure rate and the modifying factors.

Examples of the detailed stress method of reliability prediction are given in Chapter 3 of this text.

With both the parts count method and the parts stress analysis method the item failure rate can be determined directly by the summation of part failure rates if all elements of the item reliability model are in series. In the event that the item reliability model consists of nonseries elements (for example, redundancies or alternate modes of operation), item reliability can be determined either by considering only the series elements of the model as an approximation, or by summing part failure rates for the individual elements and calculating an equivalent series failure rate for the nonseries elements of the model.

The various reliability prediction methods are summarized in Table 9.8.

Table 9.8 Reliability Prediction Methods

Method	Used during	Advantages	Disadvantages
1. Similar item/ similar circuit	Feasibility	Fastest	Least accurate
2. Parts count	Initial design	More accurate than Method 1	Less accurate then Method 3
3. Parts stress	Detail design	Most accurate	Most time-consuming

9.4 AUTOMATED RELIABILITY PREDICTION PROGRAMS

A number of computer software programs are presently available to assist in performing detail stress reliability predictions in accordance with MIL-HDBK-217. These programs, marketed by a number of different suppliers, are available in a variety of formats offering a variety of different capabilities over a wide range of prices. They

Table 9.9 Reliability Prediction Program Selection Factors

1. Does the program actually perform a detail stress prediction or is it merely a generic parts count prediction?
2. If it is a detail stress prediction, can each of the application-dependent parameters such as stress level, ambient or case temperature, and environment actually be entered and modified as necessary?
3. Can each of the part-dependent parameters such as power dissipation, device complexity, and number of pins, etc., be entered in the program or are average values assumed?
4. Is it necessary to enter all of the parameters for each individual part, or does the program contain a resident part parameter data base?
5. Is it possible to communicate with an external part parameter data base and is it possible to modify or enhance a resident part parameter data base?

Table 9.9 (Continued)

6. Does the software contain default values for the various π factors if you fail to insert a specific value?
7. If so, do you know that the default value has been selected, and what the default values are?
8. What type of default values are assumed; best case, worst-case or nominal?
9. Are "global" changes or "what if?" variations possible such that a completely new prediction can be performed automatically based on one or more application-dependent changes?
10. Does the program contain any system modeling capability such that automated higher-level predictions are possible?
11. Is the program capable of handling simple forms of redundancy?
12. Are any overstressed parts automatically flagged by the program?
13. Can the program reside on a personal computer or a minicomputer or is a large mainframe computer necessary?
14. What operating system and/or programming language does the program utilize?
15. Is the program user friendly? Does it require extensive computer experience or extensive training?
16. If such training is required, how and where can such training be acquired?
17. How are subsequent changes in the MIL-HDBK-217 models incorporated in the program?
18. Is the program an outright purchase, a lease arrangement or a time-share program?

are available for installation on mainframe computers, minicomputers, and personal computers, all offering different options, and each having their own advantages and disadvantages. Time-share operation rather than outright purchase is an additional option.

In order to select properly and use the best program for a given application it is necessary for the user to understand the mechanics

Table 9.10 Automated Reliability Prediction Programs

Program Name	Organization	Comments
RADC-ORACLE	Reliability Analysis Center RADC/RAC Griffiss AFB, NY 13441	Time-share, available for military program use only
PREDICTOR	Management Sciences, Inc. 6022 Constitution Ave. N.E. Albuquerque, NM 87110	Purchase or time-share; IBM PC
RAP 217 and PC RAP 217 (parts count)	PROMPT Software Company 393 Englert Court San Jose, CA 95133	Purchase; IBM PC, or APPLE, and CP/M personal computers
Reliability prediction program	Powertronic Systems, Inc. P.O. Box 29019 New Orleans, LA 70189	Purchase; IBM PC, or CP/M personal computers
Failrate	Systems Evaluation, Inc. 31255 Cedar Valley Drive Westlake Village, CA 91362	Purchase; IBM PC
217 PREDICT	SYSCON Corp. 540 Weddell Drive, Suite 1 Sunnyvale, CA 94086	Time-share
Rel Calc 2	T-Cubed Systems 31220 La Baya Drive, Suite 110 Westlake Village, CA 91362	Purchase; IBM PC
ARPP	Director, Licensing Bell Communication Research 290 W. Mt. Pleasant Ave. Livingston, NJ 07039-2729	License purchase DEC VAX-UNIX, IBM-AT-VENIX (hard disk), or IBM-PC

of the prediction process and to recognize the strength and weaknesses of the various programs available. Some of the more important factors which should be addressed in selecting a specific reliability prediction program are addressed in Table 9.9. A sample of some of the automated reliability prediction programs presently on the market are listed in Table 9.10.

10

Environmental Considerations

10.0 DESIGNING FOR THE ENVIRONMENT

Failures are frequently associated with specific environments as shown in Figure 10.1. This chapter addresses some of the general environmental design considerations dealing with these specific environments.

Careful selection and application of materials can enhance item reliability by reducing or eliminating adverse environmental effects. The equipment environment is neither forgiving nor understanding; it methodically surrounds and attacks every component, and when a weak point exists, equipment reliability suffers. Design and reliability engineers must therefore understand the environment and its potential effects, and then select designs or materials that counteract these effects or provide methods to alter or control the environment within acceptable limits.

In addition to the obvious environments of temperature, humidity, shock, and vibration, the design engineer will create microenvironments within the equipment by his choice of designs and materials. For example, a gasket or seal under elevated temperatures or at reduced pressures may release corrosive or degrading volatiles into the system. Teflon may release fluorine, and polyvinylchoride (PVC) may release chlorine. These examples illustrate the fact that internal microenvironments which are inadvertently designed into the system can seriously affect reliability.

Materials and components selection aids are available to design and reliability engineers. For example, the text *Deterioration of Materials, Causes and Preventive Techniques* by Glenn A. Greathouse

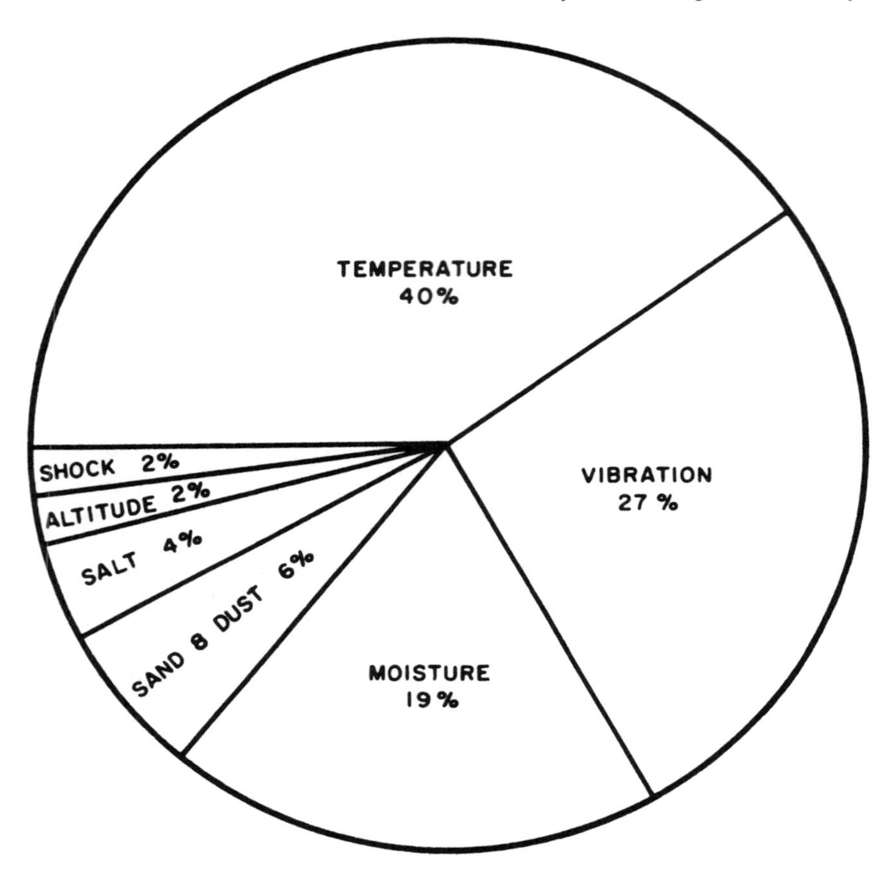

Figure 10.1 Failure distribution versus environment.

and Carl J. Wessel, as well as military specifications, standards, and handbooks provide both general and specific guidance on this subject.

Environmental Resistance

To realize fully the benefits of a reliability-oriented design, consideration must be given early in the design to the required environmental resistance of this equipment. The environmental resistance, both inrinsic and that provided by specific design features, will determine the ability of the equipment to withstand the deleterious stresses imposed by the environment in which the equipment

Table 10.1 Environmental Stresses, Effects, and Reliability Improvement Techniques in Electronic Equipment

Environmental stress	Effects	Reliability improvement techniques
High temperature	Resistance, inductance, capacitance, power factor, dielectric constant, etc., will vary; insulation may soften; moving parts may jam due to expansion; finishes may blister; devices suffer thermal aging; oxidation and other chemical reactions are enhanced; viscosity reduction and evaporation of lubricants are problems, structural overloads may occur due to physical expansions.	Heat-dissipation devices, cooling systems, thermal insulation, heat-withstanding materials
Low temperature	Plastics and rubber lose flexibility and become brittle; electrical constants vary; ice formation occurs when moisture is present; lubricants gel and increase viscosity; high heat losses; finishes may crack; structures may be overloaded due to physical contraction.	Heating devices, thermal insulation, cold-withstanding materials
Thermal shock	Materials may be instantaneously overstressed causing cracks and mechanical failure; electrical properties may be permanently altered; Crazing, delamination, ruptured seals.	Combination of techniques for high and low temperatures
Mechanical shock	Mechanical structures may be overloaded, weakened or collapse; items may be ripped from their mounts; mechanical functions may be impaired.	Strengthened members, reduced inertia and moments, shock-absorbing mounts

Table 10.1 (Continued)

Environmental stress	Effects	Reliability improvement techniques
Vibration	Mechanical strength may deteriorate due to fatigue or overstress; electrical signals may be mechanically and erroneously modulated; structures may be cracked, displaced, or shaken loose from mounts; mechanical functions may be impaired; finishes may be scoured by other surfaces; wear may be increased.	Stiffening, control of resonance
Humidity	Penetrates porous substances and causes leakage paths between electrical conductors; causes oxidation which leads to corrossion; moisture causes swelling in materials such as gaskets; excessive loss of humidity causes embrittlement and granulation.	Hermetic sealing, moisture-resistant material, dehumidifiers, protective coatings
Salt atmosphere and spray	Salt combined with water is a good conductor which can lower insulation resistance; causes galvanic corrosion of metals; chemical corrosion of metals is accelerated.	Nonmetal protective covers, reduced use of dissimilar metal in contact, hermetic sealing, dehumidifiers
Electromagnetic radiation	Causes spurious and erroneous signals from electrical and electronic equipment and components; may cause complete disruption of normal electronic equipment such as communication and measuring systems.	Shielding, material selection, part type selection

Nuclear/cosmic radiation	Causes heating and thermal aging; can alter chemical, physical, and electrical properties of materials; can produce gases and secondary radiation; can cause oxidation and discoloration of surfaces; damages electrical and electronic components, especially semiconductors.	Shielding, component selection, nuclear hardening, hardening
Sand and dust	Finely finished surfaces are scratched and abraded; friction between surfaces may be increased; lubricants can be contaminated; clogging orifices, etc.; materials may be worn, cracked, or chipped; abrasion, contaminates insulations, corona paths.	Air filtering, hermetic sealing
Low pressure (high altitude)	Structures such as containers, tanks, etc., are overstressed and can be exploded or fractured; seals may leak; air bubbles in materials may explode causing damage, internal heating may incrase due to lack of cooling medium; insulations may suffer arcing and breakdown; ozone may be formed; outgassing is more likely.	Increased mechanical strength of containers, pressurization, alternate liquids (low volatility), improved insulation, improved heat-transfer methods

Source: *Electronic Reliability Design Handbook* (MIL-HDBK-338), Naval Publications and Forms Center, Philadelphia.

will operate. Identification of and detailed description of the environments in which the equipment must operate is the initial step in the determination of the required environmental resistance. Determining the performance of the components and materials when they are exposed to the degrading stresses of these environments is the next step. When their performance is found to be inadequate or marginal, corrective measures such as derating, redundancy, protection from adverse environments, or selection of more resistant materials and components will be necessary to fulfill the reliability requirements of the equipment.

Environmental Factors

Reliability is strongly dependent upon the operating conditions that are encountered during the entire life of the equipment, therefore it is important that such conditions be accurately identified at the beginning of the design process. Environmental factors which exert a strong influence on equipment reliability are summarized in Table 10.1 and discussed in greater detail in the following pages.

10.1 TEMPERATURE AND THERMAL SHOCK

The thermal design of electronic equipment is as important to reliability and performance as is the circuit design. Excessive temperature is the primary cause of both performance and reliability degradation. Therefore, each design must be thoroughly evaluated to establish that its thermal characteristics are consistent with the equipment reliability requirements.

Reliability Implications of Temperature

Before discussing the theory and implementation of thermal control systems, a brief summary of the various ways in which temperature can influence the reliability of an electrical or electronic device is in order.

Thermal energy, quantified as temperature, may be directly linked to a number of physical and/or chemical processes which are detrimental to the reliability and/or performance of electrical devices and equipment. These processes may be grouped into two categories: those attributed to static thermal stresses and those attributed to dynamic thermal stresses.

Failure mechanisms attributable to static thermal stresses usually involve a threshold temperature which must be exceeded before any detrimental effects occur. Change-of-state failures such as internally melted parts are usually the result of overstress. Organic materials are often vulnerable to thermal overstress due to their relatively low glass transition temperatures. It is the responsibility of both the electrical designer and the thermal designer to insure that no part will ever be subjected to a temperature in excess of the maximum rated temperature of that part, even under worst-case conditions.

Reduction in the operating temperatures of components is a primary method for improving reliability. The thermal design should reduce heat input to minimal levels and provide low thermal resistance paths from heat-producing elements to an ultimate heat sink at a reasonably low temperature. Effective thermal design maintains equipment and parts within their permissible operating temperature limits under all operating conditions. Detailed thermal design is an engineering discipline in itself, and as such as beyond the scope of this text. However, the thermal design and the electrical design must be performed together in a synchronized manner.

Specific Thermal Effects

Chemical and physical deterioration are accelerated by both heat and cold for two basic reasons.

1. The physical properties of materials are greatly modified by temperature changes.
2. The rate at which chemical reactions occur is a function of the temperature of the reactants. A simple rule of thumb is that the rate of many reactions doubles for every 10°C rise in temperature; this is equivalent to an activation energy of about 0.6 eV (see Arrhenius reaction rate model Chapter 2.2).

Heat is transferred by three methods: (1) radiation, (2) conduction, and (3) convection. One or more of these three methods must, therefore, be used to protect against excessive temperatures. High-temperature degradation can be minimized by both passive and active techniques. Passive techniques use natural heat sinks to remove heat, while active techniques use devices such as heat pumps or refrigeration to create heat sinks.

Thus the thermal characteristics of every item of equipment should be evaluated from two viewpoints:

1. Is a substitute item available that will generate less heat?
2. Can the item be located and positioned so that its heat has minimal effect on other items?

To maintain a constant temperature, heat must be removed at the same rate at which it is generated. Thermal systems such as conduction cooling, forced convection, blowers, direct or indirect liquid cooling, direct vaporization or evaporation cooling and radiation cooling must be capable of handling both natural and induced heat sources.

Thermoelectric cooling is another possible cooling method. It makes use of the phenomenon of reversible flow of heat and electricity to pump heat from a colder region (cold junction) to a hotter region (hot junction). Heat pumping results from the Peltier effect: when a voltage is applied across the junction of two dissimilar materials, heat is absorbed or evolved at the junction. Thermoelectric coolers are attractive because they are compact and require no moving parts.

Some of the disadvantages of thermoelectric cooling include:

1. Low-power efficiency.
2. Single stage coolers are limited to a 70°C temperature difference.
3. The power efficiency drops drastically when using multiple stages.
4. They frequently fail when the hot junction temperature exceeds 130°C.

Passive sinks require progressive heat transfer from intermediate sinks to ultimate sinks until the desired heat extraction has been achieved. Thus, when heat sources have been identified, and heat removal elements selected, they must be integrated into an overall heat removal system, so that heat is actually removed and not merely redistributed within the system. Efficiently integrated heat removal techniques can significantly improve item reliability.

Materials expand or contract in accordance with their thermal coefficient of expansion as their temperature is changed. This expansion and contraction causes problems with seals and the fit between parts. It also generates internal stresses. Local stress concentrations due to nonuniform temperature are especially damaging. A simple example is that of hot water-glass which shatters when immersed in cold water. Metal structures, when subjected to cyclic heating and cooling, may ultimately collapse due to the induced stresses and fatigue caused by flexing. The thermocouple effect at the junction of two dissimilar metals produces an electric current that can cause electrolytic corrosion. Plastics, natural fibers, leather, and both natural and synthetic rubber are all particularly sensitive to temperature extremes as evidenced by the brittleness at low temperatures and high degradation rates at high temperatures.

Table 10.2 summarizes some of the basic precautions to enhance reliability at low temperatures. However, there is always the risk that compensating for one failure mechanism will aggravate another failure mechanism.

Table 10.2 Low-Temperature Protection Methods

Effect	Preventive measures
Differential contraction	Careful selection of materials. Provision of proper clearance between moving parts. Use of heavier skin material.
Lubrication stiffening	Proper choice of lubricants: Use greases compounded from silicones, diesters, or silicone-diesters thickened with lithium stearate. Eliminate liquid lubricants wherever possible.
Ice damage caused by freezing of collected water	Eliminate moisture by: Providing vents. Ample draining facilities. Eliminating moisture pockets. Suitable heating. Sealing. Desiccation of air.
Degradation of material properties and component reliability	Careful selection of materials and components with satisfactory low-temperature capabilities.

Source: *Electronic Reliability Design Handbook* (MIL-HDBK-338), Naval Publications and Forms Center, Philadelphia.

Thermal Performance Analysis

The failure rates of electronic system components vary significantly with temperature. Table 10.3 illustrates the reliability improvement potential associated with the operation of circuit elements at reduced temperatures. The cost of designing and implementing adequate thermal performance is usually more than offset by savings in equipment maintenance costs. Suitable thermal design also minimizes temperature excursions of components when environmental temperatures or power dissipation vary, thus further enhancing reliability.

Parts stress analysis (addressed in Chapter 3.1) in accordance with MIL-HDBK-217 is the preferred method for evaluating thermal performance of electronic equipment with respect to reliability. It determines the maximum safe temperatures for each constituent part.

Table 10.3 Reliability Improvement at Reduced Temperatures

Part description	Base failure rates[a] (per 10^6 hr)		Δt (°C)	Failure rate improvement
	Reduced temp. (°C)	High temp. (°C)		
PNP silicon transistors	0.0012 at 40°	0.0091 at 160°	120	8:1
NPN silicon transistors	0.0008 at 40°	0.0048 at 160°	120	6:1
Glass capacitor	0.00024 at 40°	0.0073 at 125°	85	30:1
Transformers and coils	0.0025 at 40°	0.0666 at 85°	45	27:1
Resistors	0.0003 at 40°	0.0054 at 120°	80	17:1

[a]From MIL-HDBK-217D assuming a 10% stress level.

Source: *Reliability Design Handbook* (RDH-376), copyright 1976 by IIT Research Institute, Reliability Analysis Center, RADC/RAC, Griffiss Air Force Base, N.Y.

Once the maximum allowable component temperatures are determined and the power dissipated by each component is ascertained, a heat-flow network can be established from each component to available heat sinks or coolants for analysis for the system's thermal performance. In situations where surface temperatures must be related to maximum allowable internal temperatures such as junction temperatures of semiconductor devices a knowledge of the internal thermal resistance of these components is required to calculate the corresponding surface temperatures for the particular operating conditions of the component.

Further specifics of the parts stress thermal analysis and design techniques are described in MIL-HDBK-251 "Reliability/Design Thermal Applications." This document provides a very comprehensive review of most aspects of thermal design.

Thermal Design

Electrical components are never 100% efficient; they always generate some waste heat. The difference between the input power and the output power is the amount of waste heat that must be dissipated by the device.

Thermal problems may be reduced by judicious component selection and the use of efficient circuit designs. Switching-regulated power supplies are an excellent example. In high-power applications a transistor operates much more efficiently as a switch than as a linear device. Thus, switching-regulated power supplies operate at considerably greater efficiencies and dissipate less thermal energy than their linear counterparts.

The optimization of circuit design is seldom sufficient to eliminate all thermal problems. It is still necessary to remove the heat from the source and dissipate it to a thermal sink.

Thermal design is based upon optimization of one or more of the three basic heat transfer methods:

1. Conduction
2. Convection
3. Radiation

Conduction

Heat conduction is effected through molecular oscillations in solids and by elastic impact in liquids and gases. Conduction heat-flow relationships are analogous to Ohm's law for electrical current flow, that is, the rate of heat transferred is analogous to current flow, temperature differences are analogous to voltage drops, and thermal resistance to heat transfer is analogous to electrical resistance.

Conduction cooling is capable of handling most thermal design problems by providing a very low thermal impedance path from the heat source to a thermal reservoir. The thermal resistance of the conductive path may be closely controlled by appropriate material selection. Thermal conductivity is a bulk material property; it is relatively immune to degradation. This is in contrast to convective and radiative techniques which are strongly dependent on the surface conditions and therefore subject to degradation with time.

Thermal control within a circuit is achieved by minimizing the resistance in one or more of the paths to thermal ground or to the ultimate heat sink. This may be accomplished by: providing high thermal conductance paths to ground; maximizing heat transfer at interfaces by use of polished surfaces, thermal grease, and so on; and by minimizing the number of thermal interfaces.

Convection

Heat transfer from the surface of a solid to moving masses of fluids, either gaseous or liquid, is known as convection. This mode of heat transfer is implemented by circulation of the fluid.

Convection cooling is often adequate where thermal densities are moderate. The most common convective medium is air, with the air flow resulting from either forced air or natural convection currents. In natural convection the flow of air is created by the existence of thermal gradients. The efficiency of natural convection cooling may be optimized by proper selection of air flow paths. When natural convection is insufficient to achieve proper thermal conditions, fans or electrostatic wind generators may be used to increase the air flow and consequently increase the amount of thermal energy transferred to the air per unit time.

Radiation

All bodies continuously emit thermal radiation in the form of electromagnetic waves ranging in wavelength from long infrared to short ultraviolet. Radiation emitted from a body can travel undiminished through a vacuum or with relatively little absorption through gases. When radiation is intercepted by a second body, part of the radiation may be absorbed as thermal energy, part may be reflected from the surface, and part may be transmitted, still in electromagnetic wave form, through the body as in the case of glass.

Radiation-based techniques are seldom used except in space applications where convective and conductive techniques are impractical. Control of thermal radiation may be achieved by the use of radiation shields and by appropriate surface coatings. For most other applications radiative heat transfer is seldom a significant factor in the thermal characterization of equipment.

Table 10.4 Steps in Effective Thermal Design

1. Establish the maximum and minimum environmental temperatures of anticipated heat sinks and coolants.
2. Characterize the available cooling techniques, that is, forced air convection, liquid or vaporization cooling, and so on.
3. Develop a heat-flow network using electrical analog techniques for the conditions of maximum allowable component temperatures and maximum environmental heat sink or coolant temperatures. Then determine the thermal resistance from the parts to their respective heat sinks. (Computer programs are available to greatly simplify this task.)
4. Select packaging approaches and component placements that will fulfill the thermal resistance requirements in terms of the available and permissible cooling techniques.
5. Determine the suitability of simple cooling techniques such as free or forced air cooling to satisfy the heat concentration and thermal resistance requirements of the proposed design. If necessary, proceed to higher-level cooling techniques until the optimum cooling method is identified.
6. Evaluate the penalties associated with the selected cooling method. Perform tradeoff analyses to identify alternative approaches and refinements if possible.

Frequently it is practical or necessary to employ a thermal design based on a combination of methods. The heat sink is a good example. With a heat sink, a low resistance thermal path is provided from the component to the air. Heat is conducted from the component to the heat sink. The heat sink is then cooled by convection, with air serving as the thermal reservoir.

Equipments exhibiting very high thermal densities often require special cooling techniques, such as heat pipes, cold plates, refrigerants, and others. Each of these techniques has specific strengths and limitations that must be considered before the optimum technique can be selected. Some basic steps used in an effective thermal design are shown in Table 10.4.

10.2 MECHANICAL SHOCK AND VIBRATION

Protection against mechanical abuse is generally achieved by the use of suitable packaging, mounting, and structural techniques. In most cases, tradeoff situations between the level of protection and reliability improvements are not as pronounced as in the case of thermal protection. One notable exception is that of fatigue damage. Where fatigue is the primary failure mechanism, the level of protection will have a significant impact on reliability.

Mechanical Design Techniques

Basic structural design techniques, such as the selection of suitable materials and the proper location of components can aid in protecting an item against failure caused by severe shock or vibration stresses.

Two possible approaches may be considered when shock or vibration are present. The first approach is to isolate the equipment from the shock or vibration, and the second is to build the equipment to withstand the shock or vibration. With isolation, the problem is that effective simultaneous control of both shock and vibration is difficult due to the conflicting requirements which they impose. When only one or the other is present, special mountings may be used. The effectiveness of protective measures against both shock and vibration stresses is generally determined by an analysis of the deflections and the mechanical stresses produced by these environment factors. This generally involves determination of the item's natural frequencies and an evaluation of the mechanical stresses within the components and materials produced by the shock or vibration. If the mechanical stresses produced are below the allowable safe working stress of the materials involved, no direct protection methods are required. If, on the other hand, the stresses exceed safe levels, then corrective measures such as reduction of inertia and bending moment effects, stiffening, and the incorporation of additional support members are indicated. If such approaches still do not reduce the stresses below the safe levels, further reduction is usually possible by the use of shock-absorbing mounts.

When using shock mounts, one factor which is frequently overlooked is the possibility of a collision between two adjacent components or separately insulated subsystems if maximum excursions and sympathetically induced vibrations are not evaluated by the designer.

Fatigue failures (the tendency for a metal to break under cyclic-stressing loads considerably below the metal's tensile strength) is another area of reliability concern relating to shock and vibration. This includes low-cycle fatigue, acoustic fatigue, and accumulated fatigue under combined stresses. The possible interaction between multiaxial fatigue and other environmental factors such as temperature extremes, temperature fluctuations, and corrosion requires careful study. Stress-strength analysis of components and parameter variation analysis are particularly suited to the evaluation of these effects. Nondestructive testing methods are also very useful in this area. Several different nondestructive evaluation (NDE) methods such as X-ray, neutron radiography, and dye-penetrant can be effectively used to locate fatigue cracks. The development of a simple design that is reliable is much more cost-effective than elaborate fixes and subsequent testing to redesign for reliability.

In addition to using proper materials and configuration, the shock and vibration experienced by the equipment should be controlled. In some cases even though an item is properly insulated and isolated against shock and vibration damage, repetitive forces may loosen the fastening devices. Obviously, if the fastening devices loosen enough to permit additional movement, the device will be subjected to increased forces, and may fail. Many specialized self-locking fasteners are commercially available, and fastener manufacturers can provide valuable assistance in selecting the best fastening methods.

An isolation system can be used at the source of the shock or vibration, in addition to isolating the protected component. The best results are obtained by using both methods. Damping devices are used to reduce peak oscillations, and special stabilizers may also be employed when unstable configurations are involved. Typical examples of dampeners are: viscous hysteresis, friction, and air damping. Vibration isolators are commonly identified by their construction and the material used for the resilient elements (that is, rubber, coil spring, woven metal mesh, and so on). Shock isolators differ from vibration isolators in that shock requires stiffer springs and a higher natural frequency for the resilient element. Types of isolation mounting systems include: underneath, over-and-under, and inclined isolators.

Table 10.5 Vibration and Shock Protection

1. Protection is defined as: temporary storage and release of energy with a changed time relation.
2. Isolation reduces the amplitude of the transmitted force.
3. Vibration isolation is a controlled mismatch between the excitation and the natural frequencies of the system.
4. Vibration isolation does not usually provide shock isolation.
5. The time distribution of application energy is the primary point of difference between shock mounts and vibration mounts.
6. Shock mounts can increase vibration damage.

In summary, some basic considerations in designing for shock and vibration protection are shown in Table 10.5.

10.3 ATMOSPHERIC CONSIDERATIONS

Many different contaminants and reliability degradation factors may be encountered in the atmosphere to which our equipment is exposed. Some of the more common contaminants include: moisture, sand and dust, corrosive airborne chemicals, explosive mixtures, and reduced atmospheric pressure. Each of these potentially degrading atmospheric conditions must be recognized and adquately addressed by our design.

Moisture Protection

Moisture is a chemical, and, considering its abundance and availability in almost all environments, is probably the most important chemical deteriorative factor to deal with. Moisture is not simply water. It is usually a solution of many impurities. These impurities cause many chemical difficulties such as the corrosion of metals. Many materials that are normally pliable at low temperatures will become hard and perhaps brittle if moisture has been absorbed and then freezes. Condensed moisture can act as a medium for the interaction between many otherwise relatively inert materials. Most gases readily dissolve in moisture. Chlorine released by PVC plastic, for example, forms hydrochloric acid when combined with moisture.

Although the presence of moisture may cause deterioration, the absence of moisture may also cause reliability problems. The useful properties of many nonmetallic materials depend upon an optimum level of moisture. Leather and paper become brittle and crack when they are very dry. Fabrics wear out at an increasing rate as moisture levels are lowered and fibers become dry and brittle. Dust (due to lack of moisture) can cause friction, increased wear, and can clog filters.

Moisture, in conjunction with other environmental factors, creates difficulties that may not be the characteristic result when these factors act alone. For example, abrasive dust and grit, which would otherwise escape, can be trapped by moisture. The permeability (to water vapor) of some plastics (PVC, polystyrene, polyethylene, and so on) is directly related to their temperature. The growth to fungus is enhanced by moisture, as is galvanic corrosion between dissimilar metals.

Some design techniques that can be used to counteract the effects of moisture are: (1) the elimination of moisture traps by providing drainage or air circulation; (2) the use of desiccant devices to remove moisture when air circulation or drainage is not possible; (3) the application of protective coatings; (4) rounding the edges of an item to allow a uniform coating of protective material; (5) using materials which are resistant to moisture effects, fungus, corrosion, and so on; (6) hermetically sealing components, using gaskets and other sealing devices; (7) impregnating or encapsulating materials with moisture-resistant waxes, plastics, or varnishes; (8) separating dissimilar metals or materials that might combine or react in the presence of moisture; and (9) separating components that might damage protective coatings.

The designer should consider possible adverse effects caused by specific methods of protection. For example, hermetic sealing, gaskets, protective coatings, and so on, may aggravate moisture difficulties by sealing moisture inside of an item or by contributing to internal moisture condensation. The gasket materials must be evaluated carefully for possible outgassing of corrosive volatiles or for incompatibility with adjoining surfaces or protective coatings. MIL-STD-454 "Standard General Requirements for Electronic Equipment" may be very helpful in this area. It provides a number of common requirements for electronic equipment such as: Requirement 15 related to corrosion protection, Requirement 16 related to dissimilar metals and Requirement 31 related to moisture pockets.

Sand and Dust Protection

Sand and dust degrade equipment primarily by:

1. Abrasion leading to increased wear
2. Friction causing both heat and increased wear
3. The clogging of filters, small apertures, and delicate equipment

Equipment with moving parts requires particular care when designing for sand and dust protection. Sand and dust will abrade optical surfaces, either by impact when being carried by air, or by physical abrasion when the surfaces are improperly wiped during cleaning. Dust accumulations have an affinity for moiture and when combined with moisture may lead to corrosion, electrical shorts, or the growth of fungus.

Dust may be composed of any fine, dry particulate material including metals, combustion products, solid chemical contaminants, and so on. Some forms of dust may provide direct corrosion or fungicidal effects on equipment, since dust may be alkaline, acidic, or microbiological.

Most electronic equipment requires air circulation for cooling, and removing dust is not a simple matter. The question is not whether to allow dust to enter, but rather how much or what size dust can be tolerated. The problem becomes one of filtering the air to remove dust particles larger than a specific nominal size.

The nature of filters is such that for a given working filter area, as the ability of the filter to stop smaller and smaller dust particles is increased, the flow of air or other fluid through the filter is decreased. Therefore, the filter surface area either must be increased, the flow of fluid through the filter decreased, or the allowable particle size increased, that is, invariably there must be a compromise.

Sand and dust protection must be planned in conjunction with other environmental protective measures. For example, it is not practical to specify a protective coating against the moisture if sand and dust will be present unless the coating is carefully chosen to resist abrasion and erosion, or unless it is self-healing.

Explosion Proofing

Protection against possible explosion is both a safety and reliability problem. This type of environment requires extreme care in design and reliability analyses.

Explosion protection planning must address two different equipment categories (not necessarily mutually exclusive):

1. Items containing or immersed in materials susceptible to explosion
2. Components located near enough to cause explosive items to explode

The first category includes flammable gases or liquids, suspensions of dust in the air, hypergolic materials, compounds which spontaneously decompose in certain environments, equipment containing or subjected to high or low extremes of pressure (that is, implosions), or any other systems capable of creating an explosive reaction. The second category is fairly obvious and includes many variations on methods for providing an energy pulse, a catalyst, or a specific condition that might trigger an explosion. A nonexplosive component, for example, could create a corrosive atmosphere, mechanical puncture, or frictional wear on the side of a vessel containing high-pressure air and thereby cause the air container to explode.

The possibility of an explosive atmosphere leaking or circulating into other equipment compartments must be recognized. Lead-acid batteries, for example, create hydrogen gas that, if confined or leaked into a small enclosure, could be ignited by electrical arcing from motor brushes, by sparks from metallic impacts, or by exhaust gases. Explosive environments, such as dust-laden air, might be circulated by air distribution systems.

Explosion protection and safety are very important for design and reliability evaluations, and should be closely coordinated and controlled. Just as safe equipment is not necessarily reliable, neither is reliable equipment necessarily safe; however, with proper design analysis the two requirements can be made compatible.

Additional Considerations

Corrosive airborne chemicals such as sulfur and chlorine compounds can degrade the reliability of equipment by attacking the connector pins, exposed contact surfaces in switches or relays, and unprotected metallization runs on printed circuit boards. Gold flash or other precious metals less subject to chemical attack are often employed to prevent this type of degradation. Printed circuit boards may also be conformally coated to reduce attack by airborne chemicals. Hermetically sealed switches and relays should be specified for use in hostile atmospheric environments.

As the atmospheric pressure is reduced the dielectric strength of the air is also reduced. Thus, when high voltages are utilized in equipment and the equipment is exposed to reduced atmospheric pressure, it may be necessary to take additional precautionary measures such as increase the spacing between adjacent pins or other high-voltage points to prevent electrical breakdown. Hermetically sealing the box and pressurizing it with an inert gas would be another alternative measure to prevent arc-over.

10.4 RADIATION ENVIRONMENTS

Radiation environments can be seen from two different viewpoints: first from that of the source or the cause of the radiation, or second, the specific type of radiation encountered, that is, its location in the electromagnetic spectrum. Both viewpoints are valid and both will be addressed in this chapter.

The entire electromagnetic spectrum may be conveniently divided into several categories ranging from gamma rays at the short-wavelength end down through x-rays, ultraviolet, visible, infrared, and radio, to the long-wavelength radiation from power lines.

Solar Radiation

Solar radiation may sometimes be a reliability concern. Near the surface of the earth, solar radiation damage is usually caused by radiation in the 0.15 to 5 m wavelength range. This range includes the longer ultraviolet rays, visible light, and up to about the midpoint in the infrared band. Visible light accounts for roughly one third of the solar energy failing on the earth; the rest is in the invisible ultraviolet and infrared ranges. The solar constant (that is, the quantity of radiant solar heat received normal to the surface at the outer layer of the atmosphere of the earth) is roughly 1 kW/m^2. In some parts of the world, almost this much can fall on a horizontal surface on the ground at noon.

Solar radiation can cause physical or chemical deterioration of materials. An example of the mechanical effects is the deterioration of natural and synthetic rubber. Solar radiation can also cause functional defects, such as temporary electrical breakdown of semiconductor devices exposed to ionizing radiation.

Some considerations to include in a radiation protection analysis are: the type of material being irradiated, its absorption characteristics, its sensitivity to specific wavelengths and energy levels, ambient temperature, and the proximity of reactive substances such as moisture, ozone, and oxygen. Specific protection techniques include: shielding, utilization of exterior surface finishes that absorb less heat and are less affected by radiation effects, and minimizing exposure time to radiation. The removal of possible reactive materials by circulation of air to other fluids and careful location of system components may also be considered in some instances.

Electromagnetic Radiation

Lightning is an obvious form of natural electromagnetic radiation. It is estimated that lightning strikes the earth about 100 times each second and each stroke releases large bursts of electromagnetic energy which encircle the globe. Even cloud-to-cloud discharges can induce serious transients on data communication lines miles away. Most of this energy is concentrated at the low frequency end of the electromagnetic spectrum with the maximum power level being concentrated at about 3 kHz.

Although natural electromagnetic energy dominates all other sources in the vicinity of a lightning strike, man-made electromagnetic energy is far more prevalent. This results not only from a low probability of being in the immediate vicinity of a lightning discharge but also from the complex sophisticated use to which man is putting electromagnetic radiation. Man-made electromagnetic radiators include power distribution systems, industrial, scientific, and medical (ISM) applications, and all forms of communications. The development of lasers has introduced yet another intense source of electromagnetic radiation, and in military applications the electromagnetic pulse (EMP) associated with nuclear weapon detonations is of considerable importance.

The EMP spectrum is similar to that created by lightning. Its maximum energy appears at about 10 kHz but with smaller amplitudes spread throughout a broad region of the frequency spectrum. EMP energy is of considerably greater magnitude than that observed in lightning and extends over a much larger area of the earth. Despite the similarities between EMP and lightning and other strong sources of electromagnetic energy, it cannot be assumed that protective measures consistent with these other electromagnetic radiation sources will protect items from the effects of EMP. The rapid rise time of the pulse associated with a nuclear detonation and the strength of the resulting pulse are unique.

Electromagnetic Interference

A variety of effects of electromagnetic radiation on materials is known and probably some effects are still unrecognized. One of the most serious environmental effects of electromagnetic radiation is the electromagnetic interference (EMI) it produces on the effective use of the electromagnetic spectrum. Equipment must be designed so that it does not generate or emit electromagnetic disturbances which would interfere with other equipment, and it must also be resistant to adverse effects from other sources of EMI.

EMI protection must generally be specifically designed for the noise and interference fields against which protection is required. This usually involves the specification of shielding and filtering effective in the frequency range of concern.

Protection against the effects of electromagnetic radiation has become a sophisticated engineering field of electromagnetic compatibility (EMC) design. The direct approach in most cases, is simply to avoid the limited region in which high radiation levels are found. When exposure cannot be avoided, shielding and filtering become important protective measures. In other cases material design changes or operating procedural changes must be instituted to provide protection. Some of the more important specifications and standards dealing with EMI are shown in Table 10.6. As one example of the EMI requirements in the documents listed in this table, consider the requirements defined in the FCC rules, part 15, subpart J for two different types of digital equipments.

Class A devices are defined as those used in commercial, industrial, or business applications. Two examples are computers (for example, any electronic device using digital techniques and generating timing signals at a greater rate than 10K pulses per second) and computer peripherals. Class A devices must be tested for compliance to the FCC criteria. Documented test data must be kept on file subject to FCC audit.

Class B devices are defined as those used in residential applications. Examples include: personal computers, electronic games, and calculators. These devices must be certified (not simply be in compliance) by submitting actual test data or hardware for review.

Table 10.6 EMI Specifications and Standards

FCC part 15, subpart J:	"Digital Equipment"
FCC part 18:	"Industrial, Scientific and Medical Electronics (ISM)"
MDS-201-0004:	"FDA Electromagnetic Capability Standard For Medical Devices"
VDE[a] 0871/6.78:	"Specification For Equipment That Generates or Processes RF"
VDE[a] 0875/6.77:	"Regulation For Equipment (Including Industrial Areas) Exposed to Unintentional RF"
MIL-HDBK-253:	"Guidance For The Design of Systems Protected Against The Effects of Electromagnetic Energy"
MIL-STD-461:	"Requirements For Equipment Electromagnetic Interference Characteristics"
MIL-STD-462:	"Measurement of Electromagnetic Interference Characteristics"
MIL-E-6051:	"System Electromagnetic Compatibility Requirements"

[a]Verbande Deutscher Electrotechniker, West Germany.

A series of technical requirements must be met by these equipments as shown in Table 10.7. They are concerned with:

1. Radiated emissions—unintentional signal transmissions
2. Conducted emissions—amount of signal fed back into the AC power lines
3. Labeling and instructions manual requirements

Similar, but somewhat more restrictive requirements are contained in the German VDE documents.

Outside of the United States and West Germany, few countries have undertaken a comprehensive standards effort to control or minimize RFI from computer equipment. The IEC has sought to harmonize the EMI standards in various different countries through its proposed "International Special Committee for Radio Interference" (CISPR) regulations. These regulations, however are not binding on member countries. Thus at present much confusion, changing

Table 10.7 FCC EMI Requirements

Equipment class	Radiation type	Frequency (MHz)	Distance (m)	Field strength (MV/m)	Maximum voltage (MV)
A	Radiated	30–88	30 (3)	30 (300)	
		88–216	30 (3)	50 (500)	
		216–1000	30 (3)	70 (700)	
A	Conducted	0.45–1.6			1000
		1.6–30.0			3000
B	Radiated	30–88	30 (3)	10 (100)	
		88–216	30 (3)	15 (150)	
		216–1000	30 (3)	20 (200)	
B	Conducted	0.45–30			250

standards, and conflicting rules exist internationally in the field of EMI/RFI.

Nuclear Radiation

Natural nuclear radiation can sometimes be a problem for electronic equipments. High-density microcircuit memories are susceptible to soft errors resulting from the natural decaying alpha-particle emissions in the device package materials themselves. This is true for both the sealing glasses used on ceramic packages and the uranium and thorium content of the silicon fillers and the antimony oxide flame retardants used in epoxy encapsulants. Polyimide die coatings have been developed to provide alpha-particle shielding of the chips themselves. However, these relatively thick coatings can aggravate stress-related problems within the devices. Also, some of the alpha-particle-shielded materials are hydrophilic, and when used in epoxy-encapsulated devices can adversely affect the moisture resistance properties of the package.

Although natural background levels of nuclear radiation exist, the primary nuclear radiation usually of interest to design engineers is that associated with earth satellite and space probe applications with man-made radiation sources such as reactors, isotope power sources, and nuclear weapons. Intense nuclear radiation can produce both transient and permanent damaging effects in a variety of materials.

Charged nuclear particles emanate from nuclear reactions; however, x-rays, gamma rays, and neutrons are the types of nuclear radiation of the greatest concern. Because of their long ranges in the atmosphere these forms of radiation can irradiate and damage a variety of materials. For example, natural electron fluence can cause ionization of surface materials. It causes degradation, deterioration and charge build-up on satellite surface components. Natural proton fluence can cause permanent degradation in solar cells and other directly exposed semiconductor devices.

Among the nuclear effects of concern are those called *transient radiation effects on electronics*, referred to as TREE. These transient effects are due primarily to the ionization effects of gamma rays and x-rays. The separation of transient and permanent effects is made on the basis of the primary importance of the radiation effects. For example, a large current pulse may be produced by ionizing radiation. This current pulse may then cause permanent damage to a device by overheating. This is considered to be a transient effect because the permanent damage results from overheating due to excess current rather than to direct-radiation-induced material property change. Extensive information is available on specific electronic components, circuits, and hardening methods.

It is not possible to protect material items from nuclear radiation completely as can be done for some other environmental conditions. The variety of effects produced by nuclear radiation for different materials and components makes protective design difficult. Rather, protective procedures usually define a radiation hardness level in a given material item and then design and test the item to that level. This is the approach used in MIL-M-38510, in MIL-STD-279 "Total Dose Hardness Assurance Guidelines for Semiconductors and Microcircuits" and in MIL-STD-280 "Neutron Hardness Guidelines for Semiconductors and Microcircuits."

Nuclear radiation protection generally consists of the use of specific components having an intrinsic hardness and the incorporation of shielding features that impart the required level of hardness to the system. The provision of nuclear protection schemes is usually a go or no-go proposition since few tradeoff situations are apparent.

Some common radiation hardening methods are:

1. EMP current pulse: Shunt "out-of-band" energy to ground. Use a reset approach which allows the upset to occur, and then ignores what occurred during the upset interval.
2. Photocurrents: Use minimum base width and small geometry semiconductor devices.
3. Neutron effects: Use semiconductors with high doping levels, low-voltage devices, high current-density devices and high cutoff frequency devices.

Nuclear radiation hardening is a large and complex field where a variety of specialists is required to deal with different aspects of the problem.

11
Production and Use Reliability

11.0 INTRODUCTION

System and equipment operational reliability experienced in actual use may be significantly less than the MIL-HDBK-217 inherent predicted reliability. The predicted reliability will not represent operational reliability unless design deficiencies have been eliminated, manufacturing and quality defects have been minimized, and the operating and maintenance procedures have been optimized. Specific attention must therefore be directed towards minimizing reliability degradation throughout the system design and development stage, the production stage, and the maintenance and use stages.

Reliability prediction and analysis methods are based primarily on system design characteristics and data regarding the attributes of the constituent parts. These estimates reflect the reliability potential of equipment during its useful life period, that is, after quality defects are removed and prior to wear-out. They represent inherent reliability, that is, the reliability potential of the equipment as defined by its design configuration, stress and derating factors, application environment, and gross manufacturing and quality factors. Design-based reliability estimates do not represent the expected early life performance of the equipment particularly as initially manufactured.

An effective reliability engineering program recognizes that the achievement of a high-level of actual-use reliability is a function of design as well as of all life-cycle activities. Design establishes the inherent reliability potential of a system or equipment item. However, the transition from the paper design to actual hardware and ultimately to operation many times results in actual reliability that is far below the inherent level. The degree of degradation from the inherent level is directly related to the inspectability and maintainability features designed and built into the system. It is also related to the effectiveness of the measures taken during production, shipment, and storage prior to use to eliminate potential failures, manufacturing flaws, and deterioration factors. Lack of attention to these areas can result in actual equipment reliability as low as 10% of the inherent reliability.

The impact of production, shipment, storage, operation, and maintenance degradation factors on the reliability of typical equipments is conceptually illustrated in Figure 11.1. The figure depicts the development of a hardware item as it progresses through its life-cycle stages. The figure shows that an upper limit of reliability is established by the design. When the item is released to manufacturing, its reliability may be degraded by the production processes. However, with subsequent process improvements and manufacturing learning factors reliability will grow. When the item is released to the field, its reliability will again be graded. As operational personnel become more familiar with the equipment and acquire maintenance experience and as field operations continue reliability will again improve.

Design reliability efforts include: selecting, specifying, and applying proven high-quality, well-derated, long-life parts; incorporating adequate design margins; using cost-effective redundancy; and applying tests designed to identify potential problems. Emphasis must be placed on incorporating ease-of-inspection and maintenance

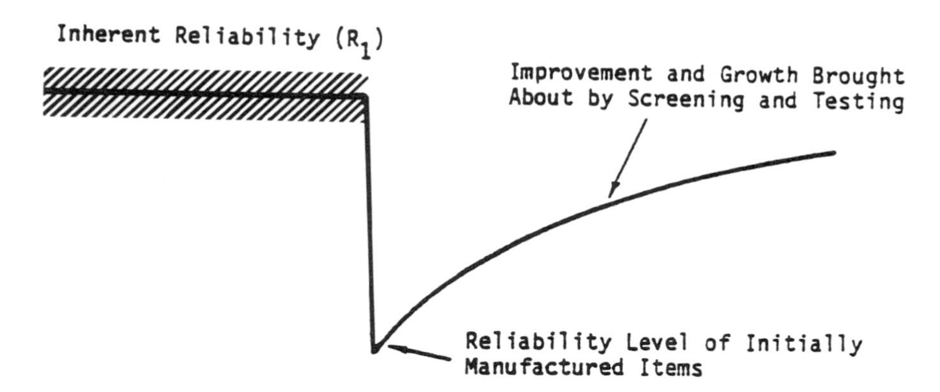

Figure 11.1 Impact of design and production activities on equipment reliability. [From *Reliability Design Handbook* (RDH-376), copyright 1976 by IIT Research Institute, Reliability Analysis Center, RADC/RAC, Griffiss Air Force Base, N.Y.]

features. This includes using easily replaceable modules (or components) and incorporating built-in testing on-line monitoring, and fault isolation capabilities. During the development phase reliability efforts address the application of systematic and highly disciplined engineering analyses and tests in order to stimulate reliability growth and to demonstrate the level of reliability that has been achieved by the establishment of an effective, formal program for accurately reporting, analyzing, and correcting failures prior to field operation.

Once the inherent or design-in reliability has been determined, engineering efforts focus on the prevention of reliability degradation. Well-planned and carefully executed inspections, tests, and reliability/quality control methods must be applied during production (as well as during storage and operation) to eliminate defects and minimize degradation. Manufacturing, transportation and storage environmental stresses as well as inspection methods and operation/maintenance procedures must be continually assessed to determine the need for better inspection, screening, and control provisions to improve reliability.

The design must also assure that any maintenance actions taken on the equipment will not degrade its reliability. It must consider the risks associated with any future maintenance and it must be friendly to the maintenance personnel. Positive identification of a failure must be assured allowing rapid localization to the general area of the fault and then restoring the function by simple replacement of a single assembly or at worst a limited number of assemblies. Most importantly, this maintenance action should not cause further degradation of the equipment or risk the introduction of any new problems such as the need for a total recalibration of the entire item.

MIL-STD-454 "Standard General Requirements For Electronic Equipment" can be a significant help to designers in this area as it brings to mind many good general design practices that may otherwise be overlooked.

The intent of this chapter then is: (1) to examine some of the basic fabrication and manufacturing processes; (2) to determine how these can be planned and implemented in order to minimize their degradation contribution to equipment reliability; and (3) to present some guidelines for implementing these tasks. Shipment and storage factors and the maintenance of the equipment while it is in use will also be examined to minimize these degradation contributions to equipment reliability.

11.1 PRODUCTION DEGRADATION FACTORS

Table 11.1 shows some of the more common manufacturing processes involved in the fabrication of electronic equipment. It identifies

Table 11.1 Production Processes and Associated Defects

Process	Induced defect	Failure mode: Intermittent	Failure mode: Short circuit	Failure mode: Open circuit	Failure mode: Value change	Failure mode: Noise	MIL-STD-454 Reference
Wire stripping	Nicked leads			x			Req. 5
	Broken strands		x				
	Short leads	x		x		x	
	Long leads		x				
Soldering	Excessive heat			x	x	x	Req. 5
	Insufficient heat	x		x		x	
	Excessive solder		x		x		
	Insufficient solder	x		x		x	
Lead cutting	Dull tools (shock)	x		x	x	x	None
Crimping	Wrong tool	x		x	x	x	Req. 19
	Wrong terminal		x	x			
	Low force			x			
	Excessive force	x		x			
Wire wrapping	Broken wire			x			Req. 69
	Loose connection	x		x		x	
Lead bending	Stress on case	x	x		x	x	Req. 5
Wire dress	Vibration Sensitivity	x	x	x		x	Req. 69
	Residual stress	x	x	x		x	

Source: *Reliability Design Handbook* (RDH-376), copyright 1976 by IIT Research Institute, Reliability Analysis Center, RADC/RAC, Griffiss Air Force Base, N.Y.

some of the potential defects associated with these processes and their resultant failure modes, together with their applicable reference in MIL-STD-454.

The inherent reliability prediction addressed in Chapter 9 assumed first of all that the item was built to the drawing or print and second that the drawing or print was correct to begin with. Unfortunately, this is not always true. Two primary tools are available to safeguard and assure that these conditions are met. Quality inspections are the first tool. They are performed to compare the product to the drawings so as to assure compliance during the manufacturing cycle. This topic is adequately addressed in the quality control literature and thus will not be addressed at length here. Design review is the second tool.

Design Review

The purpose of design review is to assure that the design and the drawings are correct to begin with. Design review may be either formal or informal. It is a synergistic, multidisciplined, detailed peer review of the paper design of the product. It is normally performed prior to the release of the design to manufacturing. In some cases the design review effort will be directed by the customer, for example, by a military agency. In other cases the design review may be directed by senior engineering staff within the company. The major goal of design review is to assure that no unique area of concern has been overlooked. Specific items of concern include: safety, potential product liability, human factors aspects of the design, and servicing and maintenance of the equipment after delivery.

Design review is not a single isolated event. It should occur at progressive stages throughout the development cycle. With military programs it is common to have a formal conceptual design review, later a preliminary design review, and finally a critical design review just prior to the final drawing release to manufacturing.

Small informal design reviews are also held periodically to assess specific aspects or elements of the design. At a design review each of the specified requirements are compared with the present design to identify potential problem areas for increased attention or for possible reevaluation of the need for that requirement. For example, one of the concerns identified at a design review may be the need to reapportion reliability to allow a more equitable distribution of the available failure rate among certain functional elements.

Formal documentation of the critical design review is very important. It may well form the legal basis of defense in the unfortunate event of a future liability suit against the product and its producer. This documentation might show, for example, that due consideration had been given as to how the product might be used or misused by the customer and the adequacy of the design under these circumstances.

The multidisciplinary aspects of the design review must be stressed. At the formal design review all affected disciplines should be represented. This would include reliability, maintainability, manufacturing, human factors, and so on, as well as the design staff. This will help ensure that all viewpoints receive adequate consideration. In small companies without this breadth of knowledge, outside consultants may be utilized to fill any gaps in areas of expertise. A thorough multidisciplinary peer review of the design prior to release to manufacturing is absolutely essential for the achievement of a highly reliable product. Design changes after the release of the design to manufacturing are extremely, if not prohibitively, expensive, particularly where retrofit of previously manufactured equipment is required.

Personnel Capabilities

Most people are generally average in their skills. This fact is often forgotten by system designers, planners, and managers. Each expects to have well-above-average people performing the tasks that they are prepared for. System designers do pay some attention to this problem when considering operators and repair personnel. The problem is frequently overlooked, however, by the equipment design engineers. We must realize the limitations of the people involved throughout the equipment's life cycle.

System and subsystem design should be straightforward, and interfaces between subsystems should be as simple as possible. The greater the complexity, the more likely errors are to occur. Checklists are a valuable aid to designers. Design reviews and other product reviews help to overcome human limitations in the system design.

The designer of an equipment needs to consider how it will be produced, for example, what kind of quality control procedures will be necessary, or what machines or operators will actually perform a given task. Reducing the number of different, but similarly appearing parts can help avoid mistakes. A design that can accept looser tolerances might be preferred over one which requires tight tolerances, even though the latter would theoretically perform better if everything were exactly right.

Production Reliability Degradation Assessment and Control

The extent of reliability degradation during production is dependent on the effectiveness of the inspection and quality engineering control program. Reliability analysis methods are applied to measure and evaluate its effectiveness and to determine the need for process improvement or corrective changes. The accomplishment of the analysis task and, more important, how well subsequent corrective measures are designed and implemented will dictate the rate at which reliability degrades or grows during production. Reliability degradation is minimized during manufacturing, and reliability grows as a result of improvements or corrective changes that:

1. Reduce process-induced defects through accelerated manufacturing learning, and the incorporation of improved processes
2. Increase inspection efficiency through accelerated inspection procedures, and the incorporation of controlled screen and burn-in tests

As process development, test, and inspection efforts progress, problem areas become resolved. As corrective actions are instituted, the outgoing reliability approaches the inherent (design-based) value.

Quality defects, however, can be overshadowed by an unknown number of latent defects. These latent defects which pass factory quality inspection are due to flaws, either inherent in the parts or induced during fabrication, that weaken the fabricated hardware such that it may fail later under conditions of stress during field operation. Reliability screen tests apply stress during manufacturing at a given magnitude over a specified time duration in order to identify these latent defects. As in the case of conventional quality inspections, screen tests designed to remove latent defects are not 100% effective.

Reliability Degradation During Production: Infant Mortality

As was discussed in Chapter 1.1 and illustrated in Table 1.1 the distribution of failures over the life-span of a large population of hardware items can be separated into quality, reliability, wear-out, and design failures. These failure distributions combine to form the infant mortality, useful life, and wear-out periods as were shown in the bathtub curve in Figure 1.1. Design and reliability defects would normally exhibit an initially high but decreasing failure rate. In an immature design these defects would initially dominate over all other defects.

A design-based approach which is limited to only the equipment's useful life is not adequate to assure product reliability. Experience has shown that the infant mortality period can vary from a few hours to well over 1000 hr. For most well-designed, complex equipment, 100 hr is a typical infant mortality period. Positive measures must therefore be taken, beginning with design, to achieve a stabilized low-level failure rate. This includes evaluating the impact of intrinsic part defects and manufacturing process-induced defects as well as the efficiency of conventional inspections and the strength of reliability screening tests.

Intrinsic defects arise from basic limitations of the parts that comprise the system or equipment. They are a function of the supplier's process maturity, and his inspection and test methods.

Process-induced defects are those which enter or are built into the hardware as a result of faulty workmanship or design, process stresses, handling damage, or test efforts. They lead to degradation of the inherent design-based reliability. Examples of failures which may occur due to manufacturing deficiencies are: poor connections; improper positioning of parts; contamination of surfaces or materials; poor soldering of parts; improper securing of component elements; and bending or deformation of materials.

These defects, whether intrinsic to the parts or introduced during fabrication, can be further isolated into quality and reliability defects. Quality defects are not time dependent and are readily removed by conventional quality control measures (that is, inspections and tests). The more efficient the inspection and test, the more defects removed. However, since no test or inspection is perfect, some defects will escape to later manufacturing states. These must be removed at a later inspection at a much higher cost or they will occur after delivery of the equipment to the customer thus resulting in reduced operational reliability and higher maintenance costs.

Stress- or time-dependent reliability defects cannot be detected by conventional quality control inspections. These defects can only be detected by the careful and controlled application of stress. Stress screen tests consist of a family of techniques in which electrical, thermal, and mechanical stresses are applied to accelerate the occurrence of potential failures. Thus, latent failure-producing defects, not usually detected during normal quality inspection and testing, are removed from the product. Included among these tests are burn-in, temperature cycling, vibration, on/off power cycling, and various nondestructive tests. A more detailed discussion of screening and burn-in is presented in the Chapter 14.

Production Degradation Summary

The reliability of initially manufactured units will be degraded; however, subsequent improvement and growth can be achieved through quality inspections, reliability screening, failure analysis, and corrective action. The extent and rigor with which the tests, failure analysis, and corrective actions are performed will determine the degree of improvement. Therefore, process defects, along with the inherent part estimates, must be evaluated in order to estimate reliability accurately, particularly during the initial manufacturing stage.

11.2 MAINTENANCE DEGRADATION FACTORS

Defects can be induced into an equipment during field operation and maintenance. Operators may stress a system beyond its design limit to meet a current operational need or constraint or may do this inadvertently through negligence or unfamiliarity with the equipment. Operational abuses due to rough handling, extended operation, or neglected maintenance can contribute to reliability degradation during field operation. The degradation is typically the result of the interaction between man, machine, and environment. Translation of factors which influence operational reliability degradation into corrective procedures requires a complete analysis of the functions performed by man and by machine plus environmental and/or other stress conditions which degrade operator and/or system performance.

The Maintenance Problem

Reliability degradation can also occur as a result of poor maintenance practices. Excessive handling brought about by frequent preventive maintenance or poorly executed corrective maintenance (for example, installation errors) may result in defects being introduced into the system. Examples of defects resulting from field maintenance, include:

Foreign objects left in an assembly
Bolts not tightened sufficiently or overtightened
Dirt injected
Parts improperly replaced
Improper lubrication

Also, during unscheduled maintenance good parts are frequently replaced in an effort to locate the faulty parts. In many cases, the good parts are reported as defective instead of being reinstalled. These parts are often returned for repair or discarded, resulting in a reported field failure rate higher than that which is actually occurring.

Extensive replacement of analog circuitry with digital circuitry, inclusion of more built-in-test features, and fault-tolerant circuitry have enhanced the field reliability of electronic equipments. However, the effects of poorly trained, poorly supported, or poorly motivated maintenance personnel on reliability degradation still require careful assessment.

Operation- and maintenance-induced defects are factors that must be carefully considered and taken into account in the assessment and control of operational reliability. Environmental factors considered in prediction techniques account for the added stress provided by operation within that environment. However, the environmental stresses imposed during field maintenance may be different than those anticipated during the original prediction. Mechanical stresses imposed on components during removal, repair, and reinstallation may exceed those designed for a given environment. Therefore, field requirements and procedures must include criteria for controlling the reliability and quality of repair action to minimize potential maintenance-induced defects in order to achieve an actual field reliability that approaches the one predicted.

The designer must also consider how the equipment will actually be repaired in the field. Field repair is often performed under pressure with understaffed maintenance personnel, many of whom are inexperienced. Also, one cannot expect that field service personnel will have the same degree of knowledge about the system that the designers have. Even where the situation is understood, the maintenance personnel under time pressures might well choose short-cut repair methods to save time. The designer should keep in mind that the equipment will be used and repaired by ordinary people who have other things in mind than "babying" the equipment. He must realize the difference between what people will actually do, and what he thinks they ought to do.

Human Engineering

This area deals largely with motor responses of operators and with varied human physical capabilities. Some typical human engineering constraints are:

1. An operation should be within the physical capabilities of 95% of the potential operators.
2. A person should not be required to do something that his coordination will not allow him to do, for example, patting his head with the left hand while rubbing his chest with the right hand.
3. Under psychological stress, people cannot easily use, read, and respond to controls and displays.

Mock-ups simulating realistic conditions are very helpful in identifying overlooked constraints. If an equipment must be used at night in extremely cold weather, someone should try using it in a freezing, poorly lit room for several hours.

Military standards and specifications contain a number of guidelines and requirements for human factors and human engineering, and thus may be helpful in this area. Design criteria, requirements, and definitions for human engineering in military systems may be found in MIL-STD-1472, "Human Engineering Design Criteria for Military Systems, Equipment, and Facilities" the MIL-STD-803 series "Human Engineering Design Criteria for Aerospace Systems and Equipment" and MIL-H-46855 "Human Engineering Requirements for Military Systems, Equipment, and Facilities." Standardization, automation, visual and auditory displays, controls, labeling, workspace design, maintainability, remote handling devices, safety hazards, and environmental requirements are some of the subjects treated in these sources.

Human Performance Reliability

The analysis of human factors recognizes that both human and machine elements can fail, and that just as equipment failures vary in their effect on a system, human errors can also have varying effects on a system. In some cases, human errors result from an individual's action, while others are a consequence of system design or manner of use. Some human errors cause total system failure or increase the risk of such failure, others merely create delays in reaching system objectives. Thus human factors exert a strong influence on the design and the ultimate reliability of all systems having a man/machine interface. A good summary of human performance reliability predictive methods is given in the IEEE Transactions on Reliability R-22 (see Chapter 15.4) titled "A Critical Review of Human Performance Reliability Prediction Methods" (August, 1973). This document contains an excellent bibliography of available predictive methods.

In the initial evaluation of a design, the man/machine interface can be put into clearer perspective by addressing the following questions:

1. In the actual environment, which of the many human performance characteristics are truly important, which must be included in the design, and under what circumstances is each characteristic important?
2. What effect will including or excluding particular characteristics have on the system design?

The reliability of a system is affected by the allocation (not necessarily quantitative) of system functions to either the man, the machine, or both. Table 11.2 lists some of the salient characteristics of humans and of machines which are pertinent to the allocation

Table 11.2 Characteristics of Humans and Machines

Characteristics tending to favor humans	Characteristics tending to favor machines
1. Ability to detect certain forms of energy	1. Monitoring men or other machines
2. Sensitivity to a wide variety of stimuli within a restricted range	2. Performance of routine, repetitive, precise tasks
3. Ability to perceive patterns and generalize about them	3. Responding quickly to control signals
4. Ability to detect signals (including patterns) in high noise environments	4. Exerting large amounts of force smoothly and precisely
5. Ability to store large amounts of information for long periods, to remember relevant facts at the appropriate time	5. Storing and recalling large amounts of precise data for short periods of time
6. Ability to use judgment	6. Computing ability
7. Ability to improvise and adopt flexible procedures	7. Range of sensitivity to stimuli
8. Ability to handle low probability alternatives (i.e., unexpected events)	8. Handling of highly complex operations (i.e., doing many different things at once)
9. Ability to arrive at new and completely different solutions to problems	9. Deductive reasoning ability
10. Ability to profit from experience	10. Insensitivity to extraneous factors
11. Ability to track in a wide variety of situations	
12. Ability to perform fine manipulations	
13. Ability to perform when overloaded	
14. Ability to reason inductively	

Source: *Human Resources Research Program*, AR 70-8.

choice. It is evident from studying Table 11.2 that the prediction of human reliability is more difficult than the prediction of machine reliability. The machine's insensitivity to extraneous factors (item 10 in Table 11.2) versus a human's sensitivity to these factors is one consideration leading to human performance variability and the subsequent capability to predict machine reliability much more precisely. In fact, a human's response can be sufficiently influenced to vary from 0.0001 to 0.9999 reliability within conditions that would not affect a machine. The machine, for example, does not react to severe psychological stress environments which may cause breakdown in human. The tradeoff depends partly on the nature of the system and human functions and partly on the way the allocation is approached; each design situation requires a separate human factors analysis. Such variables as cost, weight, size, hazard levels, adaptability, and state of technology must be considered for each system.

One approach to the choice between man and machine is to compare the predicted reliabilities of each. This approach, however, should not be based solely on failure rates, since humans are sufficiently adaptable to recover quickly and correct some human-induced malfunctions. Similarly, humans have the flexibility to handle unique situations that might cause system failure if an unadaptable machine were assigned the task. An approach based on reliability comparisons should use failure rates in conjunction with an analysis of man/machine characteristics and the desired task accomplishments.

Man/Machine Interactions and Tradeoffs

The principal determinant of man/machine performance is the complexity of human tasks within the system. A system design that requires frequent and precise adjustments by an operator may create reliability problems associated with wear-out or maladjustment of the control device, or maintainability problems from repeated replacement of the worn control. On the other hand, a design providing an automatic adjusting mechanism may cause problems of cost, weight, size, reliability, maintainability, or safety due to the control's complexity. Similarly, for the same level of effectiveness, a system that through design, location, or environment is difficult to repair must necessarily be made more reliable than a system with a less complex man/machine interface. Thus, the man/machine interaction can contribute to or detract from the effectiveness of other disciplines depending upon tradeoffs and interactions selected during the system engineering process.*

**Human Engineering Guide for Equipment Designers* by W. E. Woodson and D. W. Conover (University of California Press, Berkeley, CA, 1966) gives additional design guides and approaches for solving human factors problems and tradeoffs with other disciplines.

11.3 BIT, BITE, AND TESTABILITY

Built-in test equipment (BITE) and the built-in test (BIT) are two of the most significant aids for the maintenance man. Of the two, BITE is usually less sophisticated in that it is normally manually operated. It contains switches, lights, panel meters, and so on. In contrast, the BIT is automated and utilizes software or firmware. It offers on-line or off-line diagnostic capability and may incorporate fault printout capability to aid the maintenance personnel. In the design of BIT six basic questions should be addressed. These questions are summarized in Table 11.3 and are then each studied in greater depth.

Purpose of the BIT

The basis purpose of the BIT must be clearly defined. First it must determine that a failure or an imminent prefailure condition exists and then it must alert the operator to the fact that a failure has occurred or will occur in the near future unless steps are taken immediately to avoid the failure. Ideally the BIT should also perform rudimentary fault isolation and indicate to the operator the subsystem or that general portion of the system in which the problem lies. For example, in a large computer complex it would determine that the phone line or the modem is the problem, not the terminal or the central processor.

In a highly automated system the BIT might also perform initial fault correction such as switching to a redundant power supply, reconfiguring the system, or calling up a standby module to bypass a defective unit.

Other BIT methods for correcting a failure are error-correcting codes and majority-voting circuits. Specific data correction codes

Table 11.3 Factors to Be Considered in Design of the BIT

1. What will be the basic purpose of the BIT?
2. At what functional level will the BIT operate?
3. Will the BIT function on-line, off-line, or will it interleave the BIT test data with the existing system data?
4. Will the design of the BIT be based upon inductive logic or deductive logic?
5. Will the BIT utilize a single centralized computer or will the computation be handled by distributed processors?
6. Will the BIT be under hardware or software control?

such as Hamming codes are available that are capable of detecting and correcting a single-bit error or of detecting a two-bit error

Active Stimulus Versus Passive Monitoring

Active BIT functions by injecting a test signal at a given location and then monitoring this signal farther downstream. The interrogation of the modules by a microprocessor that then evaluates the response is an example of active BIT. Passive BIT monitors system performance without the use of an injected test signal, but utilizes the characteristics of the system already present to assess the status of the equipment. Signature analysis and parity checks are examples of passive BIT.

The determination of a meaningful test set (that is, of a set of input test vectors) for each module in a system is a significant task. However, once that test set has been determined, it is generally not very difficult to develop hardware to input the test-set conditions.

One possible disadvantage of active BIT as compared with passive BIT is the fact that the injected test signal may be disruptive to the data already being processed by the equipment.

BIT Functional Level

The actual level at which BIT is employed is a function of the detailed maintenance plan for the system. However, a typical BIT scenario for a complex system may be as follows. It may be desirable to monitor the system continuously to assure that it is performing satisfactorily on a continuous basis. Each separate subsystem may then be monitored periodically in a sequential manner. The individual printed circuit cards would not normally be monitored but could be isolated and monitored in the event of a system malfunction.

An example of using BIT to enhance preventive maintenance is the use of a dedicated microprocessor in a radar transmitter to record periodically the critical parameters of the Klystron or the magnatron tube. This would allow remote monitoring of the condition of the tube and thus predict the proper time to schedule replacement of the tube since these types of tubes exhibit well-defined degradation characteristics prior to failure.

On-Line Versus Off-Line BIT

Passive BIT is usually on-line BIT in that it operates continuously without interfering with the normal system function.

Off-line BIT need not operate continuously. It may be used intermittently and be controlled by either the computer or by the operator. It could be called into play only when there is evidence of a malfunction or if not all modules are operated continuously the BIT can also take advantage of times when a module is not needed to run a specific test sequence. This could also include the addressing of a diagnostic program to troubleshoot a specific subsystem or element.

Interleaved BIT utilizes existing "dead time" to process the BIT test data. Inverleaving can be a powerful means for using active BIT to maintain confidence in a system without disrupting the normal signal data being processed. An example of the use of interleaved BIT may be a radar system with a rotating antenna. If the entire area of rotation is not of interest, portions of the scan could be blanked out and at this point the BIT test signal could be inserted and interleaved.

One popular approach is to use continuous on-line BIT to monitor the general well-being of the hardware, and to use initiated BIT to assist in locating a specific malfunction. Initiated BIT is also very useful in testing sections of the unit which, if tested continuously or periodically, could disrupt the normal flow of operation.

BIT Logic

Deductive BIT logic assumes that if a certain function is within its stated tolerance limits, then all the variables involved in generating that function must also be within their stated tolerance limits.

In contrast, inductive BIT logic concludes that if a specified set of measured functions are found to be within their stated tolerance limits, then a single unmeasured (and perhaps unmeasurable) function must also be within its stated tolerance limits.

The BIT in each module can use either deductive or inductive logic. A parity check used as BIT deduces from a successful test that the module is working properly. An active BIT check of a memory successfully reading out a previously entered test pattern inductively assumes that the memory will function properly with the normal operating pattern.

Centralized Versus Distributed BIT

With distributed BIT each BIT circuit has as its only function the test of the one module in which it is contained. In centralized BIT the BIT circuitry handles BIT information from various different sources. It should be noted that the distribution of BIT is not necessarily the same as the distribution of fault-indicating signals. Distributed BIT can provide a centralized fault signal and centralized BIT can trigger distributed fault indications.

Centralized BIT in a computer-controlled system can have advantages in that the computer can better interleave tests with system operation and that less total hardware is often required to implement the BIT function. Distributed BIT has the advantage that a subsystem can be taken off-line and not require the use of the system computer for diagnosis of failure and verification of restoration.

A distributed rather than centralized BIT may be more practical with a system utilizing a number of microprocessors. However, additional steps or precautions may be required to: (1) isolate data between the individual processors; (2) be able to self-test each processor; and (3) to assure synchronization between the various microprocessors.

Software Versus Hardware Controlled BIT

The actual BIT function of each system can be implemented in either hardware or in software. Software, of course, implies hardware in the form of a computer or microprocessor. If a computer or microprocessor can be shared between operational and test functions, the hardware costs associated with BIT become minimal. There may be, nonetheless, some requirements for sampling and buffering signals and for interfacing signals which are not normally accessible to the computer or microprocessor.

Some of the advantages of using software to control the BIT are:

1. It is usually easier to modify or to change the software than it is to change the hardware.
2. Software-controlled BIT will usually allow the BIT testing to be more comprehensive than hardware-controlled BIT.
3. Complete end-to-end testing and go/no-go testing are usually simpler to implement with a software-controlled system than with a hardware-controlled system.
4. It is much easier to provide diagnostic capability with a software-controlled system than with a hardware-controlled system.

Some precaution, however, may be necessary:

1. The BIT test data must be properly isolated from the signal data being processed so that they do not interfere with each other.
2. Adequate tolerances must be provided for.
3. The input stimuli level and the BIT test signal levels must be compatible.
4. Incorporating BIT data in addition to the existing signal data may require additional computer and memory capability.

Finally, a well-designed BIT system can substantially reduce the need for highly trained field maintenance personnel by permitting less-skilled personnel to locate failures, replace suspect hardware items, thereby quickly returning the system to operational status, and then send the suspect items to centralized repair facilities equipped to diagnose and repair the defective hardware item.

Testability

Testability is defined as a design characteristic which allows the status (operable, inoperable, or degraded) of an item to be determined and the isolation of faults within the item to be performed in a timely manner. Some of the main considerations regarding testability are:

1. The system must be partitioned based upon the ability to isolate faults confidently.
2. Sufficient test points must be incorporated within the system for the measurement or stimulus of internal circuit nodes so as to achieve an inherently high level of fault detection and isolation.
3. The maintenance capability must ensure that sufficient off-line automatic test and manual test capabilities are provided for consistent and complete maintenance. The degree of test automation must be consistent with the available personnel skill level and the corrective and preventive maintenance requirements.
4. The number of maintenance replaceable units which may contain faults (the ambiguity group) that result in the same fault response must be adequately defined.

Testability Design Considerations

Partitioning of the system may be either physical or functional. The ease or difficulty of fault isolation depends to a large extent upon the size and complexity of replaceable items, that is, the physical partitioning of fault isolation. Whenever possible, each function should be implemented on a single replaceable item to make fault isolation straightforward. If more than one function is placed on a replaceable item, provisions should be included to allow the independent testing of each function. Where possible, the block of circuitry currently being tested should be isolated from circuitry not being tested through the use of blocking gates, tristate devices, relays, and so on. This "divide and conquer" approach is based upon the concept that test time increases exponentially with the complexity of the circuit.

There are significant differences in the approach to testability of analog equipment versus the approach to testability of digital equipment. Digital failure modes are more straightforward, they are simply on or off. Analog circuitry can exhibit a wide variety of failure modes ranging from a minor drift to a catastrophic failure. Detection of these failures, however, is a different story. Analog failure detection is usually, but not always, quite obvious. Digital failures in contrast are usually very subtle and, due to the many combinations of functions that the circuit can perform, failures cannot be detected without complete functional operation of the circuit. Nevertheless digital failures are relatively easy to predict while analog failures are very difficult to predict. Designing for testability is a new and challenging requirement. A good reference document for the further study of testability is MIL-STD-2165 "Testability Program for Electronic Systems and Equipment."

11.4 STORAGE AND SHIPPING CONSIDERATIONS

Electronic components and equipment are subject to change. Deterioration and performance degradation can occur during shipment and while in storage. To minimize performance degradation and to assure that the designed-in hardware reliability is achieved, it will be necessary to identify the significant defects, to quantify the rate of these defects, and to analyze the deterioration influenced by shipment and storage environments, dormancy, storage testing, and environmental cycling effects. Specific inspections and analyses to predict the effects of shipment and storage may be required to assess the in-storage functional status of components and equipment items, and to control deterioration mechanisms. These are all part of the overall life-cycle reliability program.

The control efforts may include first identifying the components and equipment (and their major or critical characteristics) that deteriorate during shipment and through storage, and then preparing procedures for in-storage cyclic inspection to assure reliability and operational readiness. Inspection procedures may be required to identify the quantity of items for the test and the acceptable levels of performance for the parameters under test. The results of these efforts may be used to: support long-term failure rate predictions, evaluate design tradeoffs, and define allowable test exposures. They may also be used to support packaging, handling, storage requirements, and refurbishment plans, and may be used for retest after storage decisions.

Factors Contributing to Reliability Degradation During Shipment and Storage

Defects can be induced during shipment because: (1) the packing protection was not compatible with the mode of transportation;

(2) the container or other packaging material did not meet the specification requirements; or (3) the equipment was roughly handled or improperly loaded. Numerous different failure mechanisms are responsible for the deterioration observed in some electronic components during storage periods.

The electrical contacts of relays, switches, and connectors are susceptible to the formation of oxide or contaminant films or to the attraction of particulate matter that adheres to the contact surface, even during normal operation. During active use, the mechanical sliding or wiping action of the contacts is effective in rupturing the films or dislodging the foreign particles in a manner which produces a generally stable, low-resistance contact closure. However, after long periods of dormant storage, the contaminant films and/or the diversity of foreign particles may have increased to such an extent that the mechanical wiping forces are insufficient to produce a low-resistance contact.

The formation of contaminant films on contact surfaces is dependent on the reactivity of the control material, its history, and the mechanical and chemical properties of the surface regions of the material. Gold is normally used whenever maximum reliability is required. The primary advantage of gold is that it is almost completely free of oxide film contaminants. Even gold, however, is susceptible to the formation of contaminant films by the simple condensation of organic vapors and the deposition of particulate matter. Silver and alloys of copper and nickel are highly susceptible to the sulfide contaminants that abound in the atmosphere. Shipping and storage of equipment containing these types of contacts in kraft paper boxes should be avoided because of the effects of sulfur-containing paper. Particulate contamination can also lead to corrosive wear of the contact surfaces when the particles are hydroscopic. With this condition, water will be attracted to the contact surfaces and can lead to deterioration through corrosive solutions or localized galvanic action. The source of such particles can be directly deposited from airborne dust or wear debris from previous operations.

Another failure mode which may become significant after long-term storage is the deterioration of lubricants used on the bearing surfaces of relays, solenoids, and motors. Lubricants can oxidize and form contamination products. Lubricants can also attract foreign particles, particularly when exposed to airborne dust. This can lead to lubrication failures and excessive wear.

Over a period of time, many plastics (such as those used in the fabrication of electronic components, that is, capacitors, resistors, transistors, and so on) lose plasticizers or other constituents which evaporate from the plastic, causing it to become brittle and possibly shrink. This can cause leakage of seals, electrical breakdown of insulation, and other changes conducive to fatigue failures. Also, plastics may continue to polymerize after manufacture. That is, the structure of the molecules may change without any accompanying change in chemical composition. This will result in changes in their physical properties and other characteristics.

Many materials slowly oxidize, combine with sulfur or other chemicals, or break down chemically over a period of time. These changes may affect electrical resistivity, strength, and so on. In addition, many of these materials when exposed to condensed moisture or high-humidity conditions may, through a leaching process, lose essential ingredients such as fire-retardant additives, thereby causing additional hazards to develop slowly.

Many component parts and assemblies are sensitive to contaminants and thus are sealed during manufacture. As a result of flexing due to changing temperature and atmospheric pressure these seals will often leak, allowing air, moisture, or other contaminants to reach the active portions of the component. This leakage can be so slow that the effects may not be discernible for years, but ultimately significant changes can occur.

Finally, the methods and materials of preservation, packaging, and packing used in the storage of components and equipment, such as cardboards, plastic bags, or polystyrenes, themselves may react with the items being stores and may cause decomposition and deterioration when left dormant for long durations.

Rough handling during shipment and operation, aging, and deterioration mechanisms as previously discussed can, if uncontrolled, lead to a variety of component and equipment failure modes. A summary of some of the failure modes encountered with electronic components during storage is given in Table 11.4. Protective measures must be applied to isolate the components from these deteriorative influences in order to eliminate or reduce the failure modes.

Protective Measures

Proper protection against damage and deterioration to components and equipment during shipment and storage involves the evaluation of a large number of interactive factors and the use of tradeoff analysis to arrive at a cost-effective combination of protective controls. The major control parameters which must be addressed are: (1) the level of preservation, packaging, and packing applied during the preparation of material items for shipment and storage; (2) the actual storage environment itself; and (3) possibly the need for and frequency of in-storage cyclic inspection. These parameters must be evaluated and balanced to meet the specific characteristics of the individual equipment and material.

Preservation, packaging, and packing address the protection provided in the preparation of material items for shipment and long-term storage. The three terms are defined as follows:

Table 11.4 Failure Modes Encountered with Electronic Components During Storage

Component	Failure modes
Batteries	Dry batteries have limited shelf life. Become unstable at low temperatures, output may drop to as low as 10% at very low temperatures. Deteriorate rapidly at temperatures above 35°C.
Capacitors	Moisture permeates solid dielectrics, increasing losses, and may lead to breakdown. Aluminum electrolytic dielectric deterioration.
Coils	Moisture causes changes in inductance and loss in "Q". Moisture swells phenolic forms.
Connectors	Corrosion causes poor electrical contact and seizure of mating members. Moisture may cause shorting.
Relays and solenoids	Corrosion of metal parts may cause malfunction. Dust and sand damage to contacts. Fungus growth on coils.
Resistors	Composition resistors drift. Also they are not suitable for use above 85°C. Enamelled and cement-coated resistors have small pinholes which admit moisture, causing eventual breakdown. Precision wirewound fixed resistors fail rapidly when exposed to high humidity and to temperatures above 125°C.
Semiconductors, diodes, transistors, microcircuits, etc.	Plastic-encapsulated devices have poor hermetic seals, admitting moisture resulting in shorts or opens caused by corrosion.
Motors, blowers	Swelling and rupture of plastic parts and corrosion of metal parts. Moisture absorption and fungus growth on coils. Sealed bearings are subject to failure.

Table 11.4 (Continued)

Component	Failure modes
Plugs, jacks, dial-lamp sockets, etc.	Corrosion and dirt produce high-resistance contacts. Plastic insulation absorbs moisture.
Switches	Metal parts corrode, plastic bodies and wafers warp due to moisture absorption.
Transformers	Windings corrode, causing shorts or opens to develop.

Source: *Electronic Reliability Design Handbook* (MIL-HDBK-338), Naval Publications and Forms Center, Philadelphia.

Preservation: This is the process of treating the corrodible surfaces of a material with an unbroken film of oil, grease, or plastic to exclude moisture.

Packaging: Packaging provides physical protection and safeguards the preservative. In general, sealed packaging should be provided for equipment, spare parts, and replacement units shipped and placed in storage.

Packing: This is the process of using the proper exterior container to ensure safe transportation and storage.

Various levels of preservation packaging and packing can be applied. These range from protection only against damage under favorable conditions of shipment, handling, and storage to complete protection against direct exposure to all extremes of climatic, terrain, operational, and transportation environments (without protection other than that provided by the packaging).

Part V
Management and Testing

12

Reliability Demonstration and Qualification

12.0 RELIABILITY DEMONSTRATION TESTING

Reliability demonstration testing is an empirical measurement of time-to-failure during equipment operation. Its purpose is to prove to the customer with a certain degree of statistical confidence that the equipment design is capable of meeting the stated reliability requirement. This requirement is usually stated in terms of MTBF. Reliability demonstration testing is, by nature, a statistically based sample test. If 100% confidence were desired it would be necessary to test the entire population. This is not usually economically feasible, therefore a given amount of uncertainty or degree of risk must be accepted with the test. The degree of uncertainty is expressed in a number of interrelated terms including the *producer's risk* (α), the *consumer's risk* (β), the *discrimination ratio*, and a *confidence interval, lower confidence limit*, and so on. These terms will be defined shortly.

During the demonstration test the equipment is also subjected to stimulated-use environmental conditions such as vibration, temperature, humidity cycling, and possibly altitude.

Reliability demonstration testing is normally performed under rigorously controlled conditions in accordance with either MIL-STD-781, "Reliability Design Qualification and Production Acceptance Tests: Exponential Distribution" or IEC Specification 605-1, "Equipment Reliability Testing, General Requirements." While the two documents follow the same general procedures, there is a fundamental difference between them in terms of definitions. Therefore, the two documents cannot be used interchangeably.

Whichever specification is used, it is essential that clear definitions of both satisfactory operation and unsatisfactory operation exist prior to the start of the test and that these definitions are agreed to by both the producer and the consumer. This will help to prevent later squabbles as to whether a given failure is "relevant," that is, chargeable to the equipment, or "nonrelevant," that is, not chargeable to the equipment MTBF.

The Statistical Nature of the Test

A reliability demonstration test is effectively a sampling test. It involves a sampling of objects selected from a "population." In reliability demonstration testing, the population encompasses all failures that will occur during the life-span of the equipment. Thus a test sample consists of a number of times-to-failure, and the population is all the times-to-failure that could occur with one or more equipments on test. The test equipments (assuming more than one equipment) are assumed to be identical, that is, their part populations are identical. Also, assuming an exponential failure model (constant λ), a test of 10 devices for 100 hr each is mathematically equivalent to a test of 1 device for 1000 hr. If all possible samples of the same number of times-to-failure were drawn from the same or identical equipment, the resulting set of sample means would follow a normal distribution about the true MTBF of the equipment.

A test sample then is drawn from this population by observing those failures which occur during a small portion of the equipment's life. As in any sampling test, the sample is assumed to be representative of the population, and the mean value of the various elements of the sample (for example, the times-to-failure) is assumed to be a measure of the true mean (MTBF, and so on) of the population

Statistical sampling plans are concerned with defining the sample size upon which to base a decision as to whether the population is good or bad, and defining the acceptable number of failures per sample.

Definitions

At this point it is necessary to define a number of terms unique to demonstration testing.

Confidence Interval

A point estimate (that is, a single sample) of some parameter such as failure rate or MTBF is not meaningful without some measure of its possible error. One such measure is the confidence interval. The confidence interval is that interval in which a stated percentage of the predicted events would be anticipated to occur.

Lower Confidence Limit

In reliability demonstration the customer is primarily concerned with the consumer's risk (β) since he does not wish to accept bad equipment. Therefore, we are usually concerned with a single-sided confidence limit defining the lower confidence limit, rather than a two-sided confidence interval. This parameter may be calculated using the Chi-square (χ^2) distribution.

Operating Characteristic (OC) Curve

The operating characteristic curve is a plot of the ability or inability of a specific test plan to distinguish between good and bad equipments, that is, those that meet the requirement or fail to meet the requirement. An ideal OC curve would require an infinite number of samples and would exhibit a vertical step function, that is, an infinite slope. Due to the limitations of the sample size, actual OC curves are rounded and less-sharply defined. An ideal OC curve and an actual OC curve are compared in Figure 12.1.

To assist in differentiating between good and bad equipments and to define the OC curve two MTBF parameter values, the upper and lower limits θ_0 and θ_1, must also be defined. Unfortunately they are defined differently in MIL-STD-781 and IEC Specification 605-1. This definition difference prevents using the two specifications interchangeably. For the sake of illustration the remainder of this chapter is based upon the use of the MIL-STD-781 definitions.

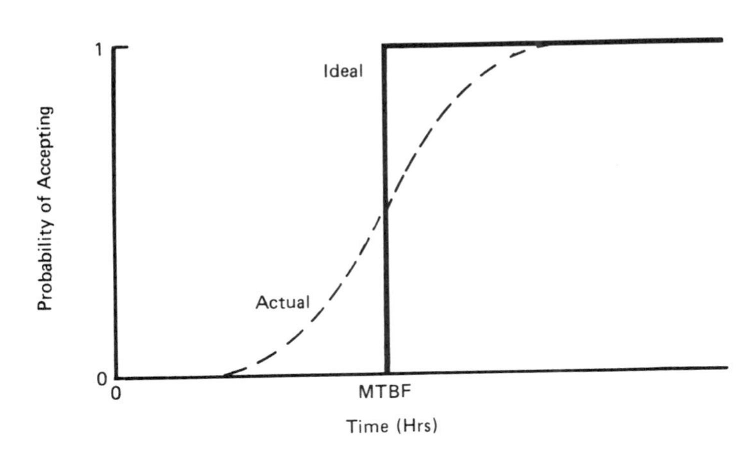

Figure 12.1 Ideal versus actual OC curve.

Lower Test MTBF (θ_1)

The lower test MTBF θ_1 is that value which is unacceptable, and an effective test plan will reject, with high probability, equipment with a true MTBF approaching θ_1. This is equivalent to noncompliance with the reliability requirement.

Upper Test MTBF (θ_0)

The upper test MTBF θ_0 is an acceptable value of MTBF equal to the discrimination ratio times the lower test MTBF θ_1. An effective test plan will accept, with high probability, equipment with a true MTBF that approaches θ_0.

Producer's Risk (α)

The producer's risk is the risk that good equipment may actually fail to pass the test. This is the probability of rejecting equipments with a true MTBF greater than the upper test MTBF θ_0.

Consumer's Risk (β)

The consumer's risk is the risk that bad equipment may actually pass the test. This is the probability of accepting equipments with a true MTBF less than the lower test MTBF θ_1.

Statistical plans have been developed to evaluate the risks α and β for given sample size, or to choose a sample size given α and β. For reliability demonstration the test plans are based on the Poisson (or exponential) distribution.

In reliability demonstration testing the size of these two risks α and β is a function of the specific test plan selected. The test plans in both MIL-STD-781 and IEC Specification 605-1 normally vary from approximately 10% to 30% for both the producer's and the consumer's risk.

Figure 12.2 illustrates the α and β risks for two different population groups, one with a 1000-hr MTBF and one with a 3000-hr MTBF.

Discrimination Ratio

The discrimination ratio establishes the capability of the specific test plan to discriminate between good and bad equipment. In other words it determines the amount of risk to be assumed based upon the limitations of sample size. The larger the sample or the longer the test time the higher the confidence will be that the results of this sample observation accurately reflect the true characteristics of the total population.

$$\text{Discrimination ratio} = \frac{\theta_0}{\theta_1}$$

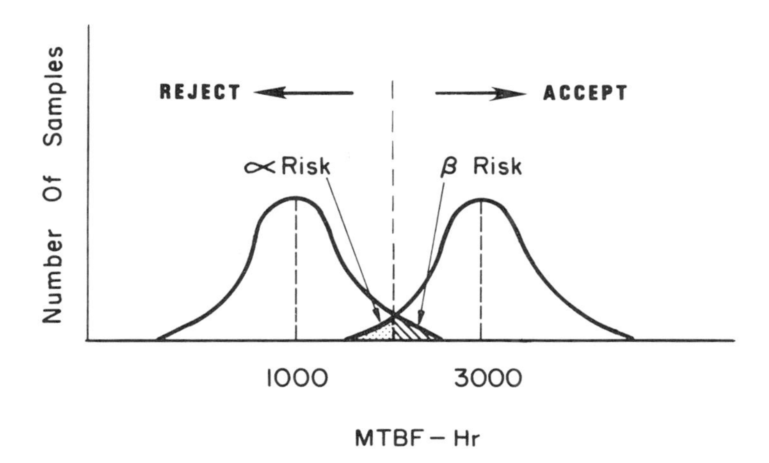

Figure 12.2 Producer's versus consumer's risk.

Fixed-Length and Sequential Tests

Two basic types of reliability demonstration test plans are defined in both MIL-STD-781 and IEC Specification 605-1. They are the *fixed-length test* plans and the *Probability Ratio Sequential Test* (PRST) plans.

The test designer thus has a choice of sampling schemes to achieve the desired α and β. In each case statistical theory will dictate the precise number of items which must be tested if a fixed sample size (fixed length) is desired. An advantage of the fixed-length tests is that they will give a numerical estimate of the MTBF of the equipment.

Alternatively, a sequential test may be selected where the conclusion to accept or reject the equipment will only be reached after an indeterminate number of observations. The major difficulty with the PRST plans is an inability to predice accurately the exact amount of time that the demonstration test will actually take. Depending upon the actual circumstances and the time of actual failures it is quite possible to remain in a never-never land for some time, (as shown in Figure 12.3) where the equipment is not bad enough to fail the test and yet it is not good enough to pass the test. Under these conditions the only logical choice is to continue testing the equipment until an accept or reject decision can be made. After 500 hr of comulative test time this equipment was accepted. PRST plans give accept/reject criteria only. They do not give a numerical estimate of the MTBF of the equipment.

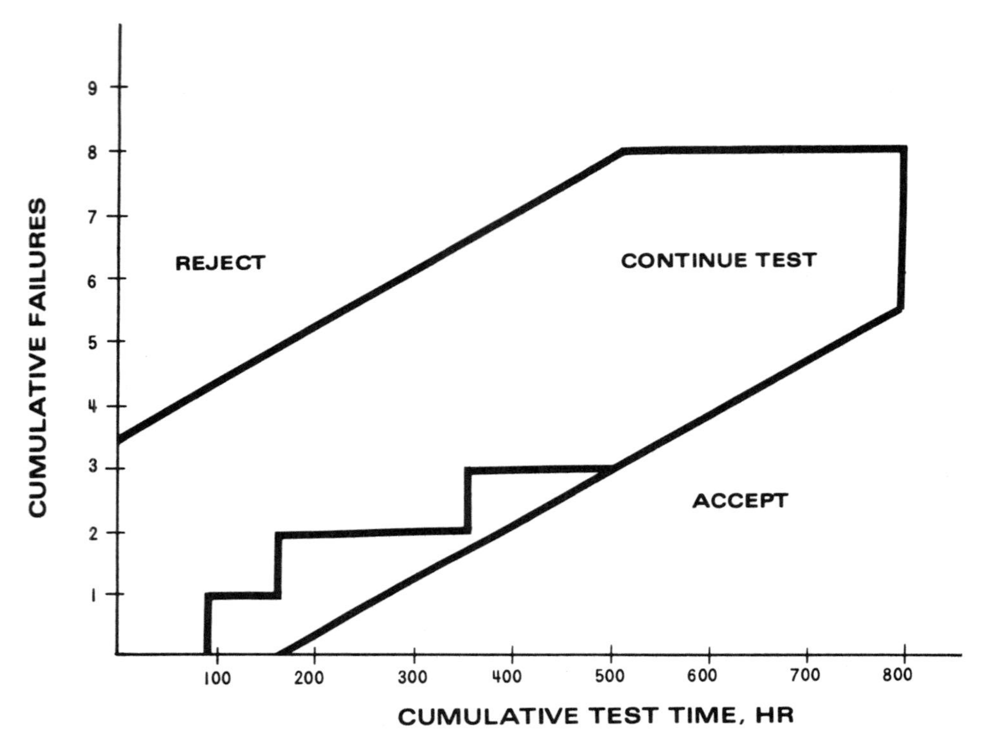

Figure 12.3 Typical PRST reliability test plan. [From *Reliability Design Qualification and Production Acceptance Tests* (MIL-STD-781), Naval Publications and Forms Center, Philadelphia.]

12.1 MIL-STD-781 QUALIFICATION TESTING

MIL-STD-781 actually addresses two fundamentally different types of demonstration testing: *initial reliability design qualification testing* and *production reliability acceptance testing* (PRAT). Both reliability demonstration tests are based on the same set of principles and both assume that the failures are exponentially distributed (this is normally the case for electronic equipment/system). Both types of demonstration testing may utilize either fixes-length or PRST test plans.

Reliability design qualification testing is performed on preproduction or initial-production hardware to determine design compliance with the specified reliability requirements.

In contrast, production reliability acceptance testing (addressed in Chapter 12.2) may be described as a periodic series of tests to indicate continuing production of acceptable equipment. They are used to assure individual item (or lot) compliance with the reliability requirement. These tests are intended to simulate in-service evaluation of the delivered item or production lot and to provide verification of the inherent reliability parameters previously demonstrated by the preproduction design qualification test.

MIL-STD-781 contains the essential procedures and requirements for specifying the applicable acceptance test plans including detailed information for test planning and for evaluation of the data. It defines the test conditions and procedures, various different test plans with their respective accept/reject criteria, and it makes extensive use of appendixes to expand and clarify the various sections of the standard. The test planning parameters θ_0 (upper test MTBF) and θ_1 (lower test MTBF) are defined, and the environmental test conditions (temperature, vibration, and moisture) based on the actual mission profile environments (that is, those encountered during the equipment's useful life) are specified.

The Operating Characteristic (OC) Curve

Since it is not economically feasible to test the complete population, we must be satisfied with a sample of the population. Based upon data from the sample, a statement can be made about the population parameter.

In essence, a statistical hypothesis is being tested. For example:

H_0 (null hypothesis): $\theta_0 \geqslant 200$ hr (accept the item)
H_1 (alternate hypothesis): $\theta_1 \leqslant 100$ hr (reject the item)

Based upon the test results, either H_0 is accepted or it is rejected and the alternate H_1 is accepted. In making this decision certain statistical (error) risks must be kept in mind:

(α) The producer risks the probability of rejecting the hypothesis when it is true (rejecting good equipment).
(β) The consumer risks the probability of accepting the hypothesis when it is false (accepting bad equipment).

Usually the sample size is set as low as possible in order to reduce costs and specify the maximum acceptable α and β risks that can be associated with θ_0 and the smallest acceptable θ_1.

The relationship between the probability of acceptance and the requirement (for example, the MTBF) is called the operating characteristic curve. An ideal OC curve as shown in Figure 12.1 would require an infinite number of samples (the producer's and the consumer's risks result from the lack of an ideal OC curve). In real life we must settle for something that gives a small probability of acceptance for MTBFs below the requirement and high probability of acceptance for MTBFs above the requirement.

Suppose, for example, that the MTBF requirement is 200 hr, the demonstration test is to last 1000 hr, and the decision rule is:

Accept if $r \leqslant 5$
Reject if $r > 5$

where r is the number of failures. From the Poisson distribution (assuming fixed-time test) the probability of rejection p(R) is:

$$p(R) = \frac{(t/m)^r e^{(-t/m)}}{r!} \tag{12.1}$$

where:

t = test time

m = MTBF

The cumulative probability of acceptance P(A) is:

$$P(A) = 1 - P(R) \tag{12.2}$$

Plotting P(A) where $r \leqslant 5$ for various values of m based upon the expected number of failures:

m	t/m	P(A) for r < 5
100	10	0.067
125	8	0.191
167	6	0.446
200	5	0.616
333	3	0.916
500	2	0.983

Thus the decision rule tends to give the right decision, but will not always result in an accept decision for m > 200. Note that there is almost a fifty-fifty chance of accepting an m of 167 hr (0.446) and a greater than 20% chance of rejecting an m of 250 hr.

Neither the producer nor consumer would be happy with this. Each would like a lower risk probability. But since P(A) = 1 − P(R), if P(A) is lowered for $m \leqslant 200$ to 0.1, P(R) must be raised for $m > 200$ to 1 − 0.1 = 0.9. What can be done?

To overcome this difficulty it is necessary to specify the reliability requirements either explicitly or implicitly in terms of two MTBF values rather than a single MTBF value.

In MIL-STD-781 the upper test MTBF θ_0 and the lower test MTBF θ_1 are used to estimate the true MTBF of the equipment and thus determine the compliance or lack of compliance with the specified requirement. The upper test MTBF (θ_0) is the minimum acceptable MTBF while the lower test MTBF (θ_1) is unacceptable based on a minimum requirement. The discrimination ratio (the ratio of θ_0 to θ_1) is used to determine to a large extent the length of time that the demonstration test will take to perform. The lower the discrimination ratio the longer the demonstration test will take. Conversely, the higher the discrimination ratio the shorter the total test time required to reach an accept or reject decision.

The consumer is only interested in the lower confidence level of the normal distribution; that is, a one-sided confidence limit or worst-case condition. The accuracy of this observation relative to the entire population may be calculated using the Chi-square (χ^2) distribution. Thus:

$$\text{MTBF lower} = \frac{2T_R}{\chi^2_{d,x}}$$

where:

T_R = total test time

d = degrees of freedom = 2(r = 1)

r = number of failures

$\chi^2_{d,x}$ = upper χ percentage point of the Chi-square distribution with d degrees of freedom

This can then be evaluated from tables of the Chi-square distribution included in a book of statistical tables.

The manufacturer of the equipment, however, must design for the upper test MTBF θ_0 to assure passing the reliability qualification test.

Defining Test Parameters

The following six characteristics must be defined for any reliability demonstration test:

1. The reliability deemed to be acceptable, θ_0 (in MIL-STD-781 the upper test MTBF)
2. A value of reliability deemed to be unacceptable, θ_1 (in MIL-STD-781 the lower test MTBF)
3. The producer's risk, α
4. The consumer's risk, β
5. The probability distribution to be used for number of failures or for time-to-failure (in MIL-STD-781 the assumed exponential distribution)
6. The sampling scheme

The discrimination ratio θ_0/θ_1, that is, the ratio of upper test MTBF to the lower test MTBF is an additional method of specifying certain test plans.

There are, of course, an infinite number of possible values for the actual reliability. In the specification of two numerical values, θ_0 and θ_1, the test achieves the producer's risk, α, and consumer's risk, β, only for those specific MTBFs. For other values, the relationship is:

Probability of acceptance $\geq 1 - \alpha$, for $\theta \geq \theta_0$
Probability of acceptance $\leq \beta$, for $\theta \leq \theta_1$
Probability of acceptance $> \beta$, for $\theta_1 \leq \theta \leq \theta_0$

Test Plan Selection

In MIL-STD-781 there are eight well-defined PRST statistical test plans which may be used to accept or reject equipment, and twelve fixed-length statistical test plans which may be used to establish the MTBF of the equipment. The primary characteristics of these test plans are shown in Table 12.1

The tests may be summarized as follows:

1. Fixed-length test plans (Test Plans IXC through XVIIC and XIXC through XXIC)
2. Probability ratio sequential tests (PRST), (Test Plans IC through VIC)
3. Short-run high-risk PRST plans (Test Plans VIIC and VIIIC)
4. All-equipment reliability test (Test Plan XVIIIC)

The accept/reject criteria are established by θ_0, θ_1, and the discrimination ratio θ_0/θ_1. Thus, by specifying any two of these three parameters given the desired producer and consumer decision risks the test plan to be utilized can be determined.

These test plans also utilize various different test levels, that is, environmental conditions, during the test.

Table 12.1 Summary of Statistical Test Plans

Test plan number	Nominal decision risks		Discrimination ratio	Time to accept decision in MTBF (θ_1) multiples		
				Minimum	Expected[a]	Maximum
PRST						
IC	10%		1.5	6.60	25.95	49.5
IIC	20%		1.5	4.19	11.4	21.9
IIIC	10%		2.0	4.40	10.2	20.6
IVC	20%		2.0	2.80	4.8	9.74
VC	10%		3.0	3.75	6.0	10.35
VIC	20%		3.0	2.67	3.42	4.5
VIIC	30%		1.5	3.15	5.1	6.8
VIIIC	30%		2.0	1.72	2.6	4.5
Fixed length	α	β				
IXC	10%	10%	1.5	45.0		
XC	10%	20%	1.5	29.9		
XIC	20%	20%	1.5	21.1		

Table 12.1 (Continued)

Fixed length	Nominal decision risks		Discrimination ratio	Time to accept decision in MTBF (θ_1) multiples		
	α	β		Minimum	Expected[a]	Maximum
XIIC	10%	10%	2.0	18.8		
XIIIC	10%	20%	2.0	12.4		
XIVC	20%	20%	2.0	7.8		
XVC	10%	10%	3.0	9.3		
XVIC	10%	20%	3.0	5.4		
XVIIC	20%	20%	3.0	4.3		
XVIIIC[b]	–	–	–	–		
XIXC	30%	30%	1.5	8.0		
XXC	30%	30%	2.0	3.7		
XXIC	30%	30%	3.0	1.1		

[a]If true, MTBF = θ_0.
[b]Test Plan XVIIIC is the all-equipment reliability acceptance test (see Chapter 12.2).
Source: *Reliability Design Qualification and Production Acceptance Tests* (MIL-STD-785), Naval Publications and Forms Center, Philadelphia.

Fixed-Sample and Sequential Tests

When θ_0, θ_1, α, and β have been specified, along with the probability distribution for time-to-failure, the test designer often has a choice of sampling schemes. To achieve the desired α and β, statistical theory will dictate the precise number of items that must be tested if a fixed sample size is desired. Alternatively, a sequential test may be selected where the conclusion to accept or reject will be reached after an indeterminate number of observations. For reliability at θ_0 or θ_1, the average sample size in a sequential test will invariably be lower than in a fixed-sample test, but the sample size will be unknown and could be substantially larger in a specific case. Usually an upper bound for sample size is known in sequential tests.

Consider, for example, a comparison between the three PRST Test Plans IIC, IVC, and VIIIC shown in Figure 12.4. These three test plans all share a common discrimination ratio of 2.0:1, but they offer increasing degrees of risk associated with shorter and shorter test times. It is quite possible, therefore, that tradeoffs may be made between the three plans at various times in the program. It is possible that Test Plan IIIC (10% risk) may be used during the

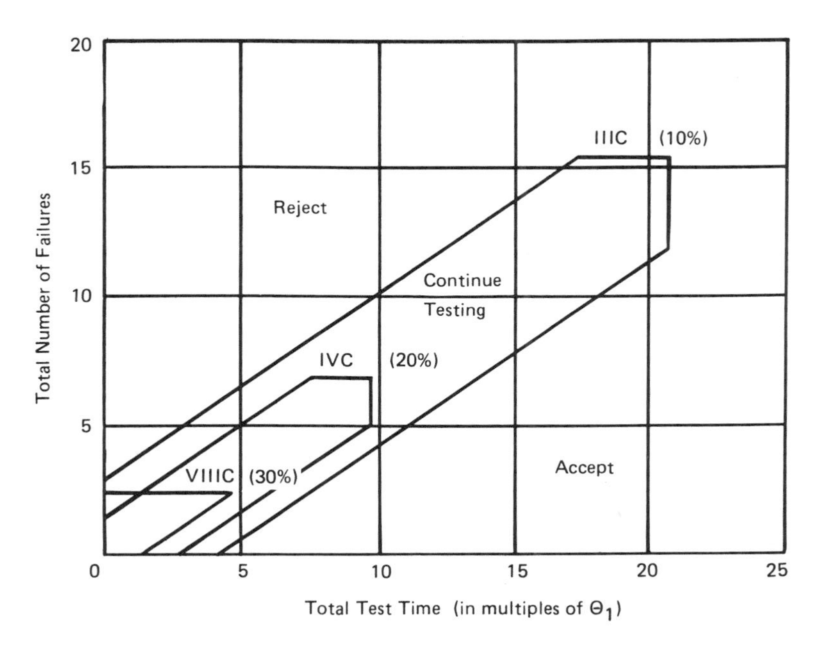

Figure 12.4 PRST test plan comparison.

initial qualification test; Test Plan IVC (20% risk) may be used during the initial PRAT test, and Test Plan VIIIC (30% risk) may be used for subsequent follow-on PRAT tests.

Determinants of Sample Size

For both a fixed-length or sequential test the number of observations required will be related to the degree of discrimination desired. In general:

The closer θ_1 is to θ_0, the larger the sample size.
The smaller α specified, the larger the sample size.
The smaller β specified, the larger the sample size.

Reliability Qualification Summary

MIL-STD-785 describes the essential elements that should be included in a reliability test program plan for development and production testing. However, it should not be invoked on a blanket basis. Each requirement should be assessed in terms of the need and the mission profile. Appendixes to the standard are designed so that they may be referenced with specific parts of the standard and invoked in the equipment specification.

MIL-STD-781 covers the detailed requirements for development and production reliability tests for equipment that experiences an exponential distribution of time-to-failure. It contains test conditions, procedures, and various fixed-length and sequential test plans with their respective accept/reject criteria. A reliability demonstration test plan should answer at least the following questions:

1. How will the equipment/system be tested? What will be the specified test conditions, for example, environmental conditions, test measures, length of test, equipment operating conditions, accept/reject criteria, test reporting requirements, and so on?
2. Where will the tests be performed and by whom? Contractor, Government, or an independent organization?
3. When will the tests be performed? During development, production, or field operation?
4. What preventive maintenance will be allowed during the test?
5. Is burn-in to be performed prior to the test, and if so, for how long?
6. Will thermal survey or a vibration survey of the equipment be required prior to the test?

12.2 PRODUCTION RELIABILITY ACCEPTANCE TESTING (PRAT) IN ACCORDANCE WITH MIL-STD-781

MIL-STD-781 covers requirements for production acceptance tests as well as preproduction qualification tests. Qualification tests are normally conducted in the development cycle, using either preproduction or initial production hardware.

Once the specified reliability has been demonstrated, production lots undergo reliability acceptance testing, usually at a level somewhat less stringent than the demonstration test level, so as to indicate continuing fulfillment of reliability requirements.

Environmental Stress Considerations

Production reliability acceptance testing per MIL-STD-781 is usually based on sampling an equipment from each lot produced as well as from all of the equipment produced. The test conditions, or stress profile, applied during the test are normally determined by the customer and incorporated into the equipment specification. If the environmental stress types and levels are not specified by the customer or are not easily estimated from a similar application, then the stress types and levels given in Table 12.2 taken from MIL-STD-781C should be applied. This table provides a summary of combined environmental test condition requirements applicable to the following categories of equipment:

Category 1: fixed ground equipment
Category 2: mobile ground vehicle equipment
Category 3: shipboard equipment (sheltered or unsheltered)
Category 4: jet aircraft equipment
Category 5: turbo-prop aircraft and helicopter equipment
Category 6: air-launched weapons and assembled external stores

Figure 12.5, also taken from MIL-STD-781, illustrates a typical test cycle for ground mobile equipment showing the timing of the various conditions.

Standard statistical test plans from MIL-STD-781 were previously summarized in Table 12.1.

All Equipment PRAT

Test Plan XVIIIC shown in Figure 12.6 may also be used for 100% production relaibility acceptance testing. This test plan is used

Table 12.2 Summary of Combined Environmental Test Condition Requirements

	Fixed ground	Ground vehicle	Shipboard	
			Sheltered	Unsheltered
Electrical stress				
Input voltage range	Nominal +5%−2%	Nominal ±10%	Nominal ±7%[a]	Nominal ±7%[a]
Voltage cycle (high, nominal, and low)	1/test cycle	1/test cycle	1/test cycle	1/test cycle
Vibration stress				
Type vibration	Sine wave, single frequency	Swept-sine log sweep	Swept-sine continuous[b]	Swept-sine continuous[b]
Amplitude[e]	–	–	–	–
Frequency range[c]	20–60 Hz	5–500 Hz	–	–
Application	20 minimum/equpment	Sweep rate 15 minimum/hr	e	e
Thermal stress (°C)				
Storage temp. (low to high)	–	−54 to 85	−62 to 71	−62 to 71
Operating temperature	g	−40 to 55	0 to 50 (controlled)	−28 to 65
Rate of change (per min.)	–	5	5	5
Maximum rate of change (per min.)	–	10	10	10
Duration (nominal)	–	–	–	–
Moisture stress				
Condensation	None	1/test cycle	e	1/test cycle
Frost/freeze	–	1/test cycle	e	1/test cycle

[a]*See* MIL-STD-1399.
[b]*See* MIL-STD-167-1.
[c]Frequency tolerance ±2% or ±0.5 Hz for frequencies below 25 Hz.
[d]*See* MIL-STD-781, Appendix B, paragraph 50.5.3.

Fighter	Transport, bomber	Helicopter	Turbo-Prop	Air-launched weapons and assembled external stores
Nominal ±10%	±10%	±10%	±10%	±10%
1/thermal cycle[e]	1/thermal cycle[e]	1/thermal cycle[e]	1/thermal cycle[e]	1/thermal cycle[e]
Random	Random	Swept-sine log sweep	Swept-sine	Swept-sine and random[c]
–	–	–	–	–
20–2000 Hz	20–2000 Hz	5–2000 Hz[d]	10–2000 Hz	20–2000 Hz
Continuous	Continuous	Sweep rate 15 minimum/hr	e	f
−54 to 71	−54 to 71	−54 to 71	−54 to 71	−65 to 71
e	e	e	e	e
5	5	5	5	5
–	–	–	–	–
3 1/2 hr	3 1/2 hr	3 1/2 hr	3 1/2 hr	3 1/2 hr
1/test cycle	1/test cycle	1/test cycle	1/test cycle	1/test cycle
1/test cycle	1/test cycle	1/test cycle	1/test cycle	1/test cycle

[e]*See* MIL-STD-781, Appendix B.
[f]*See* MIL-STD-1670.
[g]20 (heated and air conditioned); 40 (heated but not air conditioned); 60 (unoccupied tropical or semitropical). *See* MIL-STD-781, Appendix B, paragraph 50.1.4.

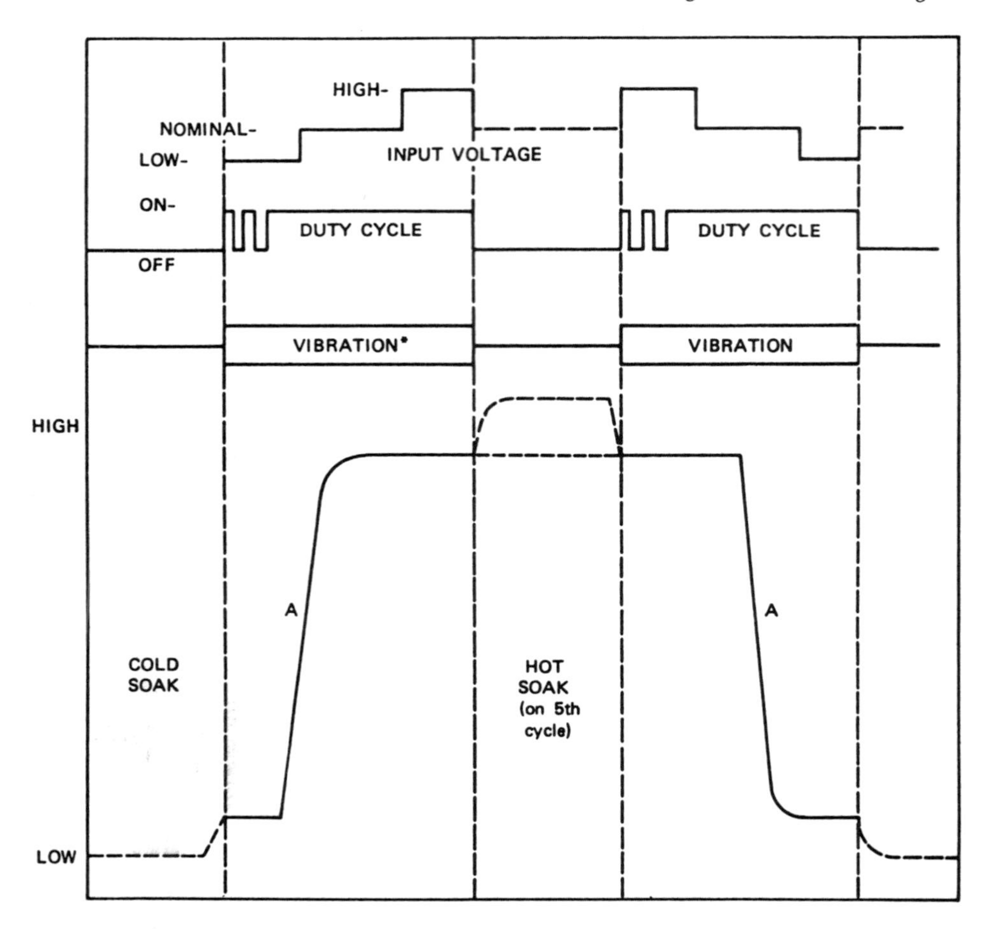

Figure 12.5 Typical environmental test cycle. [From *Reliability Design Qualification and Production Acceptance Tests* (MIL-STD-781), Naval Publications and Forms Center, Philadelphia.]

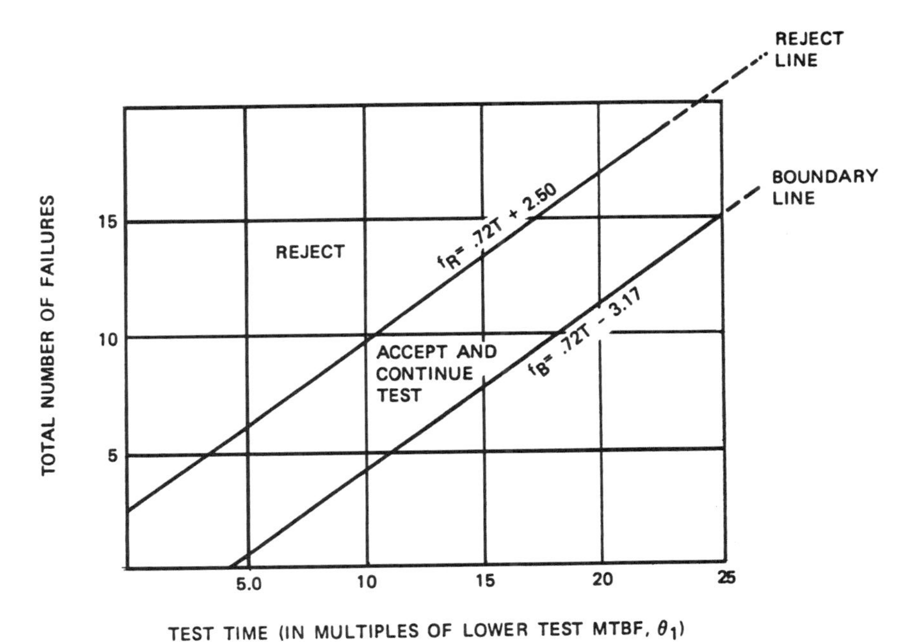

Figure 12.6 Test Plan XVIIIC. [From *Reliability Design Qualification and Production Acceptance Tests* (MIL-STD-781), Naval Publications and Forms Center, Philadelphia.]

when each unit of production equipment is to be given a reliability acceptance test. The plan consists of a reject line and a boundary line. The reject and boundary lines are extended as far as necessary to cover the total test time required for the production run. The equation of the reject line is $f_R = 0.72T + 2.50$, where T is the cumulative test time in multiples of θ_1 and f is the cumulative number of failures. The plotting ordinate is failures and the abscissa is in multiples of θ_1, the lower test MTBF. The boundary line is 5.67 failures below and parallel to the rejection line. Its equation is $f_B = 0.72T + 3.17$.

The test duration for each equipment should be specified in the test procedure approved by the customer. The maximum test duration is usually 50 hr and the minimum test time 20 hr (to the next higher integral number of complete test cycles). If a failure occurs in the last test cycle, the unit is repaired and another complete test cycle is run to verify the repair.

Optional nonstatistical plans may also be used for production reliability acceptance testing. The purpose is to verify that production workmanship, manufacturing processes, quality control procedures, and the assimilation of production engineering changes do not degrade the reliability which was found to be acceptable by the reliability qualification test. The test is applied to all production items with the item operating (for example, the power applied). The required test duration and number of consecutive, failure-free, thermal test cycles (minimum of two) which each deliverable item must exhibit is specified by the customer. The vibration, temperature cycling, and moisture environments together with any others deemed necessary may be applied sequentially. The equipment duty cycle—the sequence, duration, and levels of the environments—and the vibration option to be used in this test usually require customer approval. They are included with the test program requirements.

PRAT Test Plan Considerations

MIL-STD-785 "Reliability Program for Systems and Equipment Development and Production" incorporates the requirements of MIL-STD-781. The test criteria, including the confidence level and decision risk, should be carefully selected and tailored from MIL-STD-781 to avoid excessive cost or schedule impact without significant reliability improvement. Appendix A to MIL-STD-785 provides general guidelines for planning and implementing PRAT:

> The statistical test plan must define the compliance ('accept') criteria which limits the probability that the item tested, and the lot it represents, may have a true reliability less than the minimum acceptable reliability. These criteria should be tailored for cost and schedule efficiency.
>
> Because it is intended to simulate the item's operational environment and life profile, production reliability acceptance testing may require expensive test facilities; therefore, all-equipment production reliability acceptance (100% sampling) is not recommended.
>
> The sampling frequency may be reduced after a production run is well established; however, PRAT provides protection for the customer and motivation for the contractor's quality control program; thus, it should not be discarded by a complete waiver of the production reliability acceptance testing requirement.

Plans for performing PRAT should be prepared and incorporated into the overall reliability test plan document. The test plans should encompass the following considerations as described in Task 304 of MIL-STD-785:

1. Tests to be conducted per MIL-STD-781
2. Reliability level (that is, MTBF) to be demonstrated; the associated confidence level; and the relationship between demonstrated MTBF, confidence, test, and so on.
3. Representative mission/environmental profile
4. The number of units for test, expected test time, calendar time factors, and scheduling of effort
5. The kinds of data to be gathered during the test
6. Definition of failure (relevant, nonrelevant)
7. Authorized replacement and adjustment actions
8. Logs/data forms to be maintained that record number of units on test, test time accumulated, failures, corrective actions, statistical decision factors, and accept/reject criteria

13

Reliability Growth Management

13.0 INTRODUCTION

Experience has shown that programs that rely simply on a demonstration test by itself to determine compliance with the specified reliability requirements frequently do not achieve the reliability objectives within the available resources. This is particularly true of complex electronic systems. These systems may require new technology development and as such represent a challenge to the state of the art. Striving to meet these requirements may thus consume a significant portion of the entire development process.

In order to help ensure that equipment and systems will meet the required operational reliability requirements, the concept of reliability growth testing and growth management have been developed for equipment and system development programs.

The purpose of reliability growth management is to find and correct defects in the design and fabrication of the equipment through the systematic and permanent removal of specific failure mechanisms prior to delivery of the equipment to the customer. Achievement of reliability growth is dependent upon the extent to which testing and other techniques are used to "force out" design and fabrication flaws. Attempts are made to induce failure in the equipment so that its weaknesses can be identified and corrected by design changes. This is different from reliability demonstration where it is hoped that the equipment will not fail.

During reliability growth testing the equipment is normally stressed to some degree to help precipitate failure. It may be stressed to the extreme limits of its specified operating environment but it is not normally stressed beyond these limits, that is, it is not overstressed.

Testing of itself does not improve reliability. Only when corrective action that prevents the recurrence of failures in like equipments is taken does actual reliability growth occur.

To facilitate reliability growth it is necessary to keep accurate records of both operating time and failures. All failures should be recorded and investigated to assure that each failure mechanism involved is fully understood. If the basic failure mechanism is not understood it cannot be eliminated in future equipments. Thus a closed-loop failure reporting and corrective action system is needed to eliminate the failures identified during testing. Accurate operating-time records must be kept so that the data can be plotted and the present MTBF determined. This is done for the designer's own edification, not to meet the customer's requirement. Thus failures during a reliability growth test help by identifying potential weakness within the equipment design rather then by penalizing the designer as happens during a reliability demonstration test.

The Reliability Growth Concept

Reliability growth is defined as positive improvement of equipment reliability through the systematic and permanent removal of specific failure mechanisms. Achievement of reliability growth is dependent upon the extent to which testing and other improvement techniques have been used during development and production to force out design and fabrication flaws, and on the rigor with which these flaws are analyzed and corrected.

Figure 13.1 suggests an ideal growth process. The initial reliability of the prototype starts at some level that might be considered the state of the art at the beginning of development. Reliability grows throughout the development effort up to the point of pilot production. At that time, some loss of growth occurs due to the introduction of new manufacturing problems. During pilot production, corrective actions are continued causing resumption of growth. At the beginning of full-scale production, some loss in the achieved reliability occurs with the effects of mass production. However, growth resumes again as these problems are eliminated. Eventually, when the equipment is delivered to the customer it should have achieved the specified reliability level or, under ideal conditions, the inherent or predicted reliability level.

The rate of growth or the slope of the growth curve is affected by many variables. Therefore reliability growth must be an iterative process. As the design matures, new problems are continually identified and each of them must in turn be corrected. Thus three essential elements are involved in achieving reliability growth:

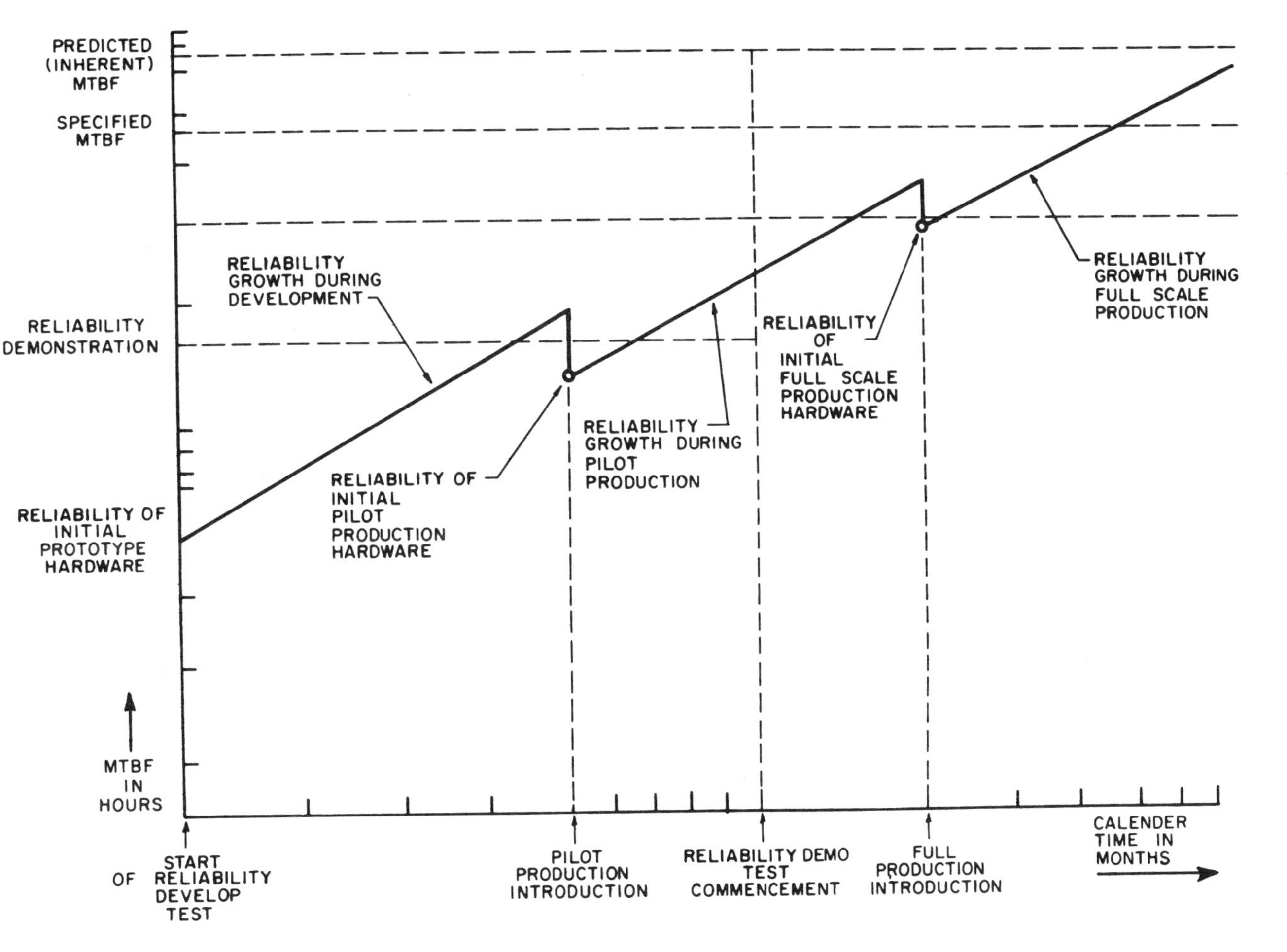

Figure 13.1 Reliability growth process.

1. Detection, by test and analysis, of failure sources
2. Feedback of problems as they are identified
3. Effective redesign to eliminate the identified problems

The rate at which reliability grows is dependent on how rapidly the activities in this iterative loop can be accomplished, how real the identified problems are, and how well the redesign effort solves the identified problems. In most cases failure sources are detected through testing, and the testing process effectively controls the rate of growth. As a consequence, the reliability growth process becomes known as one of *test, analyze, and fix* (TAAF). The reliability achieved as a result of the growth process only becomes meaningful, however, when the necessary changes developed and proven during TAAF are fully incorporated in configuration control documentation for the production hardware.

Reliability growth testing is just one aspect of a total reliability growth program. It must be accompanied by a reliability growth management program. This involves setting interim reliability goals to be met during the development testing program and the allocation and reallocation of resources to attain these goals. A comprehensive approach to reliability growth management throughout the development program consists of planning, evaluating, and controlling the growth process.

Reliability growth planning addresses program schedules, amount of testing, resources available, and the realism of the test program in achieving the requirements. The planning is quantified and reflected in the construction of a reliability-growth-program-plan curve. This curve establishes interim reliability goals throughout the program. To achieve these goals it is important that the program manager be aware of reliability problems throughout the program so that changes can be effected as necessary, for example, increased reliability emphasis. It is essential that periodic assessments of reliability be made during the test program (for example, the end of a test phase) and that they be compared to the planned reliability growth values. These assessments provide visibility or achievements and focus on deficiencies in time to affect the system design. Management can control the growth process by making appropriate decisions in regard to timely incorporation of effective fixes into the system commensurate with attaining the milestones and requirements.

13.1 GROWTH MODELING

The model most commonly used to describe the reliability growth processes for complex electronic equipments, and reliability growth testing in general, was originally published by J. T. Duane.* This model provides a deterministic approach to reliability growth such that the system MTBF versus operating hours falls along a straight line when plotted on log-log paper. Thus, the change in MTBF during development is proportional to T^{α} where T is the cumulative operating time and α is that rate of growth corresponding to the rapidity with which faults are found and changes are made to permanently eliminate the basic causes of the observed faults. The model is shown graphically in Figure 13.2, with each of the growth lines having different slopes depending upon the amount of emphasis given to the reliability growth program.

Duane postulated that as long as reliability improvement effort continues, the following mathematical expression would hold:

$$\lambda_{\Sigma} = \frac{F}{H} KH^{-\alpha} \qquad (13.1)$$

where:

λ_{Σ} = cumulative failure rate

H = total test hours

F = failures during H

K = constant determined by circumstances

α = growth rate

The original mathematical model was expressed in terms of cumulative failure rate. However, since equipment reliability is generally expressed in terms of MTBF, the following expression is more frequently used:

$$M_R = M_I \left(\frac{T_t}{t_i}\right)^{\alpha} \qquad (13.2)$$

where:

M_R = required MTBF

M_I = initial MTBF

---

*Duane, J. T., "Learning Curve Approach to Reliability Monitoring," IEEE Transactions on Aerospace, Volume 2, April 1964 pp. 363–366.

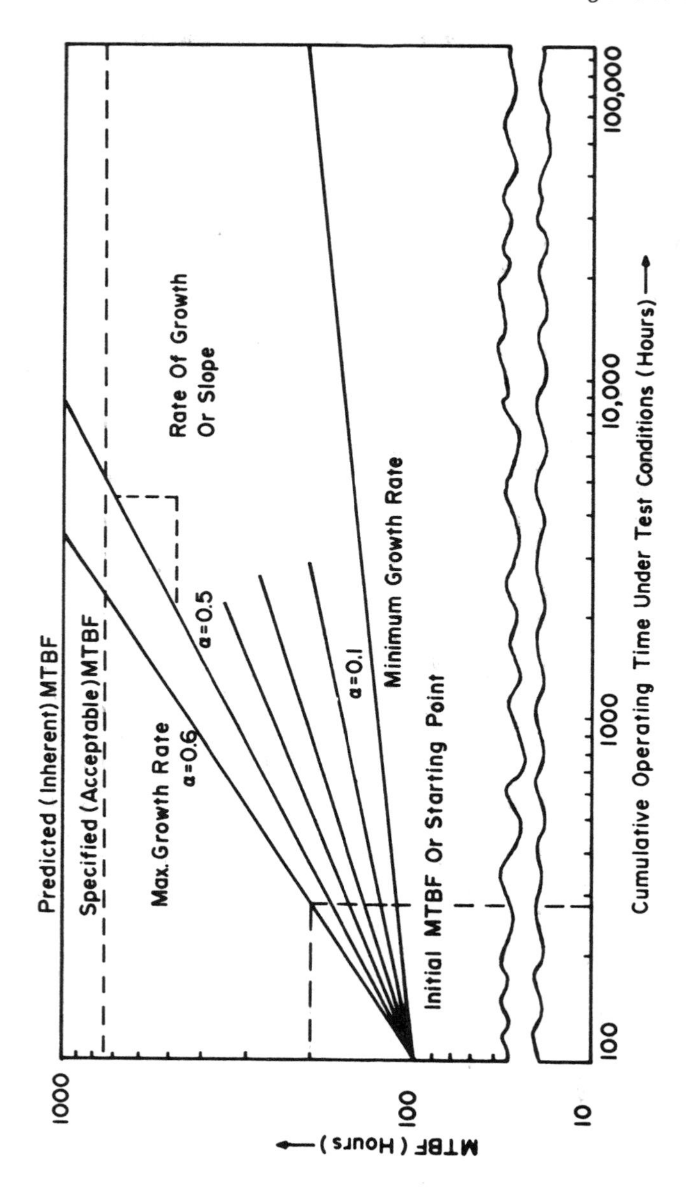

Figure 13.2 Reliability growth plot. [From *Reliability Design Handbook* (RDH-376), copyright 1976 by IIT Research Institute, Reliability Analysis Center, RADC/RAC, Griffiss Air Force Base, N.Y.]

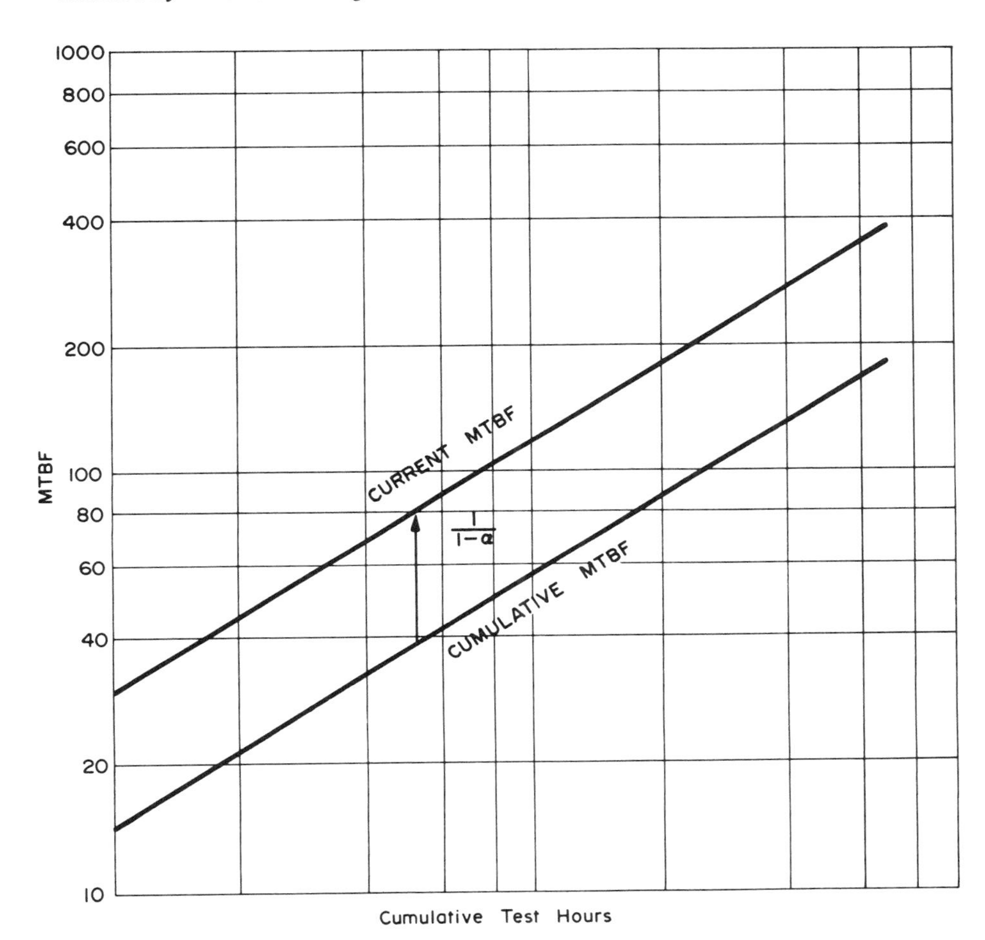

Figure 13.3 Duane plot. (From MIL-HDBK-338, *Electronic Reliability Design Handbook*, Naval Publications and Forms Center, Philadelphia.)

t_i = time at which initial datapoint is plotted (preconditioning time)

T_t = time at which the instantaneous MTBF of the equipment under test will reach the MTBF requirement

α = growth rate

Differentiating Equation (13.2) with respect to time:

$$\lambda(t) = \frac{\delta F}{\delta H} = (1 - \alpha)KH^{-\alpha} = (1 - \alpha)\lambda_{\Sigma} \tag{13.3}$$

Thus, the current instantaneous failure rate is $(1 - \alpha)$ times the cumulative failure rate, or the instantaneous MTBF is $1/(1 - \alpha)$ times the cumulative MTBF. Instantaneous MTBF may be interpreted as the MTBF that the equipment currently on test would exhibit if we stopped the reliability growth and continued testing. Thus on a logarithmic plot instantaneous or current-status curves are straight lines displaced a fixed distance from the cumulative plot by a factor $(1 - \alpha)$, as shown in Figure 13.3.

The cumulative MTBF (M_c) is normally measured in test and then converted to instantaneous (or current) MTBF (M_I) by dividing by $(1 - \alpha)$, that is:

$$M_I = \frac{M_c}{1 - \alpha} \tag{13.4}$$

Cumulative MTBF is plotted versus cumulative test time, and a straight line fitted to the data and its slope α is measured. The current MTBF line is then drawn parallel to the cumulative line but displaced upward by an offset equal to $1/(1 - \alpha)$. The corresponding test time at which this line reaches the required MTBF is the expected duration of the growth test. Much evidence has been accumulated since Duane's original report varifying the adequacy of the Duane model to representing the real world of reliability growth testing.

13.2 GROWTH TESTING

Reliability growth testing is the formal process of testing an equipment under natural and induced environmental conditions to discover and identify latent failure modes and mechanisms whose recurrence can be prevented through implementation of corrective action, thus causing the growth of equipment reliability. These tests are frequently conducted during the development phase on samples which have completed environmental tests prior to production commitment.

Structuring the Growth Test

In order to structure a growth test program (based on the Duane model) for a newly designed system, a detailed test plan is necessary. This plan describes the test-analyze-fix concept, and shows how it will be applied to the system under development. A reliability growth plan should incorporate the following features:

1. Both the specified and the predicted reliability values and the methods (model, data base, and so on) used for predicting the reliability should be described.
2. The criteria for establishing the starting points should be stated, that is, the criteria for estimating the reliability of initially fabricated hardware. Initial reliability for many newly fabricated systems has been found to vary between 10% and 30% of their predicted (inherent) values.
3. The anticipated or desired reliability growth rate should be defined. To support the selected growth rate, the rigor with which the test-analyze-fix conditions are structured must also be completely defined.
4. The calendar-time efficiency factors must be determined and the relationship of test time, corrective action time, and repair time to calendar time must be defined.

Each of these factors impact the total time (or resources) which must be scheduled to enable the reliability to grow to the specified value.

Figure 13.4 graphically depicts the four elements needed to structure and plan a growth test program. The four elements identified in the figure are:

1. Inherent reliability: This represents the value of design reliability estimated from prediction studies. Ordinarily, the specified value is somewhat less than the inherent value. The relationship of the inherent (or specified) reliability to the starting point greatly influences the total test time.
2. Starting point: This represents an initial value of reliability for newly manufactured hardware. It usually falls within the range of 10%–30% of the inherent or predicted reliability. Estimates of the starting point can be derived from prior experience, or it may be based on percentages of the estimated inherent reliability. The starting point must take into account the amount of reliability control exercised during design and the relationship

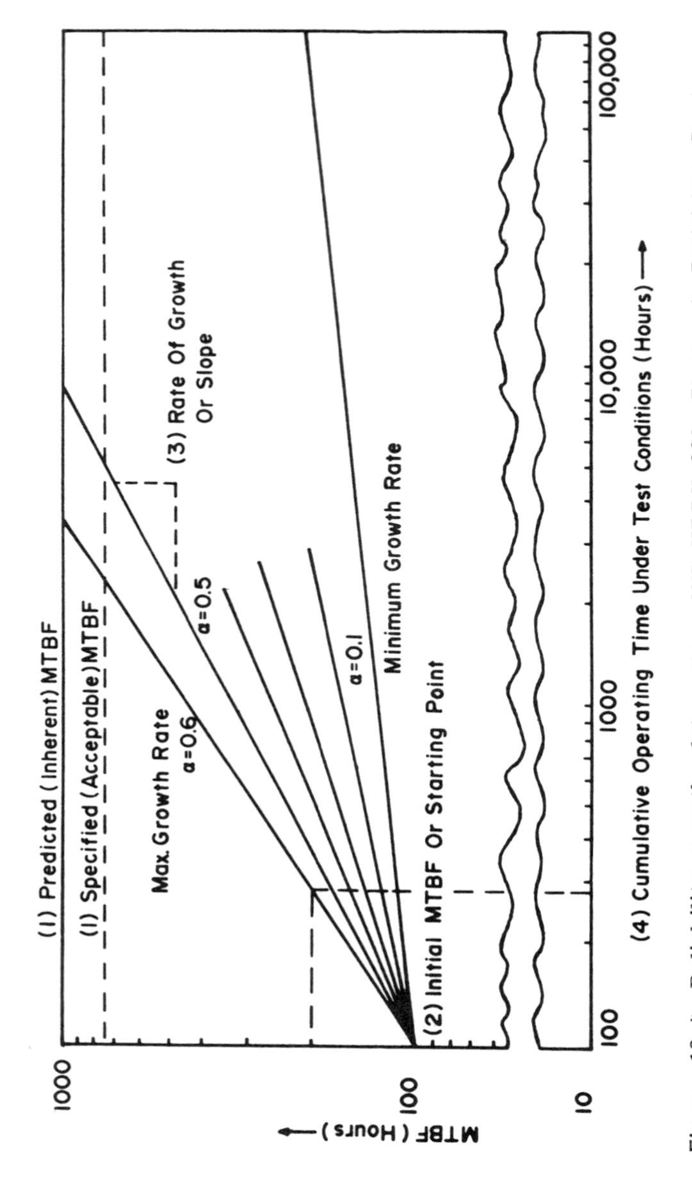

Figure 13.4 Reliability growth plot. (From MIL-HDBK-338, *Electronic Reliability Design Handbook*, Naval Publications and Forms Center, Philadelphia.)

of the system under development to the state of the art. Higher starting points, when justified, minimize test time.
3. Rate of growth: This is depicted by α, the slope of the growth curve. This is governed by the amount of control, rigor and efficiency by which failures are discovered, analyzed, and corrected through design and quality action. Rigorous test programs which foster the discovery of failures, coupled with management-supported analysis and timely corrective action, will result in a faster growth rate and consequently less total test time.
4. Calendar time/test time: This represents the efficiency factors associated with the growth test program. These factors include repair time and operating/monoperating time as they relate to calendar time. Lengthy delays for failure analysis, subsequent design changes, implementation of corrective action, or short operating periods will extend the growth test period.

Figure 13.4 shows that the value of the parameter α can vary between 0.1 and 0.6. A growth rate of 0.1 may be expected in those programs where no sepcific consideration is given to reliability. In such cases, growth is primarily due to problems impacting production, and from corrective action taken as a result of user experience. A growth rate of 0.6 may be realized if an aggressive, hard-hitting reliability program with effective management support is implemented. This type of program will include a stress-oriented test program designed to aggravate and force out defects and vigorous corrective action.

Figure 13.4 also indicates the requisite number of operating hours or test time required to obtain a given amount of reliability growth. The dramatic effect that the rate of growth has on the cumulative operating time required to achieve a predetermined reliability level is obvious. For example, for an item with an initial reliability of 100 hr, 100,000 hr of cumulative operating time is required to achieve a 200 hr MTBF when the growth rate α is 0.1, that is, when no specific attention is given to reliability growth. However, if α can be accelerated to 0.6 (by rigorous growth testing and formal failure analysis activities) then only 300 hr of cumulative operating time is required to achieve a 200 hr MTBF.

When Is Reliability Growth Testing Performed?

The formal reliability growth test is frequently performed near the conclusion of full-scale development after successful completion of environmental qualification testing and prior to reliability demonstration (qualification) testing. Although all equipment testing should be planned to contribute to reliability growth, the formal reliability growth test program is usually deferred until after environmental qualification. By this time the design of the prototype or preproduction equipment which is to be used in the reliability growth test reflects the anticipated final configuration, and it incorporates the manufacturing processes to be used in production. It is hoped that the hardware to be tested will have all significant fixes required as a result of environmental qualification testing incorporated before initiating formal reliability growth test. The reliability growth test should be successfully concluded, and all significant fixes incorporated in the test hardware prior to initiating the reliability demonstration test.

Thus, the purpose of the reliability growth test is to detect reliability problems after all performance design and environmental problems have been resolved. In contrast, the purpose of the reliability demonstration (qualification) test, discussed in Chapter 12, is to prove a specific reliability level to the customer. Some general guidance on reliability growth test time is given in MIL-STD-2068 "Reliability Development Tests":

> Fixed length test times of 10 to 25 multiples of the specified MTBF will generally provide a test length sufficient to achieve the desired reliability growth for equipment in the 50 to 2000 hour MTBF range. For equipments with specified MTBFs over 2000 hours, test lengths should be based on equipment complexity and the needs of the program, but as a minimum, should be one multiple of the specified MTBF. In any event, the test length should not be less than 2000 hours or more than 10,000 hours.

Where time is not an appropriate measurement parameter for the particular hardware, the Duane model is adaptable to other measurement parameters such as the number of cycles, events, rounds, and so on.

13.3 GROWTH PROGRAM MANAGEMENT

Reliability growth management is the systematic planning for reliability achievement as a function of time and other resources, and controlling the ongoing rate of achievement by reallocation of resources based on comparisons between planned and assessed reliability values. Reliability growth management provides a means of viewing all of the reliability program activities in an integrated manner.

A total reliability program is needed for effective reliability growth management. While it is generally recognized that reliability will grow in the presence of a reliability program, reliability growth planning provides an objective yardstick and an orderly means of measuring progress and directing resources so that the reliability requirements may be achieved in a timely and cost-effective manner. A good reliability growth plan can improve the probability of achieving the reliability program objectives; however, it is not intended to be the total reliability program.

Management of the Reliability Growth Process

There are in essence two ways to evaluate the reliability growth process.

1. Qualitative (monitoring) approach: Monitor the various reliability-oriented activities (FMEAs, stress analysis, and so on) in the growth process to assure that the activities are being accomplished in a timely manner and that the level of effort and quality of work is appropriate.
2. Quantitative (assessment) approach: Utilize assessments (quantitative evaluations of the current reliability status) based on information from the detection of failure sources.

The assessment approach is results-oriented, that is, it provides quantitative estimates of planned and achieved reliability as the program progresses. The monitoring approach, in contrast, is activities-oriented, it should be used in addition to assessments. The monitoring approach will have to be relied on early in a program before the detection of failure sources is capable of generating objective assessments.

Monitoring Management Model

Since there is no simple way to evaluate the performance of activities, management based on monitoring is less definitive than management based on assessments. Nevertheless, this method is a valuable alternative when assessments are not practical. The reliability growth program plan serves, at least partially, as a standard against which the activities being performed can be compared. Standards for level of effort and quality of work accomplished must necessarily rely heavily on the technical judgement of the evaluator.

Monitoring is intended to assure that the activities have been performed within schedule, and that they meet appropriate standards of engineering practice.

One of the best examples of a monitoring activity is design review (see Chapter 15.3). The design review is a planned monitoring of the product design to assure that it will meet the performance requirements during operational use. Such reviews of the design effort serve to determine the progress being made in achieving the design objectives. One of the most significant aspects of design review is its emphasis on technical judgements, rather than the quantitative assessments of progress.

Assessment Management Model

Assessments have been a way of life in reliability work for many years, as have the resultant decisions, and they may also be used in controlling the growth process.

The new element in reliability growth management is a formal standard against which the assessment may be compared. The fact that in the past decisions were made based on assessments implies that at least a subjective standard of acceptable reliability growth existed against which comparisons could be made. A formal, objective standard has the advantage of constancy until formally changed.

Figure 13.5 illustrates an example of a reliability growth curve, showing both the budgeted (planned) reliability growth and assessments. A comparison between the assessment and the budgeted value will suggest whether the program is progressing as planned, better than planned, or not as well as planned. Based upon the first two datapoints of assessed growth, the decision would probably be made to continue development with no changes. If reliability progress is falling short, as the two subsequent assessed datapoints may indicate, new strategies should be developed. These strategies will probably involve the reassignment of resources to work on

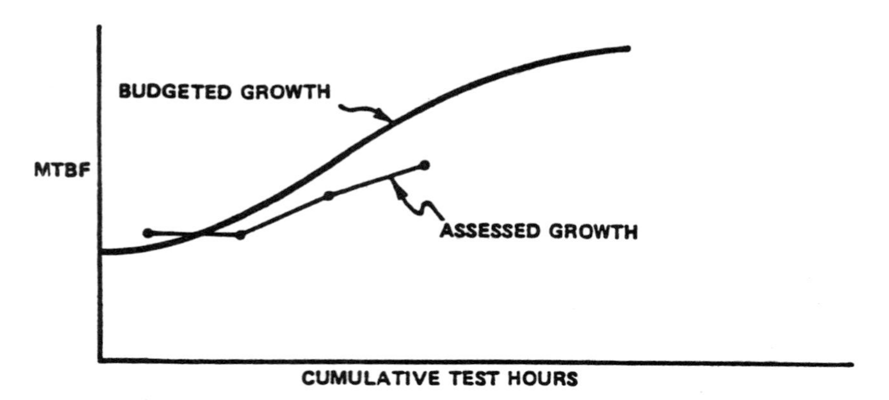

Figure 13.5 Reliability growth curve example. (From MIL-HDBK-338, *Electronic Reliability Design Handbook*, Naval Publications and Forms Center, Philadelphia.)

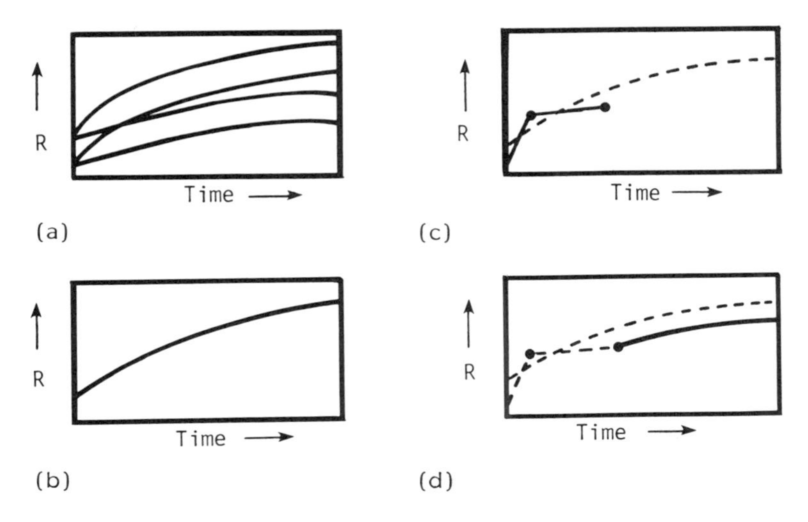

Figure 13.6 Four types of reliability growth models: (a) generic model; (b) budgeted growth; (c) assessed growth; (d) projective assessment. (From MIL-HDBK-338, *Electronic Reliability Design Handbook*, Naval Publications and Forms Center, Philadelphia.)

identified problem areas. They may, as a last resort, result in adjustment of the time frame, or relaxation of the original requirement.

Types of Models Utilized in Reliability Growth Management

In generating the reliability growth plan, the manager must predict the system's changes in reliability as the system matures, as well as track the system's progress. The manager must also project the system's status at future milestones. Reliability growth models are used to describe these changes in reliability. The majority of reliability growth models express an appropriate reliability parameter as a function of test time. These descriptions may be made either before or after the fact and are made for different purposes. The various types of growth models are illustrated in Figure 13.6.

1. Generic growth model: This model is used to depict the generalized growth pattern for a particular class of systems utilizing historical data. System characteristics that affect growth patterns include state of the art, system complexity, and the nature of the system (mechanical or electrical, and so on). Program characteristics affecting the growth patterns include external experience, analysis, levels and types of tests, failure correction, redesign effort, and resources available. The generic model may be a mathematical

model or a series of milestones depicting a typical development program for systems in the class. An example of a mathematical generic model may be: When organization X develops a system, reliability growth occurs in accordance with the Duane model. An example of a milestone-based generic model may be: When organization Y develops a system, 70% of the operational MTBF is achieved in 1 year, 100% in 3 years.

2. Budgeted growth model: This model defines the reliability expected at specific points in the life cycle. The budgeted curve has the same general shape as the generic curve, but passes through a specific set of points. Continuing the previous examples: organization Y has budgeted reliability growth during development in accordance with a specified model with a growth rate of 0.5; or organization Y has a budgeted MTBF of 700 hr at the end of 1 year, and an MTBF of 1000 hr at the end of 2 years.

3. Growth assessment model: To control technical activities a manager must have knowledge of the system's status on either a continuous or periodic basis. This knowledge is gained through assessment. Assessments can be made from test results in two different ways: the assessment may be based entirely on tests run on the current configuration, ignoring all previous information; or the assessment may be based on the statistically combined results of all tests up through the present, taking into consideration mathematically the growth that has occurred.

4. Projective assessment model: Considering where we are today, where do we expect to be at future points in time if we follow certain courses of action? A projective assessment extrapolates beyond the currently assessed value. It utilizes the generic model to establish the general shape and proposed program characteristics to determine the specific path.

Evaluating System Growth Potential

When the reliability requirement for a system has been defined, it is important to analyze, at least qualitatively, the growth potential of the system. This is necessary to give an indication of the resources required to attain the requirements.

Three factors affect the difficulty of achieving growth:

1. Reliability design effort prior to the growth effort
2. The specified reliability (MTBF) level
3. The relationship between the reliability level and the state of the art

The type of reliability design effort prior to a formal growth effort has a distinct effect on the difficulty of achieving fixes. A complete MIL-STD-785 reliability program effort prior to the growth effort may have only a small noticeable effect on the initial level of reliability. However, it affords good assurance that the reliability is capable of growth. The growth process is basically a refinement process. As such, it is very inefficient if major design changes are necessary. The rate of reliability growth is usually found to be:

1. Higher for analog hardware than for digital hardware
2. Higher in equipment of low maturity than in production hardware
3. Higher in equipment exposed to severe test conditions than in equipment undergoing bench tests
4. Higher in proportion to the hardware-oriented reliability improvement effort

The specific level of the reliability has an obvious effect on the growth process. The higher the reliability requirement, the more testing must be performed to uncover each remaining failure source.

Finally, if the reliability level approaches or goes beyond the current state of the art, fixes become very difficult since they often require minor advances in the state of the art. The baseline analysis and the analysis of technological advances can serve as an indicator of the current state of the art.

The Reliability Growth Budget

The reliability growth budget commits the manager to achieving goals, therefore it must be developed with care so as to depict accurately realistic, as well as realizable, reliability growth. While each budgeted curve must be tailored to specific program requirements, there are some considerations which apply to reliability growth budgets.

The first consideration is the general growth pattern displayed by previous similar systems. In attempting to determine the general growth pattern, or generic model, historical data must be analyzed. It is anticipated that the growth budget will be a reflection of the reliability growth pattern displayed by the historical data. A helpful exercise might be to compare the growth of different systems and programs with their respective complexities, development costs, amounts of testing program activities (FMEAs, design reviews, and so on), program design, and development time, as well as any other factors which caused the historical reliability growth to occur as it did. With this information, knowledge of the new system's requirements, program funds, and schedule, the manager is in a better position to assess his program in view of past system or program performance. Utilizing some judgement, the manager can now establish a budgeted reliability growth curve, and propose realistic tradeoffs between the schedule, resources, and requirements.

A number of methods of displaying the growth budget are in use. Reliability (or failure rate, MTBF, and so on) may be plotted as a function of test time or may be plotted against calendar time or program milestone. Generally, reliability growth tracking and its associated budget begin with the first test data for the system or component being tracked. However, in some cases it may be desirable to track system reliability that is calculated from a math model, using component test inputs.

The initial level of the budgeted reliability can be estimated in various ways. One way is to base this level on analyses of the histories of development programs of similar systems. For example, if previous, similar programs have achieved a reliability of 10% of the design predictions at the beginning of testing, the budgeted curve may be started at this point. In the case of evolutionary systems, the current reliability status may be used as a starting point.

Once a starting point has been determined, the budgeted curve may then be started at this point and extrapolated along the generic model. One of two things might happen. First, this extrapolation may meet or exceed the requirement in the allotted time frame. In this case we are ready to evaluate its cost. Second, it may fail to meet the minimum requirements in the allocated time. In this case the program will have to be reevaluated.

An alternative method to plotting the budgeted growth curve is to start with the requirement and its scheduled date and work backwards along the generic model to a starting point. However, this may cause the initial point on the budgeted curve to be unrealistically high thus requiring program reevaluation. In any case, a new starting point may have to be recalculated based on early test results.

In planning the reliability growth budget, the manager must remember that one of the purposes of a development program is to design-out failure causes. This necessarily involves time to detect and analyze failures as well as time to redesign and fabricate hardware. Early in the program system failures can be detected and corrected relatively easily. However, as test time and severity increase, the increased number and subtlety of failures can make correction of the failure causes a time-consuming activity. Unless properly planned and scheduled, a bottleneck may be created.

The question of when to start growing reliability affects the level at which reliability growth should be planned and controlled. Should it be at the system level, or should consideration be given to major subsystems? If so, what major subsystems should be considered?

The growth principles should be applied at whatever level is necessary to give the manager the information needed to manage the system affectively. If information is required at the subsystem level, the manager should not hesitate to do so. However, when using information generated at less than the system level, care must be taken to evaluate the information gained with respect to the interface problems that might occur at the system level. Also the use conditions must be evaluated to avoid a false sense of security, otherwise an expensive push may be required later on in the system's life cycle to get the system back on the right track.

Valuable information may be gained at the subsystem level if the manager is aware of these pitfalls. The timeliness of the information gained plus the generally lower cost of lower-level testing versus system-level testing can be invaluable.

Reliability Growth Assessment

Reliability growth assessment is a fundamental step in controlling the activities necessary for growth. In general, assessment of reliability should begin as soon as there is any information on the system's reliability status. A reliability prediction can be viewed as an initial assessment of the potential reliability of the system until such time as prediction data can be replaced with actual test data accumulated on the system. Growth assessments generally start at the same point that the budget does.

Assessments for reliability growth management represent an evaluation of the current system configuration, not of future configurations. Assessments may be based on the test results on the current configuration or on the statistically combined results of all tests up through the present. Special statistical techniques may also be employed to purge earlier test results to reflect improvements that have been achieved.

Assessments can give the system status, but they do not control the reliability effort. The current status, when compared to the budgeted growth curve, can indicate the need for more, less, or no change in the level of reliability growth effort. Discrepancies which exist between the budgeted growth and the assessed growth may be attributed to bottlenecks in the design-assess-redesign loop.

In some cases, it may be desirable to budget and assess reliability growth at less than system level. Such a situation might occur if the reliability apportioned to a particular subsystem, assembly, or even component is relatively high when compared to the past performance of similar items. This might also occur if it is desired to grow reliability through testing before the full system's hardware has been designed or fabricated. While testing at lower levels of assembly yields timely and credible information, perhaps at reduced costs, inferences about the system reliability will not contain concrete information regarding the effects on system reliability of interfaces between these assemblies.

13.4 FAILURE REPORTING, ANALYSIS, AND CORRECTIVE ACTION SYSTEMS (FRACAS)

Successful or satisfactory operation—the goal of all design efforts—yields little information on which to base improvements. Failures, on the other hand, contribute a wealth of data on "what to improve" or "what to design against" in subsequent efforts. The feedback of information obtained from the analysis of failures is one of the principal stepping stones of progress.

Early elimination of the causes of failure is a major contributor to reliability growth and attaining customer satisfaction after delivery. The sooner failure causes can be identified, the easier it is to implement effective corrective action. As the design documentation and preliminary hardware mature, corrective action can still be identified, but its implementation becomes more difficult and more costly with the passing of time. It is important, therefore, to employ FRACAS early in the development phase.

The purpose of FRACAS is to establish an efficient closed-loop failure reporting and corrective action system as illustrated in Figure 13.7. This includes documented procedures for the analysis of failures to determine their root causes, and then the establishment of effective corrective action to prevent future reoccurrence of this specific type of failure. Such a closed-loop system collects, analyzes, and records failures that occur on specified levels of assemblies prior to, and possibly after, acceptance of the hardware by the customer.

Essentially the FRACAS system must provide information on:

1. What failed?
2. How did it fail?
3. Why did it fail?
4. How can such failures be eliminated in the future?

FRACAS documentation should include detail procedures for: (1) initiating failure reports; (2) the analysis of failures; and (3) feedback of corrective action into the design, manufacturing, and test processes. Flow diagrams depicting failed hardware and data flow should also be included in the documentation. The analysis of failures should establish and categorize the root cause of failure.

The FRACAS system should include provisions to assure that effective corrective actions are taken on a timely basis by a follow-up audit that reviews all open failure reports, failure analyses, and corrective action suspense dates, and the reporting of delinquencies to management. The root cause for each failure shall be clearly stated. When utilized in conjunction with reliability growth testing, the method of establishing and recording operating time or cycles on equipments should also be clearly defined.

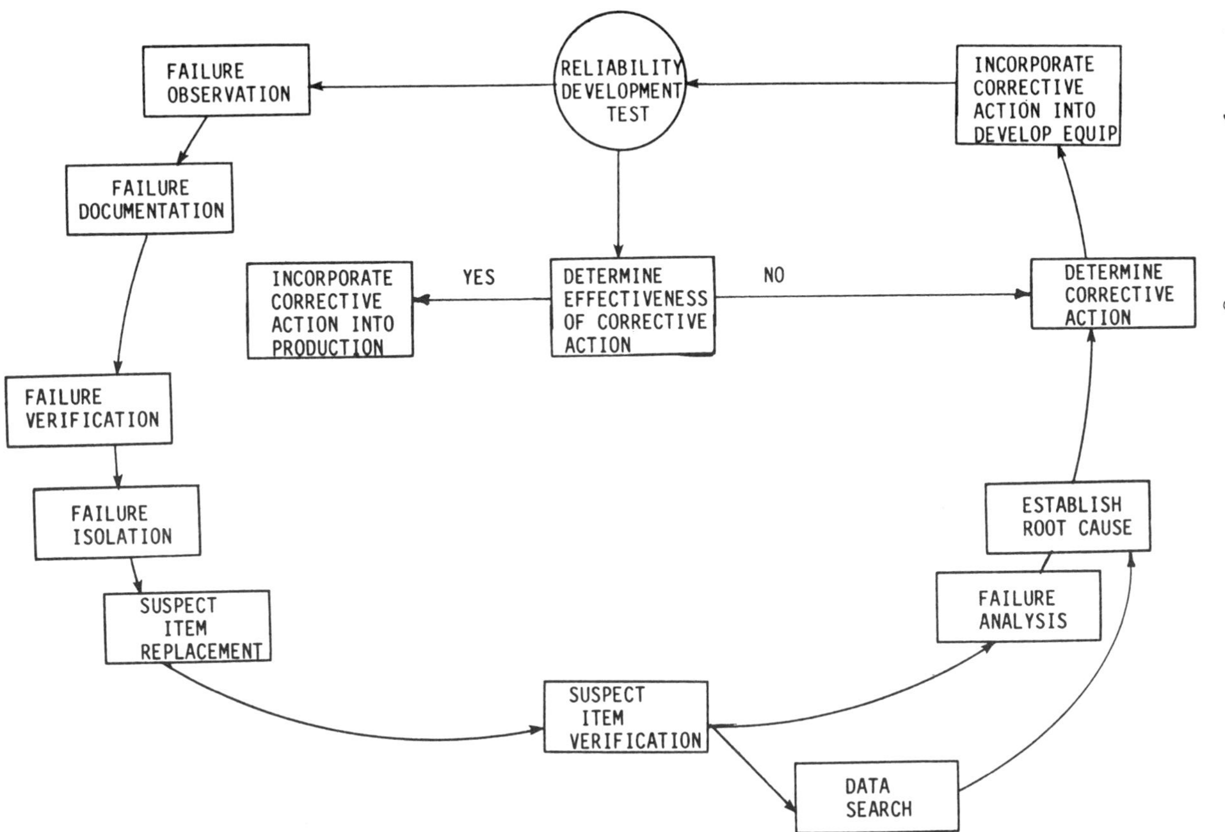

Figure 13.7 Closed-loop failure reporting and corrective action system.

Table 13.1 Steps in a Closed-Loop Failure Reporting and Corrective Action System

1. Observation of the failure
2. Complete documentation of the failure including all significant conditions which exist at the time of failure
3. Failure verification, that is, confirmation of the validity of the initial failure observation
4. Failure isolation, localization of the failure to the lowest replacement defective item in the system/equipment
5. Replacement of the suspect defective item and retest of the system/equipment to verify proper operation
6. Confirmation that the suspect item is defective
7. Failure analysis of the defective item
8. Data search to uncover other similar failure occurrences and to determine the previous history of the defective item and similar related items
9. Establishment of the root cause of the failure
10. Determination by an interdiscipline design team, of the necessary corrective action, especially any applicable redesign
11. Incorporation of the recommended corrective action in the reliability development system
12. Continuation of the reliability development test
13. Establishment of the effectiveness of the proposed corrective action
14. Incorporation of effective corrective action in the production equipments

A typical FRACAS report would consist of the steps shown in Table 13.1.

The effectiveness of FRACAS depends on accurate input data, that is, reports documenting failures/anomalies and good failure-cause isolation. Minimum documentation for the failure should include: the location of the failure, the date and time of failure, part numbers, serial numbers, model numbers, and all conditions surrounding the failure. The failure symptom and the name of the individual who observed the failure should also be documented to assist later in failure-cause determination.

These data and analyses may also be used to verify the correctness and consistency of the FRACAS. Frequently the failure causes can be determined through technical dialogue between design and reliability engineers. Occasionally, however, formal laboratory failure analyses are required to reveal failure mechanisms and provide the basis for effective corrective action. Laboratory failure analysis should be done if possible whenever the part failure mechanism is germane to the failure-cause determination.

An extremely helpful document to any organization wishing to establish a failure analysis laboratory with the capability of dissecting and analyzing defective microcircuits is MFAT-1 "Microcircuit Failure Analysis Techniques" published by the Reliability Analysis Center (see Chapter 15.4).

Another useful output product of the FRACAS is a failure summary report which groups information about failure of like items or similar functional failures. With this information the need for and extent of comtemplated corrective action and its impact can be formulated.

Key Factors in Effective FRACAS

There are several key factors that make the failure reporting and corrective action cycle effective. These are outlined as follows:

1. The discipline of the failure report writing itself must be maintained so that an accurate description of failure occurrence and proper identification of the failed items are assured.
2. The proper assignment of priority and the decision for failure analysis must be made with the aid of cognizant design engineers and system engineers.
3. The status of all failure analyses must be known. It is of prime importance that failure analyses be expedited as priority demands, and that corrective action be implemented as soon as possible.
4. The root cause of every failure must be understood. Without this understanding, no logically derived corrective actions can follow.
5. There must be a means of tabulating failure information for determining failure trends and the MTBFs of the system elements. There should also be a means for management visibility into the status of failure report dispositions and corrective actions.
6. The system must provide for a high-level technical managment approval concurring in the results of failure analysis, the soundness of corrective action, and the completion of formal actions in the corrective and recurrence-prevention loop.

Relationship Between FRACAS and FMECA

Although the respective FRACAS and FMECA (see Chapter 7) efforts are designed and capable of operating independently of each other, there is a synergistic effect when the two efforts are coupled together. An FMECA gives analytically derived identification of conceivable hardware failure modes of an item and the potential adverse effects of those failure modes on the system. The FMECA's primary purpose is to influence the system and item design to either eliminate or minimize the occurrences of hardware failure or the consequences of the failure. The FRACAS represents the real-world experience of actual failures and their consequences. An FMECA benefits the FRACAS by providing a source of comprehensive failure-effect and failure-severity information for assessment of actual hardware failures. Actual failure experience reported and analyzed in FRACAS provids a means of verifying the completeness and accuracy of the FMECA. There should be agreement between the real-world experience as reported and assessed in the FRACAS and the analytical-world experience as documented in the FMECA. Significant differences between the two worlds are cause for a reassessment of the item design and the differing failure criteria that separates the FRACAS and the FMECA.

Additional information regarding the FRACAS closed-loop system may be found in MIL-STD-2055 "Failure Reporting, Analysis and Corrective Action System."

14

Environmental Stress Screening

14.0 INTRODUCTION

Continuing advances in electronics' state of the art plus increasing emphasis on reliability and early development testing have increased the potential for providing a basically sound and inherently reliable design. As this potential has increased so has the complexity and the density of contemporary equipment packaging. This complexity amplifies the ever-present problems of detecting and correcting latent manufacturing defects. Equipment malfunction after many hours of field operation has often been attributable to something as simple as an improperly soldered wire.

Application of Screening and Burn-in During Production to Reduce Degradation

The use of screening and burn-in procedures have become an important element in an effective production reliability assessment and control program. The purpose of reliability screening and burn-in is to compress a system's early mortality period to reduce its failure rate to acceptable levels as quickly as possible. The rigor of the applied screens and the subsequent failure analysis and corrective action efforts determine the extent of degradation in reliability and hence the extent of possible improvement. A thorough knowledge of the hardware to be screened and the effectiveness and limitations of the various available screen tests is necessary to plan and implement an optimized production screening and burn-in program.

Screening is a 100% process performed on all items in their population. It utilizes stress during hardware operation to reveal inherent design defects as well as workmanship and process-induced defects. The application of stress must not weaken the product, yet it must reveal defects which ordinarily would not be apparent during normal quality inspection and testing. A number of different stresses and stress sequences could be applied to remove defects induced at the various levels of assembly fabrication. Therefore, each specific stress program must be optimized relative to the individual hardware by considering the technology, complexity, and end-item application characteristics as well as the production volume and cost constraints of the particular product being manufactured. Planning a stress screening program is an iterative process that involves tradeoff analysis to define the most cost-effective program.

Screening stresses can be applied at various manufacturing levels: (1) part, (2) intermediate (module or printed circuit board), and (3) system (unit or equipment). To detect and eliminate most of the intrinsic part defects, initial screening is conducted at the part level (this was addressed in detail in Chapter 2). Some part defects, however, are more easily detected by higher assembly-level screens; drift measurements and marginal propagation delay problems are examples. Assembly defects such as cold solder joints, missing solder joints, and connector contact defects can be detected only at the assembly or higher level.

Unfortunately, at higher assembly screening levels the item's tolerance for stress is lower; thus, the applied stresses must be lower. As a general rule, screens for known latent defects should be performed as early as possible in the assembly process as they are then the most cost-effective. A simple rule of thumb is that the cost of fixing a defect or failure rises by an order of magnitude with each higher assembly level at which it is found. For example, if it costs x dollars to find and replace a defective part in receiving inspection, it will cost 10x to find and replace that part if the defect is found at the printed circuit board level, 100x if found at the equipment level, and so on.

The idealized manufacturing process starts with screened parts procured to predetermined quality levels. Stress screens tests are then also applied as required at the different levels of assembly. Stress screening rejects should be analyzed for failures. The results of these analyses are then used to: (1) identify appropriate product design modifications and changes and manufacturing process modifications and (2) reduce where possible the overall stress burden. Stress screening data, including reject rates, failure modes, and time-to-failure data, can be incorporated into a dynamic real-time database to determine the effectiveness of the screening program. This experience database can be a major tool in the design of effective screening programs as new systems are developed and introduced into manufacturing.

Initial planning and tradeoff studies should take into account the screening effectiveness and the economic choices among part, intermediate, and final equipment- or system-level screens as well as the specific parameters that should be monitored.

Part-Level Screening

Part-level screening was addressed in Chapter 2; therefore, this will be only an overview of the topic. Part-level screening is relatively economical and can be incorporated into supplier specifications. It has the potential for maximum cost avoidance, particularly when applied to complex microcircuits and other high-technology devices where reliability is largely dependent on fabrication techniques and process control. Screen stress levels should be matched to part requirements, thus enabling the safe application of higher and more effective stress levels to remove known part defects. Part-level screens offer the advantage of simpler procedures and the ability to pass a great deal of the burden for corrective action back to the part vendors. Part-level screens, however, have no impact on the control of defects introduced during subsequent phases of manufacture and assembly.

There are three methods by which an equipment manufacturer may implement part-level screening. The screening may be performed by the part manufacturer, that is, (1) by buying the JAN part or by incorporating standard military screening requirements directly into the procurement specification; (2) by buying commercially processed parts and having an independent test laboratory perform the screening; or (3) by the equipment manufacturer doing the screening himself in his own facility. Advantages of the first two approaches are that no special training, burn-in facilities, or automatic test equipment are required. The third approach, performing the screen using in-house facilities, provides more flexibility, responsiveness to fabrication demands, and schedule control, however, it generally requires a major capital investment.

For microcircuits the governing military documents are MIL-M-38510 and MIL-STD-883. The former details the general requirements for manufacturer certification and qualification, while the latter describes the screens and the sequence in which they must be performed in order to achieve class S and class B reliability/quality levels. Method 5004 in MIL-STD-883 lists the various screens required for each class of microcircuits and describes in detail each of the test methods and test conditions.

Commercial-grade microcircuits are generally not subjected to any special screening procedures, inspections, or burn-in. They generally receive only a minimum form of visual and electrical parameter measurement at 25°C.

The governing military documents for discrete semiconductors are MIL-S-19500 and MIL-STD-750. The first details the general requirements for manufacturer certification and qualification and also lists the screens that must be applied to guarantee a given reliability/quality level. MIL-STD-750 then provides the details for each test method and test condition.

MIL-S-19500 documents the specific screens required for each reliability/quality class of discrete semiconductors JAN S (most reliable), JANTXV (intermediate), and JANTX (least reliable). There is also a fourth class, JAN, which, although subject to the certification and qualification requirements, is not subjected to 100% screening.

Screening and inspection tests for resistors, capacitors, and other passive components typically include high-temperature conditioning, visual and mechanical inspections, DC resistance measurement, low-temperature operation, temperature cycling, moisture resistance, short-time overload, mechanical shock, vibration, solderability, and life test at full-rated power.

Intermediate-Level Versus Unit- or System-Level Screening

The use of environmental stress screening at both the intermediate-level and the equipment-level has increased significantly in the past few years among both military and commercial and industrial electronic equipment manufacturers.

The concept of environmental stress screening originated with the Advisory Group on Reliability of Electronic Equipment (AGREE) testing program, specifically temperature cycling and vibration of avionics "black boxes." Current industry consensus is that temperature cycling is the most effective stress screen, with random vibration the second most effective. The vibration done in the past in AGREE testing utilized a single-frequency sine wave at a relatively low level (2.2 g's). Later research indicates that random vibration is considerably more effective than either swept-sine or single-frequency sine wave vibration.

At the intermediate-level (that is, module or assembly) thermal cycling can also be an effective screen for both part and workmanship defects. The rate of change of temperature, however, is an important parameter. High rates of change of temperature are generally more effective. Twenty to forty temperature cycles are generally recommended. Two opposing schools of thought exist on whether power should be applied during the thermal cycling. There is also a lack of general agreement on the effectiveness of vibration at the intermediate level.

At higher levels of assembly (that is, units, equipments) thermal cycling and random vibration are both effective screens. Fewer thermal cycles are necessary at these levels; 4 to 12 cycles are usually considered adequate. Power "on" is generally considered to be more effective, and a performance verification test (PVT) at each temperature extreme is recommended. Random vibration with power "on" together with continuous monitoring has been successfully used to detect intermittent failures. Low-level single-frequency vibration is not usually considered to be an effective screen.

14.1 INTERMEDIATE-LEVEL SCREENING

Intermediate-level (module, printed circuit board, assembly, and so on) screening is more expensive than part-level screening but it can remove defects introduced at the board level as well as those defects intrinsic to the parts. Because of the variety of part types incorporated into a board, somewhat lower stress levels must be applied. For example, the maximum screen temperature is limited to that of the part having the lowest maximum temperature rating of all the parts on the board. Special burn-in/temperature cycling facilities are generally required as well as special automatic test equipment (ATE). In fact, some amount of ATE is employed in virtually all large-scale screening programs. Automatic testing can perform rapid functional testing after screening or burn-in of complex boards (or other assemblies) and can help find marginal performance problems and defects arising from part interactions during operation. The extent of the facilities and equipment required is dependent on the screen conditions specified. The potential for cost avoidance is not as high with intermediate-level screens as with part-level screens. Also, the necessity to employ a lower stress level generally reduces the effectiveness of these screens to some extent.

Temperature Cycling

Temperature cycling is highly effective at the module or printed circuit board level. It reveals workmanship and process-induced defects as well as those intrinsic part defects which may have escaped detection at the part-level screen.

Specifically, temperature cycling will reveal assembly defects such as:

1. Printed circuit delamination
2. Part-to-board bond separation
3. Solder problems (fractures, cracks, opens, and so on
4. Part defects which escaped earlier part screens and inspection tests
5. Part tolerance drift

The anticipated types of latent part defects detected will depend on several factors, including:

1. The types of parts comprising the assembly (that is, microcircuits, discrete semiconductors, passive parts, low population parts, microwave parts, and so on)
2. The quality grade of the parts
3. The extent to which the parts were previously screened
4. The testability of the parts (for example, microprocessor and other LSI devices are difficult to test completely; therefore, precipitated defects may go undetected)

Figure 14.1 illustrates the environmental conditions and temperature profile of a typical temperature-cycling screen. The actual number of cycles employed is dependent upon board complexity and part density. For low complexity/density boards, three cycles are typical; for high complexity/density boards, ten cycles or more may be necessary. As shown in Figure 14.1 a functional test is usually recommended at ambient conditions after each cycle.

The number of failures should be recorded during each cycle. An analysis can then be conducted on the failed assemblies or boards to determine the underlying failure mechanisms as well as the possibility of earlier detection or the application of more stringent inspections and screens at the part level. If appropriate, the manufacturing process may be altered as well. Following analysis, repairs are made and the screen is continued. However, any failure occurring during the last cycle(s) usually requires repetition and completion of additional cycle(s) following repair. The number of cycles may be increased beyond those originally set if (1) the repair action is complex or difficult to inspect, (2) unscreened parts were used as replacements, or (3) it is likely that the repair action could induce new defects into the board.

The number of cycles initially applied represents a baseline for the temperature-cycled stress screen. Temperature-cycled screening, like any quality inspection test, should be dynamic where the number of cycles is adjusted, depending on the results of subsequent higher-level tests or field performance. The number of temperature cycles may be increased if, for example, a large number of failures are observed and reported in subsequent testing. Conversely, the number of cycles may be decreased if few failures are reported. The extent and nature of any changes, however, should be determined only through careful review and analysis of subsequent failures.

Intermediate-Level Thermal Screens

For an intermediate-level temperature-cycled thermal screen, five screening parameters must be specified: the maximum temperature,

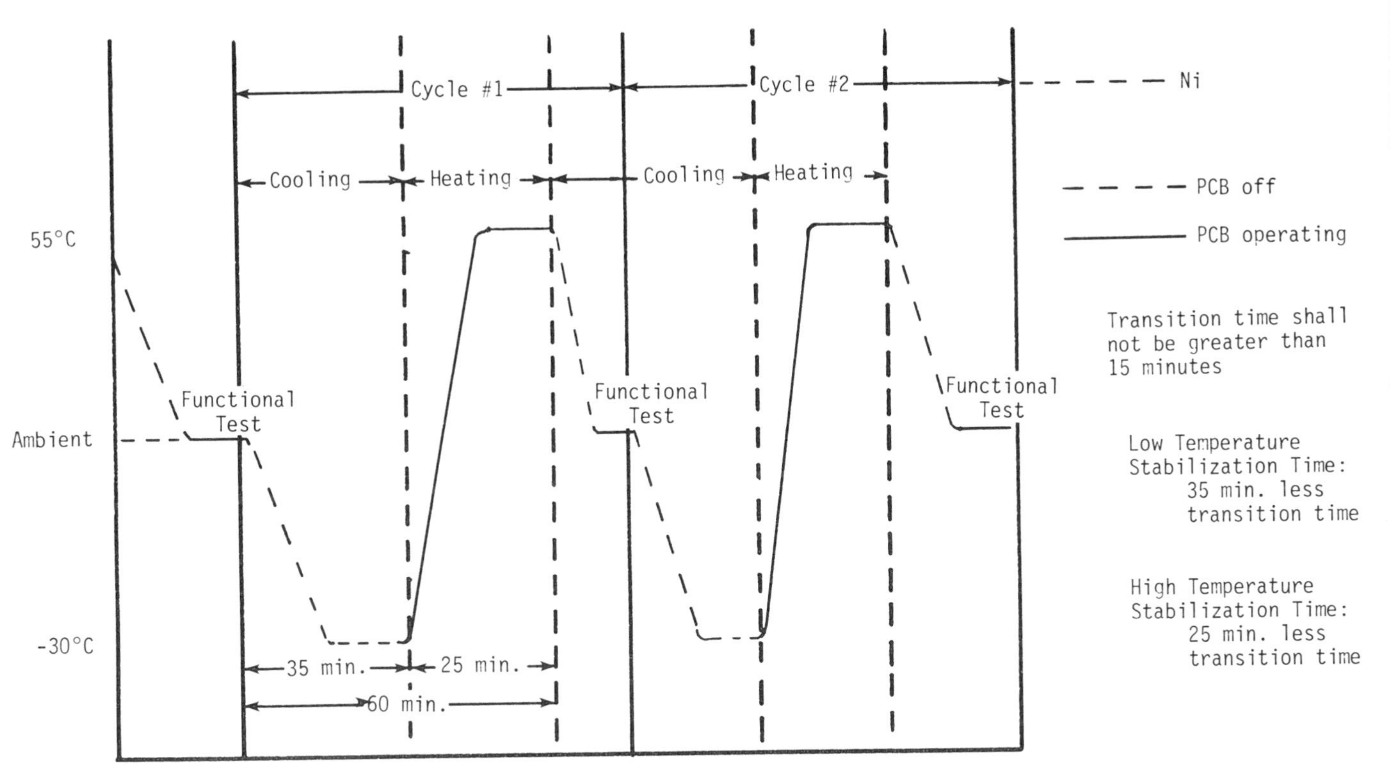

Figure 14.1 Typical temperature-cycling test. (From MIL-HDBK-338, *Electronic Reliability Design Handbook*, Naval Publications and Forms Center, Philadelphia.)

the minimum temperature, the rate of change of temperature, the length of dwell at temperature extremes, and the number of cycles.

1. Maximum temperature: The maximum temperature to which the assembly will be exposed should not exceed the lowest maximum rating of any part or material utilized in the assembly. (Nonoperating ratings for parts are normally higher than the operating ratings.)

2. Minimum temperature: The minimum temperature to which the assembly will be exposed should not exceed the highest minimum rating of any part or material utilized in the assembly.

Maximum and minimum temperatures must be carefully selected to assure that the maximum screening effectiveness is achieved. Exceeding time maximum ratings may degrade reliability by damaging nondefective parts or materials. If the operating temperature for a power-on screen cannot be readily determined analytically, a thermal survey of the item to be screened should be performed to determine the maximum and minimum screening temperatures.

3. Rate of change of temperature: Screening effectiveness increases with higher rates of change of temperature. The maximum rate of change is dependent on the thermal chamber characteristics and the thermal mass of the items to be screened.

4. Length of dwell at temperature extremes: During a temperature cycle it is usually necessary to maintain the chamber temperature constant once it has reached the maximum (or minimum) temperature. This is referred to as the "dwell" time. Dwell may be required to allow the item being screened to stabilize at the chamber temperature. The thermal lag of the item depends on its thermal mass. Most printed circuit boards have low thermal mass and thus require a relatively short dwell time.

5. Number of cycles: Twenty to forty thermal cycles are frequently recommended for screening at the assembly or module level.

Application of Power During Thermal Screening

The determination of whether to apply power to assemblies during screening and whether to perform a functional test during the screen requires consideration of the following factors:

1. Predominant type of defect present: If the predominant type of defect is expected to be a weak interconnection which is transformed to an open circuit by the screen (cold solder joint, weak wire bond, and so on), then the postscreen electrical test will detect the open circuit and a power-on screen is not required. If, on the other hand, the predominant type of defect is expected to be of an intermittent nature, then power-on screening with continuous performance monitoring is necessary.

2. Economic considerations: The fixtures and associated test equipments to house assemblies, apply power, provide stimuli, and

monitor assembly performance can be costly. Thus, the tradeoff between fixture and test-equipment cost potential benefits may prove difficult.

Intermediate-Level Vibration Screen

For intermediate-level vibration screen, the type of vibration (for example, random, swept-sine, or fixed-sine) must be selected, and also two additional parameters must be determined; the vibration level and the duration of the vibration.

Vibration Level

Random vibration at a level of 0.04–0.045 g^2/Hz is usually recommended, assuming that the assembly can withstand that level without damage. If the assembly's dynamic response characteristics to the vibration are not known, a vibration survey should be conducted first to establish properly the acceleration spectrum and level. If random vibration cannot be performed, swept-sine would be the

Table 14.1 Intermediate-Level Defects Precipitated by Thermal and Vibration Screens

Defect type detected	Thermal	Vibration
Defective part	x	x
Broken part	x	x
Improperly installed part	x	
Solder connection	x	x
PCB etch	x	x
Loose contact		x
Wire insulation	x	
Loose wire termination	x	x
Improper crimp	x	
Contamination	x	
Debris		x
Loose hardware		x
Mechanical flaw		x

second choice. Single-frequency vibration at the assembly level is generally considered to be ineffective.

Vibration Duration

Ten minutes of vibration in each of three axes is the usual suggested starting point. The need for multiaxis excitation may vary from one assembly to another. Therefore, it is desirable to determine fallout per axis during initial screens to allow later screen adjustments.

Some other factors to consider in determining the desirability of a printed circuit board vibration screen are the size and stiffness of the board. Larger boards will flex more and precipitate latent defects such as cracked metal runs, cold solder joints, and embedded conductive debris. Smaller boards, particularly if conformally coated, are usually quite stiff and thus do not benefit greatly from vibration screening.

Table 14.1 indicates the type of module or assembly defects which can be precipitated by thermal and vibration screens.

14.2 HIGHER LEVEL (UNIT, EQUIPMENT, AND SYSTEM) SCREENING

Equipment- or system-level screening is more expensive but it can remove defects introduced at all levels of fabrication. At this point in the manufacturing cycle the potential for cost avoidance is low and the permissible stress level may not adequately exercise certain specific parts. However, higher-level assembly screens are important, even if the lower-level screens have eliminated all defective parts and board defects, since the assembly of the remaining components and the boards into larger assemblies and into the final item can produce defects. Good parts can be damaged in final assembly, workmanship errors can be introduced, and product-level design defects may be present.

Equipment-level screens are primarily intended to precipitate workmanship defects and assembly-level errors. The types of equipment-level defects vary with the type of construction, but they typically include interconnection defects such as:

1. Printed circuit board connector problems (loose, bent, cracked, or contaminated contacts and cracked connectors)
2. Back-plane wiring (loose connections, bent pins, damaged wire insulation, debris in the wiring)
3. Input/output connectors (loose, cracked pins, damaged connectors, excessive, inadequate or lack of solder on wire terminations, inadequate wire stress relief)

4. Intraunit cabling (improperly assembled coax connectors, damaged insulation)

Equipment items may also contain wired assemblies integral to the item not previously screened, such as power control and BIT panels, modular low-voltage power supplies, and other purchased assemblies.

Latent defects associated with those assemblies should be considered in the selection of screens. Typical equipment-level screens include:

Thermal: both temperature cycling and fixed-temperature burn-in
Vibration: random, swept-sine, or fixed-sine wave

Thermal screens are more effective than vibration screens in precipitating latent part defects. Thermal cycling and vibration screens are both effective in precipitating latent workmanship defects, although one may be more effective than the other for a given type of defect. The equipment composition and a knowledge of prior screening will dictate the expected types of defects and aid in screen selection.

Equipment-Level Thermal Screens

For an equipment-level thermal screen the same process as described for the intermediate would be followed. Some essential differences, however, between intermediate-level and equipment-level screens are as follows:

1. Equipments have greater thermal mass; therefore, high rates of change of temperature may be more difficult to achieve. Extended dwell at the temperature extremes is usually required.
2. Power-on screening is usually more easily accomplished at the equipment-level. A functional test (PVT) at temperature extremes may be very effective in finding defects not detectable at room ambient temperature.
3. Power temperature cycles are required at the unit level; from 4 to 12 cycles is a commonly recommended range.

Equipment-Level Vibration Screens

If a vibration screen is selected, it is important that competent engineering personnel first evaluate the item to be vibrated to determine the appropriate vibration type, level of excitation, and whether or not a vibration survey should be performed. For large, massive items, low levels of vibration may be an effective screen.

If an item is subjected to an unpowered screen, subsequent testing may reveal part or workmanship defects requiring correction. However, if the item was not tested prior to the screen, it cannot be determined even with a detailed failure analysis if the defects found were precipitated by the screen or were present in the item before the screen. Stress screening has not yet advanced to the point where the quantity and type of latent defects can be accurately predicted. Therefore, some degree of experimentation is necessary to derive reasonable defect rates.

Testing prior to a screen establishes the baseline upon which postscreen testing results can be used to measure the screening effectiveness. Prescreen testing should be done immediately before the screen. This is to eliminate the possibility of introducing latent defects during processes such as cleaning, conformal coating, handling and storage which may follow the initial item testing.

Once the screening effectiveness has been established, the value of both prescreen and postscreen testing has diminished, and it may prove cost-effective to perform only postscreen testing. When major perturbations have occurred in the production cycle such as productionline changes, fabrication or assembly process changes, personnel changes, or changes in the stress-screening process, it may be advisable to reinstitute prescreen testing until the process has again stabilized.

In long-term production programs, the normal learning curve itself results in process improvements; thus, the quantity and distribution of latent defects will change accordingly. Workmanship and manufacturing process-related defects will predominate in early production, while component-related defects dominate in mature production. Stress screens have different degrees of effectiveness for different types of defects; therefore, screens that may have been effective during early production should be periodically reevaluated to assure their continued effectiveness.

14.3 STRESS SCREEN PLANNING

An effective reliability screening program requires careful planning starting during early development. Tradeoff studies should be performed and a complete test specification prepared. The effectiveness of the proposed screen could possibly be verified on prototype hardware.

A key step in specifying an effective screening program is the identification of the kinds of failures that can occur and the assembly level at which they may be uncovered. Appropriate screens are those which are most effective in accelerating the identified modes, whether intrinsic to the part or induced by the manufacturing process.

Table 14.2 Screening Effectiveness

Temperature cycling	Extremely effective at all levels of assembly; reveals part and PCB defects, solder problems, bond separations, tolerance drifts, thermal mismatches, and changes in electrical characteristics.
Power-cycled, high-temperature burn-in	Effective at all levels of assembly; will reveal time/stress-dependent part and process defects
Random vibration	Effective primarily at equipment level; reveals solder problems, part and PCB defects, connector contact problems, intermittents, loose hardware, and structural problems
High-temperature storage	Relatively inexpensive screen that can be applied at any level of assembly to reveal time/dormant stress- (nonelectrical) dependent defects
Thermal shock	Relatively simple screen that can be applied at the part or module level to reveal cracking, delamination, and electrical changes due to moisture or mechanical displacement
Fixed-frequency, sine-wave vibration	Applied at final assembly level to reveal loose hardware, connector contact problems, and intermittents

Table 14.2 lists some of the more common screens and gives an indication of their effectiveness.

It is difficult to standardize on a particular screening approach due to the varied nature of electronics equipments and their associated design, development, and production program elements. Thus, tailoring of the screening process to the unique elements of a given program is required. Screens such as temperature cycling and random vibration are the most effective. In fact, for electronics equipment, temperature cycling is generally 2 or 3 times more effective a screen than vibration. Random vibration is more effective than swept-sine, and swept-sine is more effective than fixed-sine. However, exposure levels, number of cycles, and screen durations differ widely among users.

Other less costly screens such as sinusoidal vibration, power-cycled burn-in at ambient temperature, and high-temperature soak are also used, but they are generally considered to be less effective than the former screens. Precise knowledge of the effectiveness of the various available screens is not currently available. Screens, therefore, should be selected based upon estimates of cost- and effectiveness, early development program data, equipment design, manufacturing, material, and process variables. This will at least narrow the consideration to the most cost-effective choices. The screening process should then be continuously monitored and the results analyzed so that changes in the process can be made as required to optimize the cost-effectiveness of the screening program.

Stress Screening Guidelines Matrix

The stress screening guidelines matrix (Table 14.3) is a summary developed by the Institute of Environmental Sciences from their *Environmental Stress Screening Guidelines* (1981). These guidelines are intended to act as a working document that can be used as quick reference for planning a stress screening program.

Ideally, the parameters of a stress screening program should be optimized for the specific equipment on which it is to be implemented, since each equipment has its own population of failure mechanisms peculiar to its parts, processes, packaging, worker skill levels, and so on. If preliminary studies for purposes of stress screening planning cannot be performed, these guidelines may prove helpful.

The matrix addresses the following stress screening parameters:

Thermal cycling: temperature range; temperature rate of change; number of cycles; operating versus nonoperating
Vibration: type of vibration (random, swept-sine, sine); vibration level in g's and the spectrum; duration; number of axes
Thermal cycling and vibration combined: applied sequentially; applied simultaneously
Assembly level of screening: module; unit; system

For each parameter, the matrix contains the following information:

1. Recommended application: A brief statement of the form, level, and so on, of the parameter found to provide optimum screening effectiveness.
2. Expected reduction in failure rate: A statement of the expected reduction in failure rates achievable in house or in the field as a result of performing the screen as recommended.
3. Tradeoff considerations: Identification and a brief discussion of implementation and cost tradeoffs related to optimizing the screening factor as recommended.

Table 14.3 Stress Screening Guidelines Matrix

Stress environment	Recommended application	Expected failure-rate reduction	Tradeoffs
Thermal cycling, module level			
Temp. range	Max: −55 to +125°C (180°F) Nom: −40 to +95°C (135°F) Min: −40 to +75°C (115°F)	In house: 0 to 50% Field: 20 to 75%	In-house failure rates may in some cases be increased at next assembly level; hence, equipment behavior under proposed stress-screening environment should be evaluated prior to implementation.
Temp. rate	Max: 20°C/min Nom: 15°C/min Min: 5°C/min		Temperature rates of change are as measured by thermocouple on components mounted on modules.
No. cycles	Max: 40 Nom: 30 Min: 20		Power-on screening may be continued into early production until latent design problems are exposed and production processes and test procedures are proven.
Power	On (development phase) Off (production phase)		Power-off screening is considerably cheaper and is effective on mature production hardware.
Thermal cycling, unit and system level			
Temp. range	Max: −55 to +125°C (180°F) Nom: −40 to +95°C (135°F)	In house: 0 to 75% Field: 20 to 98%	In-house failure rate may in some cases be increased at next assembly

Table 14.3 (Continued)

Stress environment	Recommended application	Expected failure-rate reduction	Tradeoffs
Temp. range (continued)	Min: −40 to +75°C (115°F)		level; hence, equipment behavior under proposed stress-screening environment should be evaluated prior to implementation.
Temp. rate	Max: 20°C/min Nom: 15°C/min Min: 5°C/min		Higher temperature rates may require open-unit exposure with higher air flow rate to overcome slower temperature response of higher mass.
No. cycles	Max: 12 Nom: 10 Min: 8		Functional testing at high and low temperature increases failure detectability.
Power	On		
Vibration, module level			
	Not recommended for noncomplex modules		Marginal payoff for noncomplex modules whose confirguations are not susceptible to vibration environmental screening.
	For complex modules, use recommendations for unit and system level	(See vibration, unit and system level)	For complex modules, refer to unit- and system-level tradeoffs.

Vibration, unit and system level			
Vibration type	Random (preferred) Swept-sine (acceptable)	In house: 0 to 25% Field: 10 to 30%	Techniques for simulating random vibration may be considered, such as two excitors to produce diagonal force vector excitation, or use of pneumatic vibration methods to provide excitation in three axes.
Vibration	Spectrum and level customized for specific equipment; 0.045 g^2/Hz recommended initial starting level with scaling up and down depending on structural response of test specimen; frequency range approximately 100 to 1000 Hz		Generalized envelope provides guideline boundaries for acceleration spectra (see Figure 14.2); for large mass, frequencies below 500 Hz disclose large number of defects; for stiff hardware with low resonant frequency modes above 500 Hz, upper frequency limit may approach 1000 Hz. Hardware responses must be large enough for screening to be effective while not exceeding hardware capability; initial response survey required. For some equipment, higher levels of random vibration (e.g., 6 g's RMS) may introduce degradation.
Vibration level and spectrum (swept-sine)	Spectrum and level customized for specific equipment	In house: 0 to 15% Field: 10 to 20%	See Figure 14.3 for recommended spectra for swept-sine vibration.

Table 14.3 (Continued)

Stress environment	Recommended application	Expected failure-rate reduction	Tradeoffs
Vibration duration and no. of axes	10 min/axis, 3 axes		If a particular preference of the equipment for failure modes in one or two axes can be defined, 3 axes may not be required; vibration survey results may be useful in such identification.
Thermal cycling and vibration combined			
Applied independently or simultaneously	Use optimized parameters presented above for thermal cycling and vibration	In house: 0 to 75% Field: 20 to 90%	Independent application of thermal cycling and vibration will result in effective screening; order of application not found significant insofar as screening effectiveness; screening time may be reduced with simultaneous application; some failure mechanism types may be more sensitive to simultaneous application of the two environments.

Applied level for screening			
Module		In house: 0 to 50% Field: 20 to 75%	Tradeoff factors include level at which failure mechanisms are detectable; percentage of defects detectable at a specific level; feasibility of implementing screening at a specific level; achievable failure-rate reduction versus reliability requirements; comparative cost savings.
Unit	See tradeoffs	In house: 0 to 75% Field: 20 to 90%	
System		In house: 0 to 75% Field: 20 to 90%	

Source: *Environmental Stress Screening Guidelines* (1981), The Institute of Environmental Sciences, reprinted with permission.

Screen Plan Summary

The guidelines in Table 14.3 are applicable to both development and production phases. The recommended approach is to:

1. Implement a screening program during development based on the guidelines
2. Refine the screening program parameters during development for use during production
3. Monitor the screening program effectiveness during production making adjustments as needed to tighten it where more effective screening is needed, or to reduce it when warranted by maturing of system reliability

After evaluating the various options, applying the previously discussed tools and guidelines, and establishing the type and level of screens to be performed, a complete screen test specification should be prepared. This specification should encompass the following:

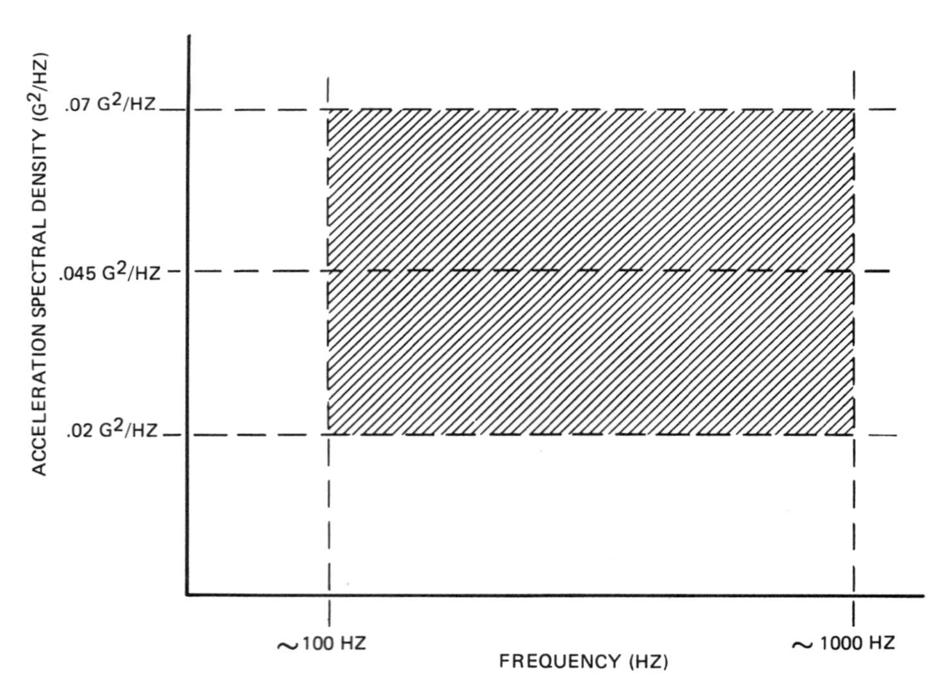

Figure 14.2 Generalized random vibration envelope. (Adapted from *Environmental Stress Screening Guidelines*, copyright 1981 by The Institute of Environmental Sciences, Mount Prospect, Illinois.)

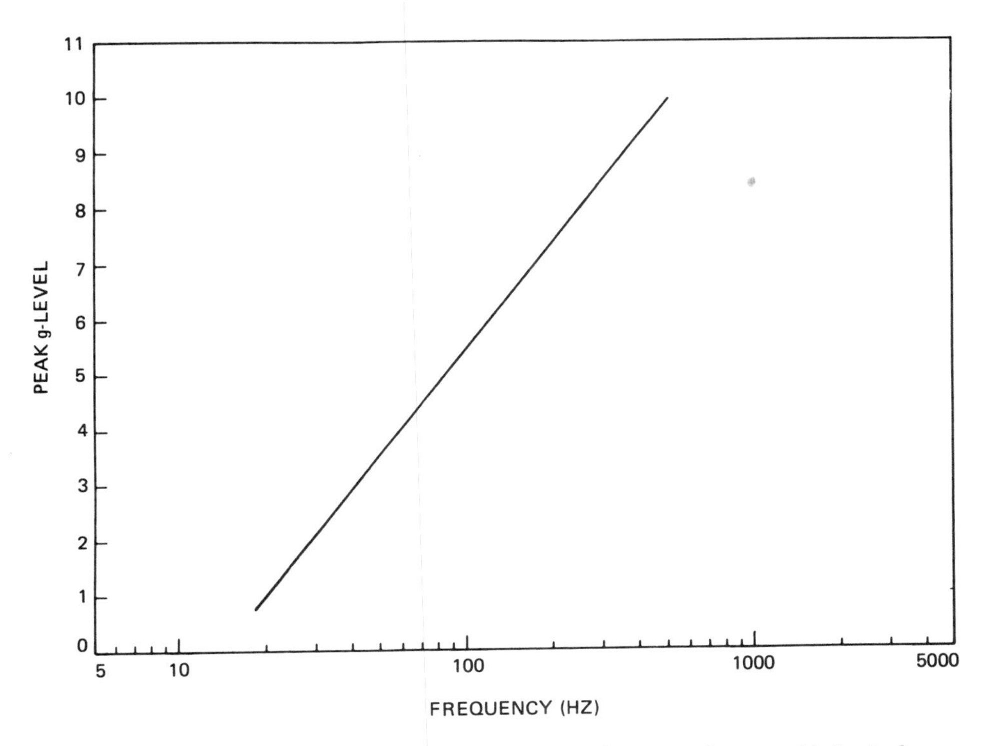

Figure 14.3 Generalized swept-sine vibration envelope. (Adapted from *Environmental Stress Screening Guidelines*, copyright 1981 by The Institute of Environmental Sciences, Mount Prospect, Illinois.)

1. The screen sequence and application levels
2. Screen conditions including duration, number of cycles, failure-free criteria, cumulative operating time, and critical electrical parameters
3. The anticipated reject or fallout rates
4. Test facilities required
5. Any special automatic test equipment (ATE) required
6. Data recording requirements and methods
7. Failure reporting analysis and corrective action procedures
8. Manpower and training requirements

In addition, if possible, provisions should be included for studies or experiments during the development phase so that the production stress-screening plan can be based on the established behavior of the specific hardware. The failure-free operation criteria should be made part of acceptance rather than stress-screening criteria. Opportunity should also be provided to perform preliminary stress-screening studies to determine costs related to the requirement. Finally, it is important to maintain visibility into the effectiveness of the overall screening program. This is accomplished via the failure reporting, analysis, and corrective action system which discussed in Chapter 13. Adjustments to the screening plan and to each screen should be made as necessary to minimize costs and maximize screening effectiveness.

15

Reliability Program Management

15.0 INTRODUCTION

This final chapter is in essence a summary or review of the material presented in earlier chapters in the text. It reviews each of the design reliability tools presented and wraps them up together into a unified, planned package for more effective usage. It closes with a brief summary of where additional reliability data can be obtained.

Reliability As a Management Discipline

In the final analysis, reliability is a management discipline. Without effective management support no reliability program can be fully effective. First, management must authorize adequate funding to allow sufficient staffing to implement and utilize the various reliability tools available to assure a given level of product reliability. Without adequate funding the necessary analysis will not be performed, the necessary tradeoffs will not be made, and optimum equipment reliability will not be achieved.

All reliability programs do not demand the same level of attention. Programs do differ, possibly quite drastically, and the reliability control elements must each be tailored to meet the specific demands of each unique program.

Once reliability has been quantitatively specified, the selection of those tasks that can materially aid in attaining the program reliability requirements is a major problem. The tasks must be judiciously selected to reflect program constraints and then be tailored to meet the specific needs of the program.

Reliability program management is well documented in MIL-STD-785, "Reliability Program for Systems and Equipment, Development

Table 15.1 MIL-STD-785 Reliability Program Tasks

Task number	Task	Text reference
	Program surveillance and control tasks	
101	Reliability program plan	Chap. 15.1
102	Monitor/control of subcontractors and suppliers	Chap. 15.1
103	Program reviews	Chap. 15.3
104	Failure reporting, analysis, and corrective action system (FRACAS)	Chap. 13.4
105	Failure review board (FRB)	Chap. 15.1
	Design and evaluation tasks	
201	Reliability modeling	Chap. 9.2
202	Reliability allocations	Chap. 9.1
203	Reliability predictions	Chap. 9.3
204	Failure modes, effects and criticality analysis (FMECA)	Chap. 7
205	Sneak circuit analysis (SCA)	Chap. 5.3
206	Electronic parts/circuits tolerance analysis	Chap. 5.1
207	Parts program	Chap. 2
208	Reliability critical items	Chap. 2
209	Effects of functional testing, storage, packaging, transportation, and maintenance	Chap. 11.4
	Development and production test tasks	
301	Environmental stress screening (ESS)	Chap. 14
302	Reliability development/growth test (RDGT) program	Chap. 13
303	Reliability qualification test (RQT) program	Chap. 12
304	Production reliability acceptance test (PRAT) program	Chap. 12.2

Source: *Electronic Reliability Design Handbook* (MIL-HDBK-338), Naval Publications and Forms Center, Philadelphia.

and Production". This document provides both general requirements and specific tasks for reliability programs. It also provides guidelines for the preparation and implementation of a reliability program plan. This military standard may be a very helpful guide to producers of both industrial and commercial systems and equipments as well as to the producers of military and aerospace systems and equipments.

There are in essence three different types of tasks addressed by the standard: reliability accounting tasks, reliability engineering tasks, and management tasks.

1. Reliability accounting tasks focus on providing the information essential to the acquisition, operation, and support management of the system/equipment.
2. Reliability engineering tasks focus on the prevention, detection, and correction of reliability design deficiencies, weak parts, and workmanship defects. An effective reliability program stresses early investment in reliability engineering tasks to avoid subsequent additional costs and schedule delays.
3. Management tasks are those tasks that relate more to the management responsibilities related to the program and less to the technical details.

A listing by number of the specific reliability tasks defined in MIL-STD-785 are shown in Table 15.1 together with their respective references in this text. Each of the tasks is then explained in the following section.

15.1 RELIABILITY PROGRAM TASKS

The following is a brief description of each of the reliability program tasks delineated in MIL-STD-785 and listed in Table 15.1

Program Surveillance and Control Tasks

Task 101: Reliability Program Plan

The reliability program plan is based upon an analysis of the specified reliability requirements. It is developed during the program conceptual design phase. The reliability program plan is a basic design tool to:

1. Assist in managing an effective reliability program
2. Evaluate the understanding of and execution of the various reliability tasks

3. Evaluate the planning to insure that documented procedures for implementing and controlling reliability tasks are adequate
4. Evaluate the adequacy of the organization to assure that appropriate attention will be focused on reliability activities, problems, or both

Task 102: Monitor/Control of Subcontractors and Suppliers

Continual visibility of subcontractors' activities is essential so that timely and appropriate management action can be taken as the need arises. It is prudent to include contractual provisions which permit the procuring activity to participate in appropriate formal prime/subcontractor meetings. Information gained at these meetings can provide a basis for follow-up actions necessary to maintain adequate visibility of subcontractors' progress including technical, cost, and schedule considerations.

Task 103: Program Reviews

Program reviews and design reviews are important management and technical tools used to insure adequate staffing and funding. Typical reviews are held to:

1. Evaluate the program progress including both technical adequacy and the reliability of a selected design and test approach (preliminary design review)
2. Determine the acceptability of the detail design approach including reliability before commitment to production (critical design review)
3. Periodically review progress of the reliability program, addressing the progress of each specified reliability task

Task 104: Failure Reporting, Analyses, and Corrective Action Systems (FRACAS)

Early elimination of failure causes is a major contributor to reliability growth. The sooner failure causes can be identified the easier it is to implement effective corrective action. Therefore, a closed-loop FRACAS should be employed early in the development phase, particularly for complex systems or equipments.

The FRACAS must assure that the disposition of failed hardware is properly controlled to preclude premature disposal. This will help ensure that the actual failed parts are subjected to the required analyses.

Task 105: Failure Review Board (FRB)

The acquisition of expensive, complex, or critical equipment or systems may require formalized FRACAS proceedings to the extent of having them controlled by a failure review board (FRB). The FRB normally consists of representatives of the procuring agency and the contractor's engineering, quality assurance, and manufacturing personnel. The task is designated to insure that FRACAS is properly implemented by providing additional assurance of tightly controlled reporting, analyses, and corrective actions taken on identified failures.

However, care should be taken with this task to prevent the duplication of quality assurance tasks which may have already been required under MIL-Q-9858.

Design and Evaluation Tasks

Task 201: Reliability Modeling

A reliability model of the system, subsystem, or equipment is required for making numerical apportionments and estimates. Reliability models are also required for evaluating the complex equipment arrangements typical of modern systems. The model should be developed as early as program definition permits, even if usable numerical input data are not yet available. Early modeling can reveal conditions where management action may be required. The model should be continually expanded to the detail level for which planning, mission, and system definition are firm.

The model is used together with duty cycle and mission duration information to develop a mathematical equation utilizing the appropriate failure rate and probability of success data to provide apportionments, estimates, and assessments of mission reliability.

Task 202: Reliability Allocations

The purpose of the reliability allocation is to convert the system reliability requirement to specific reliability requirements for each of the black boxes and lower-level items. This is one of the earliest reliability tasks to be performed and thus will probably require later updating or reallocation. Reallocation of the requirements will be performed as more detailed information regarding the design becomes known.

Task 203: Reliability Prediction

The initial prediction should be performed early in the acquisition phase to determine the feasibility of the reliability requirement. The task is then reiterated during the development and production phases to determine reliability attainability. Predictions are important in providing engineers and management with quantitative reliability information for day-to-day activities.

Predictions should be made as early as possible and updated as changes occur. Early predictions based on the parts count method are inherently unrefined; however, they do provide feedback to designers and managers on the feasibility of meeting the reliability requirements. As the system design progresses to the hardware stage, predictions mature as actual test data becomes available and is integrated into the calculations. Reliability predictions also provide essential inputs to other related activities, that is, to maintainability, safety, quality engineering, logistics, and test planning. They establish a baseline for comparing progress and performance. Predictions can also be used to detect overstressed parts and pinpoint critical areas for redesign or application of redundancy.

Task 204: Failure Modes, Effects and Criticality Analysis (FMECA)

An FMECA allows potential design weaknesses to be identified and appropriately analyzed and evaluated using engineering schematics and mission considerations. It provides a systematic identification of likely modes of failure, the possible effects of each failure, and the criticality of each failure on safety, system readiness, mission success, demand for maintenance or logistic support, or other factors.

The initial FMECA can be performed in the conceptual phase. Since only limited design definition may be available, only the more obvious failure modes will be identified. As more design definitions are developed in the validation and development phases, the analyses can be expanded to successively more detailed levels and ultimately, if required, to the part level.

FMECA results may suggest areas where the judicious use of redundancy can significantly improve mission reliability without unacceptable impact on basic reliability and where other analyses, for example, electronic parts analyses, should be made. Finally, FMECA results should be used to confirm the validity of the model used in computing reliability estimates and subsystems or functional equipment groupings, particularly where some form of redundancy is included.

Task 205: Sneak Circuit Analysis (SCA)

The purpose of SCA is to identify latent paths that may cause occurrence of unwanted functions or inhibit desired functions. It assumes that all components are functioning properly. SCA is expensive, and it is usually performed late in the design cycle after design documentation is complete. This makes subsequent changes difficult and costly to implement. Therefore, SCA should be considered only for items and functions critical to safety or mission success or where other techniques are not effective.

Task 206: Electronic Parts/Circuit Tolerance Analysis

This analysis examines the effects of parts or circuits' electrical tolerances and parameters over the range of specified operating temperatures. The analysis considers expected component value variations due to manufacturing tolerance variations and also their drift with time and temperature. In making this analysis, equivalent circuits and mode-matrix analysis techniques are used to prove that the circuit or equipment will meet specification requirements under all required conditions. Since this analysis is also expensive, its application may be limited to critical circuits only.

Task 207: Parts Program

Parts and components are the building blocks from which the system is ultimately constructed. System optimization, therefore, can be significantly enhanced by applying particular attention and resources to parts selection, control, and application. This must start early in the validation phase and continue throughout the entire life of the system.

A comprehensive parts program consists of the following elements:

A parts control program (in accordance with MIL-STD-965)
Parts standardization
Documented parts application and derating guidelines
Part testing, qualification, and screening
Participation in GIDEP in accordance with MIL-STD-1556

The objective of the parts program is to control the selection and use of both standard and nonstandard parts. An effective parts program requires knowledgeable parts engineers from both the procuring activity and the contractor.

Task 208: Reliability-Critical Items

Reliability-critical items are those whose failure can significantly effect safety, mission success, or total maintenance or logistics support costs. Reliability-critical items are identified during the part selection and application process. These items are prime candidates for detailed analyses, growth testing, reliability qualification testing, reliability stress analyses, and similar techniques to reduce the reliability risk.

Task 209: Effects of Functional Testing, Storage, Handling Packaging, Transportation, and Maintenance

Procedures must be established, maintained, and implemented to determine by test and analysis or estimation the effects of storage, handling, packaging, transportation, maintenance, and repeated exposure to functional testing on the design and reliability of the hardware. The results of this effort are used to support long-term failure-rate predictions, design tradeoffs, definition of allowable test exposures, retest after storage decisions, packaging, handling, or storage requirements, and refurbishment plans. They provide some assurance that these items can successfully tolerate forseeable operational and storage influences.

Development and Production Test Tasks

Task 301: Environmental Stress Screening (ESS)

Environmental stress screening (ESS) is a test or a series of tests specifically designed to disclose weak parts and workmanship defects requiring correction. It may be applied to parts, components, subassemblies, assemblies, or equipment (as appropriate and cost-effective). It is intended to remove defects which would otherwise cause failure during later testing or field service. ESS has significant potential return on investment during both development and production.

Task 302: Reliability Development/Growth Test (RDGT) Program

The reliability development/growth test (RDGT) is a planned prequalification test, analyze, and fix (TAAF) process in which equipments are tested under actual, simulated, or accelerated environments to disclose design deficiencies and defects. The testing is intended to provide a basis for early incorporation of corrective actions and for verification of their effectiveness, thus promoting reliability growth.

The RDGT is intended to correct failures that reduce operational effectiveness and failures that increase maintenance and logistics support costs. The RDGT should be conducted using the first prototype items available.

Task 303: Reliability Qualification Test (RQT) Program

The reliability qualification test (RQT) is intended to provide to the customer reasonable assurance that the design meets minimum acceptable reliability requirements before items are committed to production. The RQT must be operationally realistic and must provide an estimate of demonstrated reliability. The statistical test plan must adequately define successful and unsuccessful operation and define acceptance criteria which limit the probability that the true reliability of the item is less than the minimum acceptable reliability requirement. The RQT is a preproduction test, and it must be completed in time to provide management information for the production decision.

Task 304: Production Reliability Acceptance Test (PRAT) Program

The PRAT is a reliability sample testing of production hardware as delivered. Its purpose is to assure that the hardware has not been degraded as the result of changes in tooling, processes, work flow, design, or parts quality.

The PRAT is intended to simulate in-service evaluation of the delivered item or production lot. It must be operationally realistic and may be required to provide estimates of demonstrated reliability.

15.2 TASK TAILORING

It is not sufficient merely to list MIL-STD-785 as a reference document "which forms part of (the system specification) to the extent specified herein." Specific detailed direction as to which paragraphs and sections are applicable must be stated. Furthermore, MIL-STD-785 requires reliability programs to be tailored to meet specific program needs and constraints, including life-cycle cost objectives. It states that: "Tasks described in this standard are to be selectively applied to DOD contract-definitized procurements, requests for proposals, statements of work, and government in-house developments requiring reliability programs for the development, production, and initial deployment of systems and equipment."

The standard consists of basic application requirements, specific tailorable reliability program tasks, and an appendix which includes an application matrix and guidance and rationale for task selection. The document is intentionally structured to discourage indiscriminate blanket applications. Tailoring is forced by requiring that specific tasks be selected and that certain essential information relative to implementation of the task be provided by the procuring activity.

Increased emphasis is thus placed on reliability engineering tasks and tests. The thrust is toward prevention, detection, and correction of design deficiencies, weak parts, and workmanship defects. Emphasis on reliability accounting is included to serve the needs of acquisition, operation, and support management. For example, cost and schedule investment in reliability demonstration tests must be clearly identified and carefully controlled.

Both mission reliability and basic reliability must be addressed. Other areas of concern are operational readiness, demand for maintenance, and demand for logistic support. Separate requirements may be established for each reliability parameter that applies to a system, and then each of these is translated into basic reliability requirements for the subsystems, equipments, components, and parts.

Each development program is different, and each reliability program should be tailored to its specific needs. This tailoring by

Table 15.2 MIL-STD-785 Reliability Task Application Matrix

Task number	Task	Task type	Program phase			
			Concept	Validation	FSED	Production
101	Reliability program plan	MGT	S	S	S	S
102	Monitor/control of subcontractors and suppliers	MGT	S	S	S	S
103	Program reviews	MGT	S	S(2)	G(2)	G(2)
104	Failure reporting, analysis, and corrective action system (FRACAS)	ENG	NA	S	S	S
105	Failure review board (FRB)	MGT	NA	S(2)	S	S
201	Reliability modeling	ENG	S	S(2)	S(2)	S(2)
202	Reliability allocations	ACC	S	G	G	G
203	Reliability predictions	ACC	S	G	G	G
204	Failure modes, effects and criticality analysis (FMECA)	ENG	S	G(1,2)	G(1,2)	G(1,2)
205	Sneak circuit analysis (SCA)	ENG	NA	NA	G(1)	GC(1)

206	Electronic parts/circuits tolerance analysis	ENG	NA	NA	G	G
207	Parts program	ENG	S	S(2,3)	G(2)	G(2)
208	Reliability critical items	MGT	S(1)	S(1)	G	G
209	Effects of functional testing, storage, handling, packaging, transportation, and maintenance	ENG	NA	S(1)	G	GC
301	Environmental stress screening (ESS)	ENG	NA	S	G	G
302	Reliability development/growth test (RDGT) program	ENG	NA	S(2)	G(2)	NA
303	Reliability qualification test (RQT) program	ACC	NA	S(2)	G(2)	G(2)
304	Production reliability acceptance test (PRAT)	ACC	NA	NA	S	G(2,3)

Notes: Task type: ACC, reliability accounting; ENG, reliability engineering; MGT, management; FSED, pull-scale engineering development. Program phase: S, selectively applied; G, generally applied; NA, not applicable; GC, generally applied to design changes only; 1, requires considerable interpretation of intent to be cost effective; 2, MIL-STD-785 is not the primary implementation requirement, other documentation is required to define the requirements; 3, implement only to the extent necessary for this phase.

Source: *Reliability Program for Systems and Equipment Development and Production* (MIL-STD-785), Naval Publications and Forms Center, Philadelphia.

reliability specialists working with the program manager is intended to select those MIL-STD-785 tasks and requirements most suitable to the specific acquisition. These tasks and requirements are modified where necessary, to assure that each tailored task or requirement involved states only the minimum needs of the program. Table 15.2 from Appendix A of MIL-STD-785 provides tailoring guidance based upon the phase of the program. However, tailoring is not a license to specify a zero reliability program.

15.3 DESIGN REVIEW

Design review is a multidisciplined synergistic tool to assure that each design has been adequately studied to identify possible problems. It is an essential element in the reliability design process for both military and industrial or commercial equipments or systems. The intent of design review is to provide assurance that the design is capable of meeting of the specified requirements. Design reviews are critical audits of all pertinent aspects of the design conducted at critical milestones in the program. Perhaps the most significant aspect of design review is its emphasis on technical judgements rather than quantitative assessments of progress. Although the scope of the design review is normally defined in the reliability program plan it is not purely a reliability function.

Formal review of equipment design concepts and design documentation is an essential activity in any development program. Standardized procedures should be established to conduct a review of all drawings, specifications, and other design information. The review is accomplished by representatives from the various technical disciplines such as equipment engineering, reliability engineering, and manufacturing engineering. Responsible members of each reviewing department meet, generally with the customer, to consider all design documents, resolve any problem areas uncovered, and signify their acceptance of the design documentation by approving the documents for their respective departments. Major elements of design review include:

1. Analysis of the specification requirements and the operating environment
2. Formal review of all applicable engineering documentation
3. Informal design reviews conducted throughout the program as necessary

Prior to the formal design review, the requirements defined in the equipment specification and any other applicable specifications are reviewed. The expected environmental extremes are also studied

Table 15.3 Objectives of a Synergistic Design Review

1. Detect any condition that could degrade equipment reliability
2. Provide assurance of equipment conformance to all applicable specification requirements (including reliability)
3. Assure the use of preferred/standard parts as far as practical
4. Assure the use of preferred circuits as far as possible
5. Evaluate the electrical, mechanical, and thermal aspects of the design
6. Provide stress analysis to assure adequate part derating
7. Assure accessibility of all parts that are subject to adjustment or frequent servicing
8. Assure interchangeability of similar subsystems, circuits, modules, and subassemblies
9. Assure that adequate attention is given to human factors in all aspects of the design
10. Assure that the quality control effort will be effective

Source: *Engineering Design Handbook: Design for Reliability*, AMCP 706-196.

to determine possible detrimental effects on equipment performance. Checklists based on these studies are then prepared to assure that the objectives of the formal design reviews are fulfilled. Formal design review is intended to accomplish the objectives noted in Table 15.3.

Design reviews are iterative. They are repeated at appropriate stages in the design process to evaluate achievement of the reliability requirements. Reviews should include but not necessarily be limited to: current reliability estimates and achievements for each mode of operation (derived from reliability analyses or tests); potential design or production problem areas (derived from reliability analyses); and control measures necessary to preserve the inherent reliability; failure modes, effects, and criticality analyses; corrective action on reliability critical items; effects of engineering decisions, changes and tradeoffs upon reliability achievements, potential for growth within the functional model framework; status of subcontractor and supplier reliability programs; and status of previously approved design review actions. The results of all reliability design reviews should be thoroughly documented.

The review team must have sufficient breadth to handle all aspects of the items under review in order to satisfy the objectives of the review. These include both performance and reliability and the interfaces and interactions with adjacent items. Design engineering is primarily oriented towards ensuring that the design will work. In contrast, reliability engineering must identify those instances in which the design does not operate, the areas of unreliability. Systematic approaches such as the mathematical model, FMEAs, FTAs, and criticality lists assist the designer in identifying those portions of the design that he must concentrate on in order to achieve a balanced, reliable design.

In complex systems, it is often difficult for the designer to assess his design adequately without the assistance of the reliability engineering organization. Mathematical models provide for systematic functional diagramming of components as they fit into the systems and subsystems; hence, these models can aid the designer in providing an overall understanding of the complete system. The models will also aid him in creating designs that provide reliability for the total system rather than overdesign of a particular portion with little improvement of the total reliability.

Informal Reliability Design Reviews

Informal design reviews are also important elements in the design formulation process. These reviews conducted for the benefit of the equipment and system design engineers help achieve the appropriate degree of design maturity the first time around. Evaluating and guiding specified reliability characteristics and maintenance features "in process" while the design is still in the evolutionary or formative stage and still amenable to major conceptual and configuration changes is the purpose of these reviews. They are conducted on an unscheduled, as-required, informal basis at the request of the designer or the system engineer in order to: verify conformance throughout the team effort; allocate requirements and design constraints; verify the solution of problems identified in earlier design iterations; or to provide the basis for the selection of design alternatives.

Much of the reliability engineer's activity in the design support role is devoted to these reviews. He must work closely with hardware designers in developing analytical models which best represent the configuration to be verified. He must also perform real-world design assessment. Even though the informal review is a shirt-sleeve working session involving only a few selected reviewers, the results of each review should be documented in the design report. This is an important step in the scientific "analyze, then cut and try" process by which the final design configuration evolves.

During these early informal design reviews, customer participation is usually very limited since the reviews are internal, primarily for members of the design team. During formal design reviews customer specialists play a more participative role.

Formal Design Reviews

Formal design review programs for specific equipment or systems may be the subject of contractual agreement between the customer and the contractor. The recommended design review team membership and the functions of each member are briefly summarized in Table 15.4.

Preliminary Design Review (PDR)

The PDR is conducted prior to the detail design process. It should evaluate the progress and technical adequacy of the selected design approach, determine its compatibility with the performance requirements in the specification, and establish the existence of the physical and functional interfaces between the item and other items of equipment or facilities. Customer approval of the PDR is often considered as authorization to order material and may release funding for that purpose.

Critical Design Review (CDR)

The CDR conducted when detail design is essentially complete and fabrication drawings are ready for release. It should determine that the detail design satisfies the design requirements established in the specification and should establish the exact interface relationships between the item and other items of equipment and facilities. The CDR date is usually identified on the master schedule and requires final approval by the cognizant customer manager or engineer. Customer approval of the CDR is frequently considered as authorization to release drawings and specifications.

Preproduction Reliability Design Review (PRDR)

The PRDR is a formal technical review conducted to determine if the achieved reliability of a system at a particular point in time is acceptable to justify commencement of production.

The PRDR is conducted after completion of the qualification test and prior to production to ensure the adequacy of the design from a reliability standpoint. The level of achieved reliability and adequacy of design are evaluated primarily on the basis of the customer's technical and operational testing, for example, test results, failure

Table 15.4 Design Review Group, Membership and Responsibilities

Group member	Responsibilities
Chairman	Calls, conducts meetings of group, assigns action items, issues interim and final reports, and determines resolution of action items
Product Design Engineer(s)	Prepare(s) and present(s) the design and substantiates decisions with data from tests or calculations
Reliability Manager[a] or Engineer	Evaluates design for optimum reliability, consistent with the requirements
Quality Control Manager or Engineer	Ensures that the functions of inspection, control, and test can be efficiently carried out
Manufacturing Engineer	Ensures that the design is producible at minimum cost within schedule constraints
Field Engineer	Ensures that installation, maintenance, and operator considerations are adequately addressed in the design
Procurement Representative	Assures that acceptable parts and materials are available to meet cost and delivery schedules
Materials Engineer	Ensures that the materials selected will perform as required
Tooling Engineer	Evaluates design in terms of the tooling costs required to satisfy tolerance and functional requirements
Packaging and Shipping Engineer	Assures that the product is capable of being handled, stored, and transported without damage
Design Engineer(s) (not associated with unit under review)	Constructively review(s) the adequacy of the design to meet all of the customer requirements
Customer Representative	Generally voices opinion as to the acceptability of design; may request further investigation of specific items

[a]Similar support functions are also performed by maintainability, human factors, value engineering and so on.
Source: *Engineering Design Handbook: Design for Reliability*, AMCP 706-196.

reports, failure analyses reports, reports of corrective action, and other documents which could be used as necessary for backup or to provide a test history.

Design Review Checklists

A design review checklist delineates specific details to be considered for the item under review. The technical checklist should be prepared by reliability engineering to ensure that every consideration has been appropriately taken into account and furnished to the designer in the very early stages of design. The checklist should be devised for convenient use by the designer for completion along with the design, analyses, and other documentation. Table 15.5 is a typical but not all-inclusive list of details to be considered in various stages of a design review.

Table 15.5 Design Review Item Summary

Preliminary design review
Identification of critical items
Program plans
Preliminary test plans
Progress of the design
Reliability allocations and predictions
Maintenance concept
Critical design review
Subsystem and component specifications
Test plans and procedures
Critical component evaluations
Final design configuration
Reliability analyses
Test Results
Informal design review
Reliability allocations
Reliability predictions
FMECA
Failure data
Growth test data
Production assurance data

Technical checklists are often oriented in a question format to ensure that critical factors will not be overlooked.

15.4 RELIABILITY DATA SOURCES

Whenever one is working in the field of reliability there is a need for reliability data. What type of reliability data exists and where can it be obtained? Reliability data sources generally fall into three distinct classes: (1) those which are fully government-funded organizations and are usually restricted to some extent; or (2) those which are partially government-funded organizations and are not normally restricted but must charge for their services and publications; and (3) unrestricted privately operated organizations which charge accordingly for services rendered. Some of the better-known sources of reliability data are: the Reliability Analysis Center, GIDEP, the IEEE, NTIS, DTIC, and the various annual technical symposia dealing with reliability.

Reliability Analysis Center

The Reliability Analysis Center (RAC) is one of a number of information analysis centers scattered throughout the United States that are sponsored by the U.S. Department of Defense (DOD). These centers are funded and managed by the Defense Logistics Agency (DLA) but are operated by private industry under contract to DLA. Each of these centers specializes in a specific area of technical expertise. They are responsible for the collection, review, analysis, appraisal, summarization, and storage of available information on a subject of highly specialized technical area of concern. These data collections are usually computerized and are expanded on a continuing basis to incorporate the most current international research information. The critically analyzed information thus compiled in selected areas is then repackaged and disseminated according to the needs of the industry.

The RAC is located with the Rome Air Development Center (RADC) at Griffiss Air Force Base, New York and for the past fourteen years has been operated by the IIT Research Institute, the research affiliate of the Illinois Institute of Technology in Chicago.

In accordance with its charter the RAC operates under a "cost recovery" type of contract. This requires it to charge the user for its services. It maintains an extensive library and a number of data bases dealing with the reliability of microcircuits, discrete transistors and diodes, and other nonelectronic devices. It also has an extensive database dealing with the effects of electrostatic discharge on microelectronic circuits. All of this data is gleaned from various military and commercial sources with particular attention upon user data in contrast to supplier data. These library resources and data repositories are utilized to help solve technical problems for specific clients. These services are offered to U.S. DOD agencies, other government agencies, U.S. government contractors and subcontractors and private industry as well.

Specific services offered by RAC include:

Technical inquiries
Bibliographic inquiries
Reliability databooks
State of the art reports
Technology assessments
A quarterly newsletter
Reliability seminars and training courses
Direct technical consultation

The RAC may be contacted at:

Reliability Analysis Center
RADC/RAC
Griffiss AFB, NY 13441

GIDEP and EXACT

The Government Industry Data Exchange Program (GIDEP) is a fully government-funded source of reliability data. All U.S. government contractors are required by contract to be members of GIDEP. Civilian manufacturers may also join GIDEP. GIDEP maintains three large data banks. These include: (1) failure experience data exchange, (2) engineering data exchange, and (3) metrology data exchange. Access to these data banks may be achieved on a direct interactive basis. GIDEP, however, is responsible only for the dissemination of this data. It does not provide analysis or interpretation of the data.

Among the better known services offered by GIDEP are the *Alert* program and the *Urgent Data Request* program. The Alert program provides a public forum for the dissemination of reported product dificiencies associated with component parts. Under this program, if a deficiency is reported against a specific supplier's product the supplier is first notified and given ten days with which to respond to the complaint. If the supplier responds within that time period, a copy of his response will be included with the product Alert which is distributed to each GIDEP member.

The Urgent Data Request provides a vehicle for alerting the reliability community of the need for a unique type of reliability data and requests their assistance in obtaining such data.

GIDEP may be contacted at:

Officer in Charge (Code 362)
GIDEP Operations Center
Corona, CA 91720

A European interface with GIDEP is also maintained through EXACT (International Exchange of Authenticated Electronic Component Data). EXACT may be contacted at:

EXACT
Central Office
345 Ballards Lane
London N12 8LJ England

National Technical Information Services (NTIS)

The central source for the sale of government-sponsored research, development and engineering reports, and other analysis prepared by federal agencies, their contractors or grantees, or by special technology groups is the National Technical Information Services (NTIS). The NTIS is also operated by the U.S. Department of Commerce on a "cost recovery" basis and thus it must charge customers accordingly for its publications. An official document accession number or AD number must be referenced to identify each document ordered from NTIS. NTIS may be contacted at:

National Technical Information Services
5285 Port Royal Road
Springfield, VA 22151

Defense Technical Information Center (DTIC)

This is the central repository for the department of defense collection of research and development in the field of science and technology. DTIC is a fully funded government organization and its services are restricted to those organizations with current government contracts or their subcontractors and other approved organizations. These are usually organizations which are seeking to become government contractors or subcontractors in response to a specific "invitation to bid" or a "request for proposal." DTIC may be contacted at:

Defense Technical Information Center
Cameron Station
Alexandria, VA 22314

Naval Publications and Forms Center

This is the official distribution point for all military specifications, military standards, military handbooks, federal specifications, federal standards and qualified products lists (QPLs) and so on. This is a fully government-funded organization and its services are restricted to only those organizations holding current government contracts or subcontracts. They may be contacted at:

Naval Publications and Forms Center
5801 Talbot Ave.
Philadelphia, PA 19120

Private Reprinting Services

Any organization desiring copies of U.S. government specifications, standards and handbooks, and so on, including those without U.S. government contracts or subcontracts, may purchase these documents through officially authorized reprinting services such as:

Global Engineering Documentation Services, Inc.
3301 W. Macarthur Blvd.
P. O. Box 5020
Santa Ana, CA 92704

or:

Documentation Inc.
P. O. Box 1240
Melbourne, FL 32901

International Electrotechnical Specifications

Copies of International Electrotechnical (IEC) specifications may be purchased in the United States through:

American National Standards Institute (ANSI)
1430 Broadway
New York, NY 11018

In Europe the IEC publications may be purchased through:

Bureau Central de la Comission Electrotecknique Internationale
1, Rue de Varembe
Geneva, Switzerland

Symposium Proceedings

A number of technical symposia with documented proceedings are held annually which are of significance to the reliability community.

These symposiums together with their major sponsors and/or the source for their proceedings are:

1. The International Reliability Physics Symposium (held annually in April) and the Reliability and Maintainability Symposium (held annually in January):

   IEEE Service Center
   Conference Publication Sales
   445 Hoes Lane
   Piscatway, NJ 08854

2. EOS/ESD Symposium (held annually in September):

   Reliability Analysis Center
   RADC/RAC
   Griffiss AFB, NY 13441

3. Environmental Sciences Symposium (held annually in May):

   Institute of Environmental Sciences
   940 E. Northwest Highway
   Mount Prospect, IL 60056

4. Society of Reliability Engineers (SRE)
   P. O. Box 131
   Crum Lynne, PA 19022

Bibliography

Introductory Remarks

All Department of Defense (DOD) and Military (MIL) Specifications, Standards (STD), and Handbooks (HDBK) listed here may be obtained from:

Naval Publications and Forms Center
5001 Talbot Ave.
Philadelphia, PA 19120

In the United States International Electrotechnical (IEC) Specifications may be purchased from:

American National Standards Institute (ANSI)
1430 Broadway
New York, NY 10018

In Europe IEC Publications may be purchased from:

Bureau Central de la Comission Electrotecknique Internationale
1, rue de Varembe
Geneva, Switzerland

International Reliability Physics Symposium Proceedings, Reliability and Maintainability Symposium Proceedings and other IEEE Conference Proceedings listed here may be purchased from:

IEEE Service Center
Conference Publication Sales
445 Hoes Lane
Piscatway, NJ 08854

Government-sponsored research, development, and engineering reports may be purchased from National Technical Information Services (NTIS). These documents are identified in the bibliography with accession (AD) number which must be used when ordering the document from NTIS.

National Technical Information Services
5285 Port Royal Road
Springfield, VA 22151

Chapter 1

MIL-HDBK-338, Electronic Reliability Design Handbook.

Anderson, R. T. Reliability Design Handbook (RDH 376), (IITRI), March 1976, Reliability Analysis Center, RADC/RAC, Griffiss AFB, NY 13441.

Engineering Design Handbook: Reliability Measurement, AMCP-786-198, January 1976, AD#A027371.

Lloyd, R. K., Lipow, M., Reliability: Management Methods, and Mathematics, second edition, TRW, Redondo Beach, CA, 1977.

Mann, N., R. Schafer, Singpurwalla, N., Methods of Statistical Analysis of Reliability and Life Data, John Wiley and Sons, New York, 1974.

O'Connor, P., Practical Reliability Engineering, Heyden & Son Ltd., London, Philadelphia, 1981.

Quality Assurance Reliability Handbook, AMCP 702-3, Oct. 1968, U.S. Army Material Command, Washington, DC 20315, AD#702936.

MIL-STD-105, Sampling Procedures and Tables for Inspection by Attributes.

MIL-STD-721, Definitions of Effectiveness Terms for Reliability, Maintainability, Human Factors and Safety.

IEC 271, List of Basic Terms, Definitions and Related Mathematics for Reliability.

Bazovsky, I., Reliability Theory and Practice, Prentice-Hall, Englewood Cliffs, NJ, 1961.

Dey, K. A., Practical Statistical Analysis for the Reliability Engineers, (SOAR-2), Spring 1985, (IITRI), Reliability Analysis Center, RADC/RAC, Griffiss AFB, NY 13441.

Chapter 2

MIL-HDBK-338, Electronic Reliability Design Handbook.

MIL-M-38510, General Specification for Microcircuits.

MIL-STD-883, Test Methods and Procedures for Microelectronics.

MIL-S-19500, General Specification for Semiconductor Devices.

MIL-STD-750, Test Methods for Semiconductor Devices.

MIL-STD-198, Capacitors, Selection and Use of.

MIL-STD-199, Resistors, Selection and Use of.

MIL-STD-202, Test Methods for Electronics and Electrical Component Parts.

MIL-STD-965, Parts Control Program.

MIL-STD-1286, Transformers, Inductors and Coils, Selection and Use of.

MIL-STD-1346, Relays, Selection and Use of.

MIL-STD-1353, Connectors, Selection and Use of.

Denson, W. K., Turner, T., Impact of Electrostatics on IC Fabrication, RAC-TR-83-09-E01, Sept. 1983, Prepared by the Reliability Analysis Center for the Naval Sea Systems Command, NAVSEA code 06C31, Washington, DC 20362.

Rickers, H. C., Microcircuit Screening Effectiveness (TRS-1) 1978 (IITRI), Reliability Analysis Center, RADC/RAC, Griffiss AFB, NY 13441.

Doyle, E., Morris, W., Microelectronics Failure Analysis Techniques, A Procedural Guide, available from the Reliability Analysis Center, RADC/RAC, Griffiss AFB, NY 13441.

Michael, K. W., New Hybrid Electronic Moulding Compounds, Proceedings of the 26th Electronics Component Conference, IEEE, 1976, pp. 382-385.

Priore, M. G., IC Quality Grades: Impact on System Reliability and Life Cycle Cost, (SOAR-3) Winter 1984/85, (IITRI), Reliability Analysis Center, RADC/RAC, Griffiss AFB, NY 13441.

Integrated Circuit Engineering Basic Technology, ICE Corp., 15022 N. 75th St., Scottsdale, AZ 85260-2476, 1975.

Microcircuit Manufacturing Control Handbook, Volume I, ICE Corp., 15022 N. 75th St., Scottsdale, AZ 85260-2476, 1975.

Chapter 3

MIL-HDBK-338, Electronic Reliability Design Handbook.

MIL-STD-454, Standard General Requirements for Electronic Equipment, Requirement 18, Derating of Electronic Parts and Materials.

Brummett, S. L., et al., Reliability Parts Derating Guidelines, RADC TR 82-177, June 1982, AD#A120-367.

Part Derating Guidelines, AFSC P 800-27, Dec. 1983 Dept. of the Air Force, Headquarters Air Force Systems Command, Andrews Air Force Base, DC 20334.

Parts Application and Reliability Information Manual for Navy Electronic Equipment, NAVSEA TE000-AB-GTP-010, Sept. 1985, Naval Sea Systems Command, Department of the Navy, Washington, DC.

MIL-STD-975(NASA), NASA Standard (EEE) Part List, Appendix A Standard Parts Derating Guidelines.

Eskin, D. J., Mccanless, C. R., Reliability Derating Procedures, RADC TR 84-254, AD#A153268, 1984.

Chapter 4

MIL-HDBK-338, Electronic Reliability Design Handbook.

DOD-STD-1686, Electrostatic Discharge Control Program for Protection of Electrical and Electronic Parts, Assemblies and Equipment (Excluding Electrically Initiated Explosive Devices).

DOD-HDBK-263, Electrostatic Discharge Control Handbook for Protection of Electrical and Electronic Parts, Assemblies and Equipment (Excluding Electrically Initiated Explosive Devices).

Fuqua, N. B., ESD Protective Materials and Equipment: A Critical Review, (SOAR-1) Spring 1982, (IITRI), Reliability Analysis Center, RADC/RAC, Griffiss AFB, NY 13441.

Denson, W. K., Turner, T., Impact of Electrostatics on IC Fabrication, RAC-TR-83-09-E01, Sept. 1983, Prepared by the Reliability Analysis Center for the Naval Sea Systems Command, NAVSEA code 06C31, Washington, DC 20362.

NASA ESD Control Information Manual, D-TM-82-1A, Jan. 1982, Office of Chief Engineer, NASA Headquarters, Washington, DC.

Denson, W. K., Electrostatic Discharge Susceptibility of Electronic Devices, (VZAP-1), Spring 1983, (IITRI), Reliability Analysis Center, RADC/RAC, Griffiss AFB, NY 13441.

EOS/ESD-1, through EOS/ESD-7 Electrical Overstress/Electrostatic Discharge Symposium Proceedings, 1979 thru 1985, available from Reliability Analysis Center, RADC/RAC, Griffiss AFB, NY 13441.

Requirements for Electrostatic Discharge Protection of Electronic Components and Assemblies, NAVSEA OD 46363, March 1977, Naval Sea Systems Command, Washington, DC 20363.

Chapter 5

MIL-HDBK-338, Electronic Reliability Design Handbook.

Anderson, R. T., Reliability Design Handbook (RDH 376), (IITRI), March 1976, Reliability Analysis Center, RADC/RAC, Griffiss AFB, NY 13441.

Fink, D. G. and Cristiansen, D., editors, Electronic Engineers' Handbook, 1982, McGraw-Hill Book Co., New York.

Engineering Design Handbook: Reliable Military Electronics, AMCP706-124, Headquarters U.S. Army Material Command, Jan. 1976, 5001 Eisenhower Ave., Alexandria, VA 22333, AD#A025665.

Practical Reliability, Vol. 1 - Parameter Variations Analysis, NASA CR-1126, July 1968, Research Triangle Institute, Research Triangle Park, North Carolina, 27709.

Engineering Design Handbook: Design for Reliability, AMCP706-196, January 1976, AD#A027370.

Jensen, R. W., McNamee, L. P., Handbook of Circuit Analysis Languages and Techniques, Prentice-Hall, Inc., Englewood Cliffs, NJ 07632.

Morrison, G. N., et al., Thermal Guide for Reliability Engineers, June 1982, RADC TR *2-172, ADA#118839.

Clardy, R. C., Sneak Circuit Analysis Development and Application, 1976 Region V IEEE Conference Digest, 1976, pp. 112–116.

Hill, E. J., Bose, L. J., Sneak Circuit Analysis of Military Systems, Proceedings of the 2nd International System Safety Conference, July 1975, pp. 351–372, System Safety Society, 14252 Culver Dr., Suite a-261, Irvine, CA 92714.

Godoy, S. G., Engels, G. J., Sneak Circuit and Software Sneak Analysis, Journal of Aircraft, Vol. 15, August 1978, pp. 509–513.

Contracting and Management Guide for Sneak Circuit Analysis (SCA) NAVSEA TE001-AA-GYD-010/SCA, September 1980, Naval Sea Systems Command, SEA 6151, Washington DC 20362.

Sneak Analysis Application Guidelines, RADC TR 81-179, ADA#118479.

Chapter 6

MIL-HDBK-338, Electronic Reliability Design Handbook.

Anderson, R. T., Reliability Design Handbook (RDH 376), (IITRI), March 1976, Reliability Analysis Center, RADC/RAC, Griffiss AFB, NY 13441.

Klion, J., A Redundancy Notebook, December 1977, RADC-TR-77-287, AD#A050837.

Shooman, M., Probabilistic Reliability: An Engineering Approach, 1968, McGraw-Hill Book Co., New York.

Barrett, L. S., "Reliability Design and Application Considerations for Classical and Current Redundancy Schemes," September 1973, Lockheed Missiles and Space Co., Inc., Sunnyvale, CA.

Chapter 7

MIL-HDBK-338, Electronic Reliability Design Handbook.

Engineering Design Handbook: Design for Reliability, AMCP706-196, January 1976, AD#A027370.

MIL-STD-1629A, Procedures for Performing a Failure Mode Effects and Criticality Analysis, Nov. 1980.

Yurkowsky, W., Nonelectronic Reliability Notebook, March 1970, RADC-TR-69-458, AD#A868372.

Augus, J. E., Arno, R. G., Mann, N. R., Study and Investigation to Update the Nonelectronic Reliability Notebook, April 1985, RADC TR 85-66.

Dussault, H., The Evolution and Practical Application of Failure Modes and Effects Analyses, RADC-TR-83-72, ADA#131358.

Nonelectronic Parts Reliability Data (NPRD-3), (IITRI), Winter 1985/86, Reliability Analysis Center, RADC/RAC, Griffiss AFB, NY, 13441.

Chapter 8

MIL-HDBK-338, Electronic Reliability Design Handbook.

Haasl, D., F., et al., Fault Tree Handbook, NUREG-0492, Jan 1981 Division of Technical Information and Document Control, U.S. Nuclear Regulatory Commission, Washington, DC 20555.

Reliability Data Analysis and Interpretation, Volume 4 of Reliability Guides, 1974 Bird Engineering-Research Assoc. Inc., Vienna, VA 22180.

IEEE STD 352/ANSI N41.4-1976, IEEE Guide for General Principles of Reliability Analysis of Nuclear Power Generating Station Protective Systems, January 1976.

Barlow, R., Fussell, J., Singpurwalla, N., (Editors), Reliability Fault Tree Analysis, 1975, Society for Industrial and Applied Mathematics (SIAM), 117 S. 17th St., Suite 1400, Philadelphia, PA 19103-5052.

MIL-STD-882, System Safety Program Requirements.

Goddard, P. L., Davis, R., Automated FMEA Techniques, 1984, RADC TR 84-244, AD#A154161.

Chapter 9

MIL-HDBK-338, Electronic Reliability Design Handbook.

MIL-STD-756, Reliability Modeling and Prediction.

MIL-HDBK-217, Reliabi.ity Prediction of Electronic Equipment.

Engineering Design Handbook: Design for Reliability, AMCP706-196, January 1976, AD#A027370.

Reliability Engineering Handbook, NAVWEPS 00-65-502, 1970, Bird Engineering-Research Assoc. Inc. Vienna, VA 22180.

Von Alvin, W. H., editor, Reliability Engineering, Prentice-Hall Inc., Englewood Cliffs, NJ, 1964.

Winter, F., ORACLE and Predictor Computerized Reliability Prediction Programs, RADC TR-83-172, AD#B077240, 1983.

Colt, D. W., Priore, M. G., Impact of Nonoperating Periods on Equipment Reliability, RADC TR-85-91.

Chapter 10

MIL-HDBK-338, Electronic Reliability Design Handbook.

Anderson, R. T., Reliability Design Handbook (RDH 376), (IITRI), March 1976, Reliability Analysis Center, RADC/RAC, Griffiss AFB, NY 13441.

Engineering Design Handbook: Environmental Series, Part One, Basic Environmental Concepts, AMCP 706-115, AD#784999.

Engineering Design Handbook: Environmental Series, Part Two, Natural Environmental Factors, AMCP 706-116, AD#A012648.

Engineering Design Handbook: Environmental Series, Part Three, Induced Environmental Factors, AMCP 706-117, AD#A023512.

Engineering Design Handbook: Environmental Series, Part Four, Life Cycle Environments, AMCP 706-118, AD#A015179.

Engineering Design Handbook: Environmental Series, Part Five, Glossary of Environmental Terms, AMCP 706-119.

Engineering Design Handbook: Design for Reliability, AMCP706-196, January 1976, AD#A027370.

MIL-HDBK-251, Reliability/Design Thermal Applications.

Arsenault, J. E., Roberts, J. A., Reliability and Maintainability of Electronic Systems, Computer Science Press, 9125 Fall River Lane, Potomac, MD 20854, 1980.

Morrison, G. N., et al., Thermal Guide for Reliability Engineers, Juee 1982, RADC TR 82-172, ADA#118839.

MIL-STD-810, Environmental Test Methods.

EMC Technology and Interference Control News, Vol. 3, No. 4 Oct.-Dec. 1984, P. O. Box 1145, Dover, NJ 07801.

Chapter 11

MIL-HDBK-338, Electronic Reliability Design Handbook.

Anderson, R. T., Reliability Design Handbook (RDH 376), (IITRI), March 1976, Reliability Analysis Center, RADC?RAC, Griffiss AFB, NY 13441.

MIL-STD-454, Standard General Requirements for Electronic Equipment.

MIL-E-5400, Electronic Equipment Airborne, General Specification For.

MIL-STD-1388, Logistics Support Analysis.

McCormis, E. J., Human Engineering, McGraw-Hill Publishing Co., New York, 1967.

Leuba, H. R., The Impact of Policy on Systems Reliability IEEE Transactions on Reliability R-18, August 1969, pp. 137–140.

Hecht, H., Hecht, M., Kim, K. H., Fault Tolerance, Reliability, and Testability of Distributed Systems, 1983, RADC TR-83-36, AD#A129438.

MIL-STD-1472, Human Engineering Design Criteria for Military Systems, Equipment and Facilities.

MIL-H-46855, Human Engineering Requirements for Military Systems, Equipment and Facilities.

Engineering Design Handbook: Design for Reliability, AMCP706-196, January 1976, AD#A027370.

Human Resources Research Program, AR 70-8.

Woodson, W. E., Conover, D. W., Human Engineering Guide for Equipment Designers, University of California Press, Berkeley, CA, 1966.

Davis, B. P., Cordoni, C. N., People Subsystem Measurement for Total Reliability, Proceedings of the 1970 Annual Symposium on Reliability, 1970, p. 394.

Coppola, A., A Design Guide for Built in Test, RADC-TR-78-224 April 1979, ADA#069384.

Byron, J., Deight, L., Stratton, G., RADC Testability Notebook, RADC-TR-82-189, June 1982, ADA#11888.

MIL-STD-2165, Testability Program for Electronics Systems and Equipment, 1985.

Meister, D., et al., The Effect of Operator Performance Variables on Airborne Electronic Reliability, July 1970, RADC-TR-70-140, AD#A873368.

Regulinski, T. L., editor, Special Issue on Human Performance Reliability: IEEE Transactions on Reliability R-22, No. 3, August 1973.

MIL-E-17555, Electronic and Electrical Equipment, Accessories, and Repair Parts, Packaging and Packing of.

Chapter 12

MIL-HDBK-338, Electronic Reliability Design Handbook.

MIL-STD-781, Reliability Design Qualification and Production Acceptance Tests: Exponential Distribution.

IEC 605-1, Equipment Reliability Testing, Part 1 General Requirements 1978.

MIL-HDBK-781 (Aug 1983 DRAFT), Reliability Test Methods, Plans and Environments for Engineering Development, Qualification and Production.

Chapter 13

MIL-HDBK-338, Electronic Reliability Design Handbook.

Reliability Engineering Handbook, NAVAIR 01-1A-32, July 1977, Naval Air Systems Command, Washington DC.

Arsenhault, J. E., Roberts, J. A., Reliability and Maintainability of Electronic Systems, 1980, Computer Science Press, 9125 Fall River Lane, Potomac, MD 20854.

MIL-STD-2068(AS), Reliability Development Tests, March 1977.

Duane, J. T., Learning Curve Approach to Reliability Monitoring, IEEE Transactions on Aerospace, Volume 2, April 1964, pp. 363-366.

Crow, L. H., On Tracking Reliability Growth, Proceedings 1975 Annual Reliability & Maintainability Symposium, pp. 438-443.

MIL-HDBK-189, Reliability Growth Management, February 1981.

MIL-STD-1635 (EC), Reliability Growth Testing, February 1978.

Bezat, A. G., Montague, L. L., The Effect of Endless Burn-In on Reliability Growth Projections, Proc 1979 Annual Reliability & Maintainability Symp, pp. 392-297, (IEEE Cat No 79CH1429-OR).

Selby, J. D., Miller, S. G., Reliability Planning and Management (RPM), Paper No. S1-471, ASQC/SRE Seminar, Niagra Falls, NY, September 26, 1970, pp. 1-7.

Codier, E. O., Reliability Growth in Real Life, Proc. 1968 Annual Symp on Reliability, January 1968, pp. 458-469.

Crow, L. H., Reliability Analysis for Complex, Repairable Systems, Reliability and Biometry/Statistical Analysis of Life Length, 1974, pp. 379-410, SIAM, 117 S. 17th St., Suite 1400, Philadelphia, PA 19103-5052.

Haase, R. W., Kapur, K. C., Lamberson, L. R., Applications of Reliability Growth Model During Light Truck Design and Development, SAE Paper No. 780240, FEB/Mar 1978, SAE Headquarters, 400 Commonwealth Dr., Warrendale, PA 15096.

Simpkins, D. J., A Reliaability Growth Management Approach, Proc., Annual Reliability and Maintainability Symp, 1979, pp. 356-360.

Crow, L. H., Confidence Interval Procedures for Reliability Growth Analysis, U.S. Army Material Systems Analysis Activity, Technical Report 197, Aberdeen Proving Ground, MD, 1977, AD#A004788.

Reliability Engineering Handbook, NAVAIR-01-1A-32, July 1977, Naval Air Systems Command, Washington, DC.

MIL-STD-2155, Failure Reporting, Analysis and Corrective Action System, July 1985.

MacDiarmid, P., Morriss, S. Relaibility Growth Testing Effectiveness, 1984, RADC TR-84-20, AD#A141232.

Chapter 14

MIL-HDBK-338, Electronic Reliability Design Handbook.

MIL-STD-785, Reliability Program for Systems and Equipment Development and Production.

Navy Manufacturing Screening Program, NAVMAT P-9492, May 1979, AD #A088705.

IES Environmental Stress Screening Guidelines, The Institute for Environmental Sciences, 940 E. Northwest Highway, Mount Prospect, IL 60056, 1981.

IES Environmental Stress Screening Guidelines for Assemblies, The Institute for Environmental Sciences, 940 E. Northwest Highway, Mount Prospect, IL 60056, 1984.

MIL-STD-2164(EC) Environmental Stress Screening of Electronic Equipment, April 1985.

Stress Screening of Electronic Hardware, RADC TR-82-87, AD#A118261, 1982.

Chapter 15

MIL-HDBK-338, Electronic Reliability Design Handbook.

MIL-STD-785, Reliability Program for Systems and Equipments Development and Production.

IEC 300, Managerial Aspects of Reliability, 1969.

Coppola, A., Sukert, A. N., Reliability and Maintainability Management Manual, RADC-TR-79-200, July 1979, AD#A073299.

Engineering Design Handbook: Design for Reliability, AMCP706-196, January 1976, AD#A027370.

Directory, DOD Information Analysis Centers, April 1985, Defense Technical Information Center, Defense Logistics Agency, Cameron Station, Alexandria, VA 22304-6145.

Reliability and Maintainability Planning Notebook, Federal Aviation Administration, Washington, DC, 1980.

MIL-Q-9858, Quality Program Requirements.

Index